ESSENTIALS OF
BEHAVIOUR GENETICS

To my parents

ESSENTIALS OF BEHAVIOUR GENETICS

DAVID A. HAY   MA PhD MAPsS
*Senior Lecturer, Department of Genetics and Human Variation,
La Trobe University, Melbourne, Australia*

Blackwell Scientific Publications
MELBOURNE   OXFORD   LONDON
EDINBURGH   BOSTON   PALO ALTO

© 1985 by
Blackwell Scientific Publications
Editorial offices:
107 Barry Street, Carlton
    Victoria 3053, Australia
Osney Mead, Oxford
    OX2 0EL, UK
8 John Street, London
    WC1N 2ES, UK
23 Ainslie Place, Edinburgh
    EH3 6AJ, UK
52 Beacon Street, Boston
    Mass 02108, USA
The Downing House,
    744 Cowper Street
    Palo Alto, CA. 94301, USA

First published 1985

Typeset in Australia by
Abb-typesetting Pty Ltd
Printed in Singapore

DISTRIBUTORS

USA and Canada
    Blackwell Scientific
    Publications Inc
    PO Box 50009, Palo Alto
    Ca. 94303

Australia
    Blackwell Scientific Book
    Distributors
    31 Advantage Road, Highett
    Victoria 3190

South-East Asia
    P G Publishing Pte Ltd
    Alexandra PO Box 318
    Singapore 9115

Other
    Blackwell Scientific
    Publications
    Osney Mead, Oxford
    OX2 0EL, UK

Cataloguing in publication data
    Hay, David A.
        Essentials of behaviour
        genetics.
        Bibliography.
        Includes index.
        ISBN 0 86793 064 0.

    1. Behavior genetics.
    2. Human genetics.
    I. Title.

155.7

Contents

Preface

This text introduces behaviour genetics to the advanced student in genetics or psychology. Twelve years of teaching these two groups of students at La Trobe has revealed their different expectations of behaviour genetics. Psychology students want an approach to behaviour genetics which adds to knowledge of the behaviours and concepts discussed in their other psychology courses. But the same does not apply in genetics. Contemporary concepts of molecular and population genetics are often far removed from the very simple level of genetic analysis currently needed for most behaviour genetic studies and except in the area of quantitative genetics, there is only a very limited role for behavioural examples. Instead genetics students can be presented with behaviour genetics as a different way of viewing genetic analysis. Behaviour is often much more difficult to measure than the characteristics to which genetics students are accustomed and they need to appreciate that a small procedural manipulation can totally alter the results of a behaviour genetic analysis. Particularly in human behaviour genetics, the student has to think beyond the results of any genetic analyses to their potential significance for society.

This text is structured around the diverse needs of these two student groups. Psychology students will find many examples where the behaviour genetic approach provides a different view of material presented in courses on abnormal, cognitive, developmental, educational and experimental psychology. For genetics students there is an introduction to the problems of behavioural measurement to alert them to the unique difficulties of behaviour genetic analysis. While a discussion of basic genetics is mandatory in any behaviour genetics text, it is presented here in terms of the role of genetics in mental retardation, so that students already having a grounding in genetics will find something new to maintain their interest.

Behaviour genetics may be a relatively new discipline, but it has already had more than its share of controversy. There is controversy among behaviour geneticists over the best approaches to particular problems and over the best methods of analysis. There are also critics of behaviour genetics who find fault with all the available data and question the motives behind those researching in the field of behaviour genetics. These disputes can only confuse the student and make him or her unclear whom to believe. Rather than ignoring these questions which simply delays the problem until the student consults the original references, they are introduced in what I hope is a fairly unbiased manner and are often returned to in the discussion questions which conclude each chapter. To more accurately capture the original ideas of the contributors to such disputes, extensive use has been made of quotations, especially in Chapter 8 which deals with contentious social and racial issues.

The introduction to every recent behaviour genetics text has included the disclaimer that the authors have had to be selective and that people should not be offended if their favourite experiment is omitted. The present text is no exception. To keep this text to a length which does not daunt the student, it has been necessary to pick representative examples from different areas rather than trying for a comprehensive coverage. Also for the needs of a student, a limited list of references at the end of each chapter replaces a full bibliography. While not ideal in many ways, it seems better to direct the student to a few key references than to leave him or her alone in a veritable sea of information.

Many people have contributed to this book although I must take responsibility for the final form. I should like to acknowledge: Professor Peter Parsons of La Trobe University who gave the initial impetus to this text and Robert Campbell and Mark Robertson of Blackwell's whose advice and patience helped see it through to fruition; Pauline O'Brien and Marena Ross who kept my research projects functioning while I was occupied with the task of writing; Lindon Eaves (Medical College of Virginia), Lynn Feingold (Oxford University), Norman Henderson (Oberlin College), and Glayde Whitney (Florida State University) for helpful discussions and useful comments about a draft of this manuscript; Professor Walter Nance and the Medical College of Virginia where I held an A. D. Williams Distinguished Visiting Professorship while completing Chapters 6-8; graduate students, past and present, particularly Jeffrey Cummins, Karen Lavery and Grant Singleton, but also Richard Rosewarne, Tommy Solopotias and Theresa Theobald, who provided advice from their particular fields of expertise; Judi Bolton, Marianne de Ryk, Debra Duckworth, Marlene Forrester, Toni McElhenny and Ann Monkman for their efforts in typing the text and preparing the diagrams; Sally Collett, Carol Johnston and Lesley Tan for their help in checking references and preparing the index; Wayne Singleton whose cartoons enliven the text; the many former students whose reactions to the examples and to the discussion questions have shaped the text — perhaps I learned more from them than they did from me.

May 1984   David A. Hay

Acknowledgements for material cited

Permission to reproduce copyright material was given by the following:

AAAS (Figs 3.4, 4.4, 7.4; Table 5.2b, extract on p. 301), American Medical Association (Fig. 2.13), American Psychological Association (Figs 3.18, 7.7, 8.4; Tables 6.7, 7.5), Annual Reviews Inc. (Fig. 4.2, extracts on pp. 4, 6, 321), Cornell University Press (Fig. 1.8), W. H. Freeman Inc. (Table 2.8), Grune & Stratton Inc. (Fig. 7.1, Table 8.1), Harvard University Press (extract on p. 24), D. C. Heath & Co. (extracts on pp. 208, 290), N. Kessel and H. Walton (Table 7.6d), McGraw-Hill Inc. (Figs 3.20, 7.2), Macmillan Inc., New York (Fig. 2.24; extract on p. 82), Macmillan Journals Ltd (Figs 1.3, 2.20, 4.1(b), 4.9, 4.13, 4.15, 4.16, 5.2, 6.11, 7.8; Tables 2.5, 5.2a, 7.9, 8.2), Munksgaard Inc. (Table 7.6a), Raven Press (Fig. 4.1a), C. C. Thomas Inc. (Figs 1.2, 2.3, 7.11), John Wiley & Sons Inc. (Figs 2.17, 2.18, 3.16; Table 7.6c).

1 The development and scope of behaviour genetics

Topics of this chapter

1 The interrelationships between genetics, psychology and behaviour genetics and the concept of a 'genetically aware psychology'.
2 'Garbage in, garbage out' — the importance of adequate behavioural measures in any behaviour genetic analysis.
3 The history of behaviour genetics and the reasons why its development has been so recent.
4 Distinctions between behaviour genetics, sociobiology and other approaches to the determinants of behaviour.

What does behaviour genetics involve?

Behaviour genetics is a new area of science growing out of two disciplines, genetics and psychology, which themselves are both products of the twentieth century. It is only since 1960 and the first behaviour genetics text (Fuller and Thompson 1960) that the area of behaviour genetics has come to be recognized as a distinct entity.

Many people associate behaviour genetics only with the well-publicized controversies over the genetics of differences in human intelligence within and between racial groups and the social and educational consequences which may follow. It is easy to see how such a misconception arose since many recent discussions of behaviour genetics (Kamin 1974; Taylor 1980; Vernon 1979) do deal solely with intelligence test performance.

Although the intelligence debate may have much relevance to our society, the approach in this text is that the whole issue of genetics and behaviour is best understood by examining the inheritance of many different behaviours in organisms ranging from bacteria to man. We shall see that some behaviours are largely inherited, while others even in the same species are influenced mainly by environmental factors.

In the case of human behaviour genetics where the analytic methods are much in dispute, this diversity of results is important. While the outcome of genetic analysis of intelligence test scores is often dismissed by postulating defects and biases in the methods of analysis, it is a lot more difficult to explain away the fact that similar analyses of other behaviours yield quite different outcomes. The strength of the methods is that they do produce different results when applied to different behaviours and do not reflect biases towards consistently finding behavioural variation to be largely inherited.In the case of experimental animals the study of different behaviours and their differing degrees of inheritance is a valuable means of understanding the organization and evolution of behaviour.

The variety of species becomes important because it is obviously much easier to carry out unambiguous genetic analyses in laboratory organisms whose breeding and environment can be rigorously controlled.If we can find parallels between the results of animal and human research, we can then put more reliance on the human data with all their imperfections.

The study of behaviours in a diversity of species leads to a better appreciation of the genetic analysis of intelligence or any other single characteristic than does an exclusive concentration on that one behaviour.

The organization of the text

Chapter 1 introduces behaviour genetics, its relationship to genetics and psychology and the reasons why behaviour genetics is such a recent discipline.

Chapter 2 presents the prerequisite genetic knowledge, by discussing the ways in which genetic factors may contribute to mental retardation and the use to which genetic knowledge may be put in helping retarded individuals and their families.

Chapter 3 describes the genetic techniques for analysing the behaviour of experimental animals and the importance of considering genetic variation in all research on animal behaviour, not just that directed specifically towards genetics.

Chapter 4 may seem far removed from human behaviour since it involves research solely on invertebrates. But techniques are available with invertebrates which provide unique information on how genes influence behaviour and on how evolution and ecology can shape behaviour.

Chapter 5 discusses results obtained with behaviour genetic analyses of nonhuman vertebrates, mainly mice and rats, where genetic diversity is an important means of analysing different components of behaviour, their relationship with biochemical variables and the effects of environmental manipulation.

Chapter 6 outlines the controversies surrounding the twin, adoption and family study methods of human behaviour genetics and introduces several newer methods which avoid some of the pitfalls. Intelligence test data provide a common example to illustrate all the methods.

Chapter 7 applies the same methods to other areas of human behaviour, particularly to the mental illnesses and to an understanding of the components of intelligence and personality and how these change during development.

Chapter 8 ventures into a discussion of the significance of genetic differences within and between human racial groups and dismisses several conventional myths such as the idea that genetics is the antithesis of social change.

A text of this length cannot discuss every aspect of behaviour genetics but can only provide illustrations of the different methods and controversies. Apart from one other introductory text (Plomin *et al* 1980), lengthier texts are available with a bias towards the student of psychology (Fuller and Thompson 1978) or of genetics (Ehrman and Parsons 1981). This text places more emphasis on the understanding of behaviour than on genetics while still stressing that we can learn much from developments in genetics. To understand why such distinctions between the 'behaviour' and the 'genetics' in 'behaviour genetics' exist, we must consider the ways in which behaviour genetics has grown from the two parental disciplines.

Behaviour genetics — part of genetics or of psychology?

Until about 1975 it was unnecessary even to ask this question. Most research in behaviour genetics was directed to providing the first unambiguous evidence that genes could influence behaviour. This required the application of conventional genetic techniques to conventional measures of behaviour, with both parent disciplines contributing more or less equally. It was possible for geneticists at that time to view behaviour as something to be analysed in the same way as any measurable physical or biochemical characteristic — usually referred to as the 'phenotype' — of animals or even plants.

My purpose is to examine the experimental and analytical consequences of accepting the premise that behaviour, like any other property of an organism, is a phenotype, determined jointly by inherent causes (usually referred to as nature or genotype) and external agencies (nurture or environment). (J. L. Jinks (1965) *Bulletin of the British Psychological Society* **18**, 25.)

At the same time, experimental psychologists were realizing that all organisms were not identical in behaviour and that individual differences could be studied systematically by genetic techniques.

It was then while overcome by feelings of disenchantment (obviously without laws, behavior study could never be science) that I embraced genetics. There was true science. My passion became even more intense when I realized that, like thermodynamics, genetics has three laws: segregation, independent assortment and the Hardy-Weinberg law of population equilibria. What a foundation they provided for my beloved individual differences. (J. Hirsch (1970) *Seminars in Psychiatry* **2**, 89)

Geneticists and psychologists worked together to accumulate a sizeable store of information on the nature and extent of genetic determination of behaviour. There was also growing cooperation with other disciplines concerning anatomical, biochemical and physiological factors underlying behavioural differences. But science cannot proceed by the simple cataloguing of more and more examples of the inheritance of behaviour. Geneticists became critical of behavioural traits, where the difficulty of adequate measurement and the apparently very complex interplay of genetic and environmental influences made behaviour a very poor choice when it came to trying to learn more about genetics. For example,

There has not been, and in the present state of developmental and neural biology cannot be, any attempt to analyze cellular and developmental mechanisms of gene action in influencing cognitive traits. Nor can it be maintained that work on the biometrical genetics of intelligence has somehow led to progress in biometrical genetics as a general approach. On the contrary, because of the impossibility of control and manipulation of human environments and mating, man is among the worst choices of experimental organisms for testing the methods of quantitative genetics. (R. C. Lewontin (1975) *Annual Review of Genetics* **9**, 401.)

While this viewpoint may hold for some geneticists, psychology and society as a whole have placed so much emphasis on intelligence, its determinants and consequences that it seems shortsighted to dismiss research on this area just because it does not lead to a greater understanding of genetics.

The dilemma facing behaviour genetics in the early 1970s is best summarized by Vale, who wrote,

Two basically different concepts of behavior genetics derive from the question of how close its formal ties to the goals and methods of genetics should be. On the one hand, behavior genetics may be seen as a genetics of behaviors, that is, a subspecies of genetics whose primary concern is the extension of the range of traits examined with the methods of genetics from morphological and physiological to behavioral. On the other hand, it may be seen as a genetically aware psychology, whose primary *raison d'etre* is the further understanding of behavior, and which therefore seeks the best use of genotype in behavioral analysis. This difference is crucial not only at the conceptual level, but, perhaps more importantly, at the operational level, because of differences in the kinds of techniques employed and information generated. A genetics of behaviors studies the functioning of genes through the use of genetic tests applied to behavior. A genetically aware psychology seeks to augment the study of behavior through the manipulation of genotype. In the first case the experimental situation is so structured as to provide the maximal genetic information, while in the second it is structured so as to provide the maximal information about behavior. (J. R. Vale (1973) *American Psychologist* **28**, 872.)

This text is directed towards the latter option, Vale's 'genetically aware psychology', explaining how both animal and human behavioural research can be aided by the consideration of genetic variation. The mere demonstration that a particular behaviour has a genetic determinant should not be seen as the final aim of research, but only the beginning, the stage from which more detailed analyses of the ramifications of genetic differences can proceed. For example, rodent strains differing in alcohol preference differ also in many other aspects of biochemistry, physiology, behaviour, reaction to other addictive drugs and the incidence of diseases related to alcohol consumption. In this way we learn far more about alcohol and its effects than just that alcohol preference is partly inherited. The same applies to human behaviour. The evidence in Chapter 7 that different intellectual skills vary in the extent to which they are inherited may be of little interest to the geneticist, but provides uniquely important information about a longstanding debate in psychology over the structure of human abilities and has possible implications for educational practice.

The approach we adopt is one where genetics is a means to an end and not an end in itself. But while psychologists can claim this approach to behaviour genetics as part of their discipline, two recent developments in genetics show that the interest of geneticists in behaviour genetics is not dead but has merely changed direction. Both these developments may also direct behaviour genetics away from what has always been a major complaint about psychology, namely the limited range of behaviours studied and the limited number of species in which these are studied. Such criticisms of psychology are more frequent nowadays (Lockard 1971; Wilcock 1972), but are not new.Figure 1.1 is from a 1950 article bemoaning the narrowness of psychology at that time.

FIG. 1.1   Beach's prediction of the future of experimental psychology. 'Unless they escape the spell that *Rattus norwegicus* is casting over them, experimentalists are in danger of extinction.' (From F. A. Beach (1950) *American Psychologist* **5**, 115.)

Neurogenetics or 'molecular ethology'

This topic discussed in Chapter 4 concerns the use of behavioural changes to indicate genetic alterations at the level of the molecules which encode genetic information. Geneticists are using behaviour as a tool in studying gene action, because altered behaviour can sometimes be a more convenient indicator of genetic changes than either altered anatomy or physiology. Now that the molecular bases of genetics are fairly well understood, some geneticists feel behaviour provides the next challenge, not

The problem of gene structure and coding was exciting while it lasted. The story of the past two eventful decades, including my own contributions, has been well told, and need not be repeated here. But molecular genetics, pursued to ever lower levels of organization, inevitably does away with itself: the gap between genetics and biochemistry disappears. More recently, a number of molecular biologists have turned their sights in the opposite direction, i.e. up to higher integrative levels, to explore the relatively distant horizons of development, the nervous system, and behavior. The problem of tracing the emergence of multidimensional behavior from the genes is a challenge that may not become obsolete so soon. (S. Benzer (1971) *Journal of the American Medical Association* **218**, 1015.)

and one to which the same methods can be applied

In principle, it should be possible to dissect the genetic specification of behaviour much in the same way as was done for biosynthetic pathways in bacteria or for bacteriophage assembly. (S. Brenner (1973) *British Medical Bulletin* **29**, 269.)

A subsidiary advantage of neurogenetics is that it introduces into behaviour genetics a greater diversity of organisms, chosen because of unique features making them suitable for particular experiments and not merely because they are readily available, a frequent criticism of psychology in general.

Of the more than one million described species of animals, fewer than five per cent possess a backbone and are known as vertebrates; the other 95 per cent comprise the invertebrates. Of the thousands of studies on animal learning published in the past decade, only some five per cent utilized invertebrates as subjects. There are hundreds of thousands of species that have never been introduced to a maze or a Skinner box nor watched at length by an ethologist — indeed, whole phyla have been totally neglected. Two of the many possible causes for this scientific bias are easy to identify: (a) man is anthropocentric, and (b) man is lazy. (J. V. McConnell (1966) *Annual Review of Physiology* **28**, 107.)

Ecological and evolutionary influences on behaviour

This approach is outlined in Chapters 4 and 5 and is emphasized in one recent text (Ehrman and Parsons 1981). While we normally think of genes influencing behaviour through a sequence of chemical and physical pathways, the questions of ecology and evolution remind us that behaviour can also influence genes or at least the extent to which particular genetic types are represented in particular populations. Genetically determined differences in behavioural responses to environmental stresses or in migration or mating propensity are all factors which influence how a population evolves and its genetic structure. Consideration of genetics and evolution can influence which behaviours are measured and in which organisms. Psychology has only recently come to appreciate that animals are biologically adapted and have evolved so that some behaviours are more likely than others to occur in given situations, a concept termed 'biological preparedness' by Seligman (1970).

A simple example of this would be the response of an animal experiencing an electric shock or some other painful stimulus. In this situation, it is much easier to 'condition', that is to train the animal to avoid the shock by jumping out of the apparatus or running down an alleyway than by running in an activity wheel or a shuttle box. In a shuttle box an animal shocked in one side of the box moves to the other side of the box. When shocked here, the animal shuttles back again and the cycle is repeated. In real life no animal would return to where it had just been hurt as in the shuttle box or put in the effort in a running-wheel which restrains it in the environment where the pain was administered. Hence the animal is contraprepared to learn in these two situations.

Another example is that many animals in the wild adopt a 'win-shift' strategy, moving on to look for food elsewhere once they have found it in a particular spot. But this strategy is the opposite of that favoured in the maze-learning experiments of traditional animal psychology where the animal is trained to return to the same spot.

Unfortunately behaviour genetics lags behind such developments in psychology. Among the experiments on rodent behaviour discussed in Chapter 5, apparatus such as the shuttle box and the maze feature widely despite their inadequacies. The same can be said for many of the measures of human behaviour described in Chapters 6 and 7.

Beyond the specific neurological and ecological genetic approaches, there is an even more fundamental reason why behaviour genetics should not become divorced from advances in genetics. It is not too harsh to say that behaviour genetics at the present time has ignored most developments in genetics since the early 1950s when the main methods for analysing complex phenotypes became established in plant and animal breeding. That is, a geneticist of that time would find few new genetic concepts in present day behaviour genetics but just some new characteristics to which his methods are applied and some extension of the methods to human data.

Since then there has been an explosion of interest in molecular genetics, in studying the biochemical means by which genetic information is stored and passed onto subsequent generations. Certainly, this knowledge has meant that we no longer think of genes for behaviour as such. Rather genes code for specific proteins which go on to determine anatomical, biochemical and physiological characteristics in association with the internal and external environment. The outcome of all these processes may include some effect on behaviour.

In one of the first symposia on behaviour genetics (Hirsch 1967), Caspari indicated some potential applications to behaviour of this knowledge. Eleven years later Fuller and Thompson (1978) did the same, suggesting that this information has had very little influence on the majority of behaviour-genetic research. For example, behaviour genetics has largely ignored the possibility that genes can be 'regulatory' rather than 'structural', and regulate the function of other genes instead of directly influencing a pathway to behaviour. The situation now arises where the expression of an individual's genetic endowment is not necessarily obvious at birth but may alter with the internal (e.g. hormonal) or external environment.

Medical genetics illustrates the significance of such a possibility. Haemoglobin in the fetus binds oxygen so strongly that there is a flow of oxygen from the mother to fetus. Postnatally this tight binding could deprive tissues of oxygen and so the production of fetal haemoglobin is diminished and genes determining the production of the adult form become active. The existence of this mechanism is confirmed by disorders that involve a persistence of the fetal form, that is, where its genes are not 'switched off'. While behavioural parallels have not yet been described, Chapter 7 provides examples from human intellectual development that may be explained by regulatory genes acting at the time of puberty.

Some potential applications of molecular techniques to mental retardation and to mental illness are discussed in Chapters 2 and 7 respectively, but one fundamental difficulty remains before behaviour genetics can fully utilize such recent advances in genetics. We must be able to measure the behaviour of interest with a sufficient degree of precision, unconfounded by other variables. As the next section illustrates, this is not as easy as it might seem.

'Garbage in, garbage out' — the need for adequate behavioural measurement

The dictum 'garbage in, garbage out' can be applied to many areas of behavioural science, and is directly relevant to behaviour genetics. It means that no matter how sophisticated and expert our genetic analysis, the results of our study will be worthless unless we have been able to measure behaviour adequately in the first place. Apart from the question of *validity* and whether a measure actually measures what it is supposed to, there are two criteria of particular relevance to behaviour genetics.

The first, *reliability*, examines whether the measure of behaviour is consistent or if the result varies when the measure is repeated or made by someone else. When we measure a behaviour which reflects underlying and constant genetic influences, we would expect the result always to be the same. If our measurement techniques are inadequate and results fluctuate, then we are more likely to conclude that the behaviour is determined largely by the environment and that this explains the variation from one test to another.

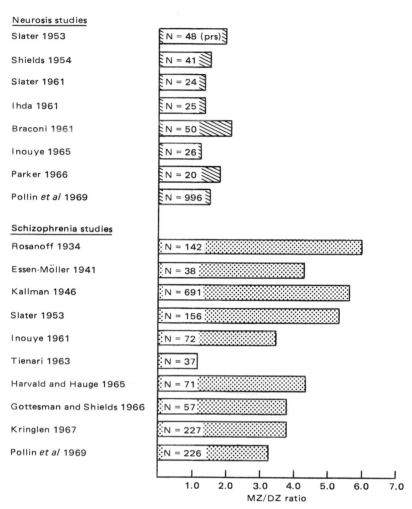

FIG. 1.2   The similarity of MZ (identical) relative to DZ (non-identical) twins
in studies of neurosis and schizophrenia. An MZ/DZ ratio of 1 means they are
equally similar and greater than 1 that MZ are more similar than DZ. (From W.
Pollin (1976) in *Human Behavior Genetics*, Kaplan A. R. (ed.) p. 279, Thomas,
Springfield, Ill.)

Figure 1.2 is an example from the area of mental health of how the
reliability of measurement sets an upper limit on how much of the
variation in a behaviour is seen to be determined by inheritance. The
twin method is discussed in Chapter 6 but basically involves comparing the similarities between identical twins (monozygotic (MZ)
twins who are genetically identical) with those between fraternal
(dizygotic (DZ) twins who genetically are only as alike as any two
brothers or sisters). If genetic factors are involved, the identical twins
will be much more similar, as seen for the data on schizophrenia in
Fig. 1.2. If it is mainly the environment that matters, the two types of
twins will show the same degree of similarity, as seen for the neurosis
data in Fig. 1.2.

But do these results mean that schizophrenia is necessarily more of a genetic disorder than neurosis? Schizophrenia is a much more severe disorder and there is a higher degree of consistency between psychiatrists in its diagnosis. Neurosis is a much milder disorder covering a wide range of symptoms and overlapping with the behavioural differences seen among the normal, non-mentally ill population and varying more with society's tolerance of people who are in some way 'different'. Thus reliability of diagnosis is potentially much lower.

To make things even more difficult, Fig. 1.2 is a compilation of twin data from studies in many different countries and there are international differences in diagnostic categories. United States psychiatrists have in the past had a far wider definition of schizophrenia so that no less than 50% of the schizophrenics classified in New York hospitals during the US–UK Cross National Project were later diagnosed by the project staff as having some other psychiatric disorder. With the more precise British classification, agreement between psychiatrists on schizophrenia can be as high as 92%. The situation is changing with the introduction in 1980 by the American Psychiatric Association of a new *Diagnostic and Statistical Manual of Mental Disorders*, abbreviated to DSM-III. While this has resulted in a much more rigorous definition of many disorders, especially schizophrenia, it has created an additional problem for geneticists who compile data on individuals diagnosed over a period of time, some under the old DSM-II, others under DSM-III.

If the psychiatric diagnosis of an individual is subject to so many variables, then what hope is there of accurately comparing twins or other relatives for the purpose of genetic analysis? We are left with a situation where the apparent differences in inheritance between schizophrenia and neurosis may be real, or may just be due to differences in the reliability of diagnoses. A satisfactory answer depends not so much on better genetic analyses, but more on improved methods of diagnosis in the first place.

The second criterion is *ease and speed of measurement*. Ever since the first genetic experiments carried out by Mendel on peas over a century ago (Chapter 2), it has been clear that any breeding experiment involves the scoring of many individuals. The laws of genetics are based on probability and accurate estimates of probabilities require large samples. Behaviour genetic analysis is no exception and imposes additional constraints. Behaviour may have to be measured in individuals of the same age to eliminate developmental differences and at the same time of day to avoid diurnal rhythm effects. In organisms such as the vinegar fly, *Drosophila*, widely used in genetics because of its short generation time, the total lifespan of 2–3 weeks may leave only a very short period during which behavioural measures can be made. The same problems do not arise with humans, but since people usually volunteer for research, the tests must again often be short as well as being limited in other ways for ethical reasons.

The dilemma imposed by the need for fast, reliable measurement is summarized below:

There can be no doubt that these requirements have influenced the choice of behavioural characters used in such studies. It has got to be something easily and quickly measured, and if measurement can be automated so much the better. The number of squares entered in an open field and faecal boli deposited there (rats), time taken to emerge from a small box, revolutions of a running wheel or speed of acquisition of a conditioned avoidance response (mice), scores of preening, walking or standing still in response to a mechanical stimulus (*Drosophila*) — these are typical measures employed.

Most ethologists would have grave doubts about the relevance of such measures to the behaviour of the animals in their natural habitats. This is not to say that behaviour genetics must always satisfy ethological criteria, but it does make arguments on the evolution of behaviour derived from some quantitative studies much more hazardous. Further, some measures, however easy to make in the laboratory, are subject to all the problems of validating behavioural units touched on earlier. (A. Manning (1975) in *Function and Evolution in Behaviour* Baerends G., Beer C. and Manning A. (eds), Clarendon, Oxford, p.77.)

Much of behaviour genetics is a compromise between what Manning sees as the approach of ethologists, namely the detailed observation of behaviour often in the field rather than the laboratory, and the limitations imposed by the numbers needed for genetic analysis. It might seem extreme to say that behaviour geneticists often do not know what they are measuring, except that whatever it is, they are measuring it quickly and reliably. The following three examples, one each from *Drosophila*, rodents and humans all deal with behaviours that have been the subject of extensive genetic analyses and suggest that this criticism is not without foundation.

Taxes in Drosophila

In 1959, Hirsch introduced a method for 'reliable mass screening of individual differences' in geotaxis (the attraction to and from gravity). *Drosophila* are introduced into the start of a vertically placed multiple-choice maze (Fig. 1.3). At each choice-point they can either go up against gravity (negative geotaxis) or go down (positive geotaxis). Their sequence of choice is reflected in the tube they reach at the end of the maze. Flies with a consistently negative geotaxis will be in the top tube (-5), those with a positive geotaxis in the bottom tube ($+5$). A system of one-way cones prevents the flies from turning back in the maze.

The system can be adapted to study phototaxis (the attraction to and from light) by covering a horizontally placed maze with a screen, which only allows light to reach one side of each choice-point. Positively phototactic flies turn towards the light, while negatively phototactic flies turn away.

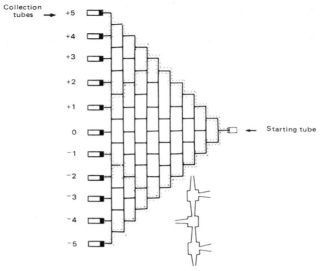

FIG. 1.3   Schematic diagram of the type of maze used to study *Drosophila* taxes. This particular maze has 10 choices, but others with 14 or 15 are routinely used. The inset shows a typical junction with the interconnecting one-way cones which prevent flies retracing their path. Dotted lines represent a modified maze used to study learning — see Chapter 4. (Derived from D. A. Hay (1975) *Nature* **257**, 44.)

This apparatus would appear to be an ideal method for behaviour genetic analysis. A large number of flies can be introduced into the maze and all the experimenter has to do is return the next day to find them neatly sorted according to their behavioural preference. In the 1960s this maze was used in many experiments that have been influential in our understanding of how genetic factors could affect behaviour (see Chapter 4). But was the maze measuring taxes?

Murphey and Hall (1969) studied flies selectively bred for 181 generations for negative geotaxis and found that in the course of the selection the flies had been bred also for:

1   Resistance to desiccation and/or starvation in a plastic environment. Flies may spend several hours negotiating the maze without any food or moisture and only the hardy ones survive.

2   Reduced locomotor activity in this environment, where flies which are too active tend to die.

3   Increased response stereotypy. To get a high taxis score, a fly must consistently turn in the same direction at each choice-point. The negative geotaxis flies do so, even in a horizontally placed maze where there are no gravity cues.

4   Low levels of 'claustrophobia', or reluctance to pass through those narrow one-way cones seen in the inset to Fig. 1.3.

None of these points affect the value of the geotaxis experiments in demonstrating how the inheritance of behaviour can be analysed, but highlight two very important aspects of validity. Firstly they limit possible comparisons between experiments measuring taxes in different ways and any extrapolations made from such data about the role of taxes in evolution. To quote from a major review of the problems and different experimental designs used in measuring *Drosophila* phototaxis:

Since designs may vary in their utility for research directed at diverse aspects of the (phototaxis) response, different designs must be used in accordance with the research interests of the investigator. However, when comparisons or generalizations are to be made, the operational nature of the measurement must be considered. In addition, when generalizations are made concerning the evolutionary or ecological significance of the response, it must be remembered that there is no *a priori* reason to suppose that one particular design provides a measure of phototaxis which is more closely akin to that which exists in nature than any other design. (R. F. Rockwell and M. B. Seiger (1973) *American Scientist* **61**, 339.)

Secondly, they demonstrate that an apparently fairly simple behaviour may involve many of what Murphey and Hall (1969) called 'task correlates' — other undetected behaviours that influence the final performance. The result is that any genetic analysis of a taxis is really the analysis of a complex of traits, quite possibly inherited to different extents and subject to different environmental influences. There is nothing amiss with such an analysis at the genetic level, but the behavioural interpretation is much more complex.

Open-field behaviour in rodents

The most widely used task in behaviour genetics research on rodents has undoubtedly been the open-field test (Fig. 1.4). The animal is exposed to a novel, featureless, brightly-lit arena and its reactions observed over a short period of time. The most usual measures are ambulation and urination plus defecation, the latter often being taken as a measure of 'emotionality'. In one of the most detailed justifications of any apparatus and procedure in behaviour genetics, Broadhurst (1960) quotes from an earlier text on the open-field, an account of an early Persian battle which is a unique validation of the concept of ambulation, urination and defecation as a reaction to stress. 'To save their lives they trampled over the bodies of their

Fig.1.4   The open-field test involves observing the rodent in an unfamiliar empty arena

soldiers and fled. Like young captured birds they lost courage. With their urine they defiled their chariots and let fall their excrement.' (p. 37.)

Despite extensive use of the open-field test and the many years spent selectively breeding both mice and rats for aspects of their performance in this situation, debate continues as to what the open-field actually measures. Archer (1973) has argued against the concept of emotionality in the open-field on the grounds that open-field performance does not relate sufficiently both to physiological measures of arousal, e.g. heart-rate, and to other behavioural measures where emotionality may be involved, e.g. learning to avoid an electric shock. His solutions are to study a wider range of behaviours in the open-field over a longer period of time or to introduce a far greater range of behavioural and physiological measures apart from the open-field, in both cases without necessarily assuming that they all reflect a single underlying concept of emotionality.

Walsh and Cummins (1976) review some of the more practical problems with the open-field, concerning the variety of ways in which measures can be made and the enormous effects these can have on the results. It should be noted that Broadhurst (1960) was at pains to spell out exactly how to conduct an open-field test, right down to the levels of illumination and of background noise. Figure 1.5 typifies how easy it is to alter open-field behaviour. The activity of mice varies both with the situation in which they are tested and with their experience prior to the test. To make matters worse the three genetically distinct strains of mice respond differently to the situation.

(a)

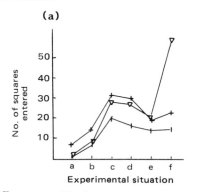

(b)

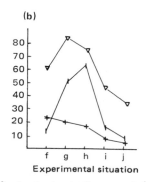

FIG. 1.5   The fluctuations in ambulation for 2 minutes in the open-field of mice from three genetic strains denoted as C57BL(▽), C3H(ı) and BALB/C(+). a–f represent six alternative variations in the format of the open-field and g–j represent four different pretest experiences. See reference for further details. (From D. A. Hay (1980) *Neuroscience and Biobehavioral Reviews* **4**, 489.)

So what use is the open-field test if it varies this much? Broadhurst (1975) demonstrated that two rat strains bred specifically for high and low defecation in a very rigidly controlled open-field situation differed also on a wide variety of other tasks where emotionality might be involved. Thus it seems that the genetic work can justify a behavioural concept which otherwise might be open to much criticism.

Readers interested in the general issue of behavioural measurement in rodents are recommended to Silverman (1978). Although based around drug research, this text is an excellent introduction to how to choose behaviours to study and the pitfalls in methodology and interpretation.

Intelligence testing

What the open-field test has been to rodent behaviour genetics, the intelligence test has been to human behaviour genetics. It has the highest reliability of any measure of human behaviour and certainly much better than the mental illnesses. Measurement can be made fairly rapidly. In fact if a *group* test is used, that is one where the person simply completes a test form — a so-called 'paper and pencil' test — then there is no limit to the number of people who can be tested at the one time. As distinct from the group test is the *individual* test, where an experienced tester deals with a single person at a time, often administering problems verbally and frequently timing how long the person takes to complete each item.

'It's about Benny, doctor. He's just come from school with an IQ of 104! Should I put him right to bed?'

FIG. 1.6   The problems of the IQ score. © 1955 Cowles Magazines, Inc. (Reproduced in *Psychological Corporation Test Service Bulletin* (1959) **54**.)

The problems of the intelligence test are epitomized in Fig. 1.6. Apart from the ethical issues of why intelligence is assessed in particular cases and why test results are released without adequate explanation, the IQ score summarizes ability into a single number. (IQ or intelligence quotient is the ratio of a child's mental age as assessed by an intelligence test to his or her chronological age. So the 10 year old with a mental age of 10 will have an IQ of 100. If he or she were mildly retarded with a mental age of only 5, the IQ would be 50. Because intelligence does not grow throughout life, a different formulation of IQ is used for adults (Brody and Brody 1976) but with equivalent meaning.)

The problems of a single IQ score are like those of a *Drosophila* geotaxis score — there are many ways of reaching the same result. For example, tests often involve items of graded difficulty with the later, more complex items carrying higher scores. So a person could hurry through all the easy items, ignore the more difficult ones and get the same score as someone who spends time solving the difficult problems and leaves the quick, easy ones. This speed versus power strategy is ignored in most formal IQ tests except the recently developed British Ability Scales. It is worth remembering that many of the IQ tests used in behaviour genetics were established quite early in the history of psychology. They predate much of our understanding of cognition, most especially the work of Piaget, whose research on the development of thought processes in the child has had enormous influences on both education and psychology.

An even more important issue than how people solve intelligence test problems is what the intelligence test measures. While scientists have argued throughout this century about data on the inheritance of intelligence, they have argued equally vehemently about how intelligence is structured. Is there a single general ability (often abbreviated to g) which underlies the diversity of items in tests or are there group factors determining specific skills and taking into account that a person can do well on verbal problems but no so well on numerical problems or *vice versa*? Brody and Brody (1976), Eysenck (1979) and Vernon (1979) all provide accounts of this issue from differing viewpoints. In particular, Eysenck discusses the extremes represented by Spearman who argued in 1927 for a two factor theory comprising g and also s factors

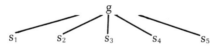

specific to each test, and Thurstone who argued in 1938 for a set of primary mental abilities or independent group factors including such things as numerical ability, memory, spatial ability and verbal relations. A more modern reconciliation is a hierarchical model of intelligence involving group factors as an intermediate level of organization

What implications do these psychological arguments have for behaviour genetics? Do we analyse a single IQ score as a measure of g

and assume that is the level at which genetics operates? Or do we assume that the group factors represent fairly independent units which can be inherited to different extents? If the latter situation is correct, then the genetic analysis of IQ scores is the analysis of a *potpourri* of quite distinct behaviours and the end result is confusion. The genetic methodology may be above reproach, but the behavioural measures to which they are applied are found wanting.

As we shall see in Chapters 2 and 7 the answer lies somewhere between the hierarchical and the group factor approaches. Genetically determined mental retardation affects most intellectual abilities, but some more than others. In the normal population detailed genetic analyses suggest that while there is a substantial genetic contribution to g, the genetic determinants of the various group factors differ to some limited extent. The situation resembles the circular argument of rodent open-field behaviour where the results of the genetic analysis help validate the behavioural measures used in the analysis.

These examples from *Drosophila*, rodents and humans illustrate solutions to the problem of how to obtain data from sufficient individuals for genetic analysis. In all cases, the measures are quick and reliable, but are they valid measures of the intended behaviour or are they obscured by Murphey's 'task correlates'? The problems with these measures are not specific to behaviour genetics. At the time they were introduced into psychology these measures were among the best available and it is a mark of their very importance that much work has gone into studying their limitations, revealing issues that could never have been anticipated. Among scientists, behaviour geneticists need to pay special attention to such examples because of the time and effort needed for their research. Other behavioural scientists may readily be able to change their experimental design but the behaviour geneticist may be locked into years of effort, breeding flies or mice or collecting data from human families and is not in a position to change half-way through an experiment.

In summary, the aim of much genetic analysis is to determine the extent to which genetic (G) and environmental (E) effects contribute to the external phenotype (P), a problem often written as $P = G + E$. All we can ever observe is P and it is from the genetic analysis of P that G and E are inferred. Everything hinges upon adequate measures of P, not ones which either (a) are confounded with other behaviours — the *Drosophila* mazes, (b) could perhaps have been measured in other ways less open to criticism and less sensitive to minor procedural changes — the rodent open-field, or (c) do not reflect current thinking in psychology and where some interpretations suggest characteristics differing in inheritance are combined in the one score — intelligence tests.

But so much effort in behaviour genetics has been devoted to these three measures that they cannot be dismissed. It may be better to think of them as models of what analyses are possible in behaviour genetics, but certainly not as models of how behaviour should be measured. To answer the obvious question of why behaviour genetics has developed this way with often less than adequate contact with changing methods and theories in psychology, we must examine its history, a topic explored fully by Vale (1980).

The history of behaviour genetics

Most philosophers of the past expressed some views on the inheritance of behaviour. Developing from the ideas of the Greek Stoics, Locke argued that the newborn was a 'tabula rasa', a blank slate on which experience wrote all the determinants of behaviour. On the opposite viewpoint, Descartes with his famous dictum 'Cogito ergo sum' (I think therefore I am) and Rousseau with his concept of man as 'the noble savage' were arguing far more for inborn influences on behaviour. It must be emphasized that these were philosophical statements based on debate not on experimental evidence. Yet they have stamped their influence on experimentation in psychology in general and in behaviour genetics in particular. Loehlin (1983) observes that Locke was far more willing to accept genetic differences in personality than in intelligence and speculates that this may be why we see so much controversy over the genetics of intelligence and much less over personality.

From Locke's view that ideas come to us through our senses and then interact with the mind grew the Associationist tradition. The Associationist philosophers with their highpoint in James Mill (1773-1836) considered thought as an association among sensations or between sensations and ideas which depended in turn on previously occurring sensations. This concept gave rise to the early psychological method of introspection where the person described his mental experiences and which became more formalized in the psychophysical methods of studying sensory perception. At this time there was no intelligence testing as such, but the view was that since everything came to us through our senses, sensory ability itself gave some indication of a person's abilities

The first person to begin using sensory tests from a genetic viewpoint was Sir Francis Galton in the latter part of the nineteenth century. Not only did Galton devise an extensive battery of sensory tests for colour vision, smell, taste, touch and auditory and visual acuity, he also developed methods of studying genetics based on twins and on pedigrees, particularly of eminent families. He did not have to look far, since one of his cousins was Charles Darwin whose ideas on evolution had caused so much consternation only a few years previously. From Galton's pioneering work in London came Karl Pearson's correlation methods for the statistical analysis of familial resemblance which are now used throughout science and indirectly also the first intelligence test.

In 1896 Binet and Henri in France criticized the emphasis on sensory function favoured by Galton and his students, such as the American, James McK. Cattell who had established a testing laboratory in Pennsylvania in 1890. Out of their ideas came the first intelligence test as we know it, the Binet–Simon test of 1905, designed to screen out Paris school children needing special education. Following the international adoption of these tests and the work of people such as Spearman, Burt, Vernon and their American counterparts Thurstone, Guilford and R. B. Cattell, has developed the field of psychometrics, the measurement of individual differences in behaviour and the analysis of their structure and determinants (Willerman 1979).

Psychometrics has had its share of controversy over the role of inheritance in intelligence and the social and political consequences arising therefrom, but it has always recognized genetic analysis as a viable research area. In complete contrast, a second influence of the philosophers led to the virtual elimination of behaviour genetic research on animals other than man for many years. The problem stemmed from the theories of the philosophical schools of psychology in existence early this century. Predominant among these were Freud's psychoanalysis which stressed such inborn determinants of behaviour as thanatos, the death instinct, and Gestalt psychology, a mainly European school concerned with perception and innate perceptual mechanisms. The most visible school in the USA was McDougall's hormic psychology which although useful in stimulating much experimental work, went too far in enumerating instincts or inherited dispositions responsible for the impetus and direction for all activities. In 1908 McDougall specified 12 instincts to explain behaviour and by 1930 this had grown to 17. Other members of McDougall's school had up to 118 instincts.

It is too easy to propose instincts to explain behaviour and not surprisingly there was a backlash spearheaded by Watson (1924) in his book *Behaviorism* (Norton, New York), from which comes the most famous quote in modern psychology.

I should like to go one step further now and say, 'Give me a dozen healthy infants, well-formed, and my own specified world to bring them up in and I'll guarantee to take any one at random and train him to become any type of specialist I might select — doctor, lawyer, artist, merchant-chief and, yes, even beggarman and thief, regardless of his talents, penchants, tendencies, abilities, vocations, and race of his ancestors'. I am going beyond my facts and I admit it, but so have the advocates of the contrary and they have been doing it for many thousands of years.

Watson was reacting against two things. One was the imprecise nature of the data provided by introspection and by the psychoanalytic interview. He preferred animal research for which interview methods are obviously impractical and where objective measurement is the only possible approach. At the same time he overreacted to the instinct concept, denying the role of genetics in behaviour. It was unfortunate that Watson's stressing of objective measurement became linked in the minds of so many psychologists with the idea that all behaviour was environmentally determined. Altering behaviour and dealing with individual differences was just a question of finding and manipulating the correct environmental factors.

The ideas of Watson and his successors in the Behaviourist movement, most notably Skinner, held sway over experimental psychology up until 1960 and the rediscovery of behaviour genetics. It is no wonder then that as we saw earlier the behaviour geneticists of the early 1960s showed an almost missionary zeal in pointing out examples of genetic influences on behaviour.

One topic for the historian of psychology is how psychology maintained two such disparate views in the period 1920–60. On the one side were the Behaviourists with a blanket denial of individual differences and possible genetic determinants, while on the other, the psychometricians and also anthropologists (Freeman 1983) were having a raging battle, the nature–nurture controversy, over the causes of individual and racial differences. Block and Dworkin (1976) review this period where the invective of the Lippmann–Terman debate makes modern scientific writing seem pallid. To take a sample of Terman's scorn for environmental effects on intelligence.

And just to think that we have been allowing all sorts of mysterious, uncontrolled, chance influences in the nursery to mold children's IQs, this way and that way, right before our eyes. It is high time that we were investigating the IQ effects of different kinds of baby talk, different versions of Mother Goose, and different makes of pacifiers and safety pins (p. 37.)

Block and Dworkin also reprint Karier's discussion of the impact at that time of the eugenics movements. Galton had initially proposed eugenics as 'the science which deals with all influences that improve the inborn qualities of a race, also with those that develop them to the utmost advantage'. Early this century groups advocating eugenics became very vocal in the USA, dedicated to improving the 'genetic quality' of Americans, positively through encouraging the most able to breed, negatively through discouraging the least able. As well as limiting entry into the USA of migrants from those ethnic groups supposedly of lower intelligence, negative eugenics extended to mass sterilization of the 'morally and sexually depraved', a category which at its most absurd included those convicted of stealing chickens or cars.

It is worth stressing the very limited evidence on which their ideas were based. The most influential work concerned the Kallikak family (Fig. 1.7). Apart from emphasizing the supposedly genetic basis to the feeble-minded descendants of the 'tavern girl' (many of whom were classified as such on the basis of anecdotal evidence), the history of this family also perpetuated the idea that mental retardation, criminality and immorality were all interconnected.

We would not accept such interpretations of the evidence nowadays and some historians have queried whether or not they were actually needed at the time. The political pressure groups trying to change US society by promoting such measures as selective migration would still have argued along the same lines irrespective of any scientific or other evidence about genetics and intellectual ability, such was the climate of the times.

Martin Kallikak

He dallied with a feeble-minded tavern girl

He married a worthy Quakeress

She bore a son known as 'Old Horror' who had ten children

She bore seven upright worthy children

From 'Old Horror's' ten children came hundreds of the lowest types of human beings

From these seven worthy children came hundreds of the highest types of human beings

FIG. 1.7   The two families of Martin Kallikak. (From H. E. Garrett (1955) *General Psychology*. American Book Company, New York.)

It is unfortunate that people such as Kamin (1974) still link behaviour genetics with the evidence, philosophies and motives of that period. The data we consider in this text are very different from the subjective reports of 60 or 70 years ago. Controversies over the objectivity of earlier data continue. In Chapter 6 we examine the work of Sir Cyril Burt on identical twins reared apart which was influential support for the genetic viewpoint but which is now recognized as hopelessly flawed and perhaps fabricated. The faults were not only on the part of the supporters of the biological viewpoint. Freeman (1983) has criticized the work of the famous cultural anthropologist, Margaret Mead and others have drawn parallels between her actions and those of Burt. When Mead in the 1920s described a very free lifestyle among Samoan adolescents in contrast to the more stressful situation in western society, her results were heralded as evidence for cultural determinism rather than for any biological factors which presumably should apply worldwide. But Freeman argues that she falsified the situation, interpreting the evidence selectively in terms of her ideology that culture was all-important.

Other disciplines and behaviour genetics

As well as being influenced by psychometrics and Behaviourism, other disciplines have influenced behaviour genetics. At the same time as the Behaviourists were becoming supreme in the USA, a very different approach to behaviour, ethology was developing in Europe. In its early days it was predominantly a topic for zoologists interested in observing behaviour of varied animal species in the field and with a considerable emphasis on 'innate' behaviour. Such innate behaviours differed greatly from the three behaviour genetic examples discussed earlier. Pioneering ethologists such as Lorenz and Tinbergen emphasized patterns of behaviour which were species-specific and, once initiated, fixed and independent of external cues. An example of such a *fixed action pattern* is the way in which grey-lag geese roll eggs into the nest — even if the egg rolls out of reach, the stereotyped movement sequence is completed minus the egg. When such behaviours are called 'innate', the implication is that they are genetically determined and develop in all members of a species irrespective of individual experience.

While this may seem like behaviour genetics there are fundamental differences. Behaviour genetics is predominantly concerned with explaining individual differences *within* species, usually by some breeding program in animals and by the study of familial similarities in humans. Traditional ethology emphasized behavioural differences *between* species and was less interested in variation within species. By definition species are groups which do not interbreed, so only very rarely was it possible to use breeding experiments. Figure 1.8 is one of the few examples of species hybrids. Whereas the Cape Teal (Fig. 1.8a) extends the neck forward to make the burp call which is part of male courtship, the yellow-billed pintail keeps the head erect (Fig. 1.8b) and the hybrid between them holds the neck diagonally. But apart from their rarity, species hybrids are often abnormal in behaviour and fertility and cannot be the basis of genetic analysis. The main method for determining whether a behaviour was 'innate' was the deprivation experiment, rearing animals in isolation from experience and seeing if the behaviour pattern still eventuated.Experience could still play a part, but its effects were limited by the innate constraints.

The distinction between ethology and behaviour genetics is typified in imprinting, a form of learning in young animals where exposure at a particular age to a particular cue such as the colour or call of their mother has almost irreversible consequences for their behaviour. While ethologists are primarily concerned with the evolutionary significance of imprinting, behaviour geneticists are more interested in whether members of the same species differ in the strength of imprinting and the ease with which an imprinted preference can be changed. Immelmann (in Royce and Mos 1979) demonstrated that zebra finches differ considerably in the age at which their imprinting becomes resistant to further social influences. Similarly, Kovach (1979) has carried out an elaborate series of breeding

(a)

(b)

FIG. 1.8   Posturing during the burp call of (a) the Cape teal and (b) the yellow-billed pintail. (From P. A. Johnsgard (1965) *Handbook of Waterfowl Behaviour*. Cornell University Press.)

experiments in quail, showing genetic variation in 'innate' preferences for particular colours and in the degree of imprinting to these colours.

Considerable conflict existed between ethologists and experimental psychologists who objected to the small scale and apparent subjectivity of some ethological research as well as to the deprivation experiment. Rearing an animal in isolation tells one a lot about the effects of solitary confinement but very little about inheritance. At about the same time that experimental animal behaviour genetics became accepted once more, a reconciliation developed between ethology and psychology with an appreciation that both genetic and experimental factors can determine a particular behaviour (Beer 1973).

One legacy of the conflict between ethology and psychology is that some scientists still believe in the either-or situation. Once some environmental effect has been found, then any genetic determinants can be totally dismissed. As we see in Chapter 8 this simplistic view is still commonly applied to human behaviour, although it also occurs in animal experiments. Consider the following from a study on the relative rates at which a food pecking response could be extinguished in parrots from temperate and semiarid regions of Australia, that is, how quickly do they stop working when not rewarded with food. The question was whether differences between the two groups reflected genetics, environment or both factors:

After manipulating the rearing environments of genetically similar samples of rats, *Rattus norvegicus*, Gluck and Pearce (1977) demonstrated group differences in resistance to extinction when the animals were tested as adults, suggesting that different early environmental experiences could be sufficient to account for the parrot data. (J. S. Watson and P. J. Livesey (1981) *Australian Journal of Psychology* **33**, 245)

Apart from the fact that information on American laboratory rats may have little to do with Australian parrots, note how the demonstration of one environmental effect is sufficient to exclude all genetic explanations.

Recently a new discipline, sociobiology, has developed which draws upon ethology, psychology, behaviour genetics, ecology and evolutionary and population biology. The theme of sociobiology is an understanding of social behaviour in an evolutionary context. Data are gathered from a wide variety of animal species and applied to the fundamental question of 'What are the best strategies in social behaviour for an individual to get his or her genes into the next generation?' Everything from the caste system in social insects to the different social structures of baboon and orangutan communities can be interpreted in this evolutionary framework.

What is surprising are the different emphases of sociobiology and behaviour genetics, given that sociobiology revolves around evolution and evolution depends on there being genetic differences within a species on which natural selection can act (see Chapter 2). Sociobiology has paid very little attention to within species differences, focussing more on differences in social systems between species. While behaviour genetics is concerned with much more than social behaviour, the different approaches of the two disciplines can be demonstrated with the example of learning. Learning is a problem for sociobiology. To quote from the text which initiated sociobiology:

Viewed in a certain way, the phenomenon of learning creates a major paradox. It seems to be a negating force in evolution How can learning evolve? Unless some Lamarckist process is at work, individual acts of learning cannot be transmitted to offspring. If learning is a generalized process whereby each brain is stamped afresh by experience, the role of natural selection must be solely to keep the *tabula rasa* of the brain clean and malleable. To the degree that learning is paramount in the repertory of a species, behaviour cannot evolve. This paradox has been resolved in the writings of Niko Tinbergen, Peter Marler, Sherwood Washburn, Hans Kummer, and others. What evolves is the directedness of learning — the relative ease with which certain associations are made and acts are learned, and others bypassed even in the face of strong reinforcement. (E. O. Wilson (1975) *Sociobiology : The new synthesis*, p. 156, Harvard University Press, Mass.)

(Lamarckism was a pre-Darwinian view of evolution and inheritance in terms of the use and disuse of body parts influencing what was passed to the next generation.) While sociobiology is content to explain learning in terms of 'biological preparedness', we see in Chapters 4 and 5 that behaviour geneticists approach learning in terms of why some animals learn when others do not and address specific questions. Are there biochemical or physiological differences underlying the genetic differences in learning? If learning is impaired on one task, is performance on other learning tasks also affected? Are all types of learning ability inherited to the same extent? If we breed animals that can or cannot learn, what other behaviours are affected?

Despite these differences, behaviour genetics cannot ignore sociobiology. The extension of sociobiology to human behaviour has led to much controversy and the emergence of a very vocal group of scientists, objecting to all research including human behaviour genetics which they see as favouring biological determinism.

The past 10 years have seen a reemergence of such particularised biological determinist explanations of social problems . . . The list is a long one. Genetics and IQ to deal with racism and unemployment, XYY — the criminal chromosome — to deal with crime, genetics and sex role differences to deal with sexism, drug 'therapies' for 'hyperactive' children, and leucotomies for enraged women and prisoners are among the wares plied by scientists of different disciplines offering explanations for the irrevocability of the ills of a society predicated upon divisions based on race, sex and class. (British Society for Social Responsibility in Science (1976) *New Scientist* **70**, 348.)

Many scientists find unconvincing those sociobiological explanations of human behaviour based on extrapolation from other animals. But at least in the case of behaviour genetics the simplest answer to such emotional attacks is to consider the data dispassionately. In Chapter 2 we see that the connnection between XYY and criminality has gone out of favour, not because of invective like that in the quotation but because more and better scientific data became available. Chapter 7 and particularly Chapter 8 quash the idea that genetics is a means of maintaining the *status quo* distribution of resources between social classes and between races. In many ways the opposite is the case and environmental explanations of behaviour are responsible for denying resources to those who would most benefit.

As was pointed out at the beginning of this chapter, the best way to approach such issues is through an understanding of behaviour genetics across a wide range of species and behaviours and not to follow some of the critics in focussing on just a small area of research in isolation.

Conclusion

Behaviour genetics is the product of many scientific developments. Human behaviour genetics grew alongside psychometrics and the study of individual differences. Although the emphases on genetic determinants of human behaviour have fluctuated during this century with changing social attitudes (Cronbach 1975), interest in research never died completely. Public debate on genetics and intelligence erupted again in 1969 with the publication of Jensen's article on the problems of compensatory education for disadvantaged minorities and the suggestion that genetically determined racial differences in ability might be involved along with environmental factors (Chapter 8). The arrival of sociobiology has ensured that the arguments will continue.

Behaviour genetic research on animals almost disappeared during the Behaviourist period but has fully recovered to overlap both with experimental psychology and with ethology. Hirsch (1963) distinguished these three disciplines by ethology being involved with differences between species, and behaviour genetics with differences within and between species, while experimental psychology overlooked all differences both within and between species. Such distinctions are no longer clear. Skinner (1966) claims that Behaviourists have always accepted species differences and genetic differences within species, but many of his arguments appear to reinforce rather than correct Hirsch's view. On the other hand, Thiessen (1972) argues in the journal *Behavior Genetics* that behaviour genetics has been too preoccupied with differences within species (which he unkindly refers to as 'genetic junk') and should pay more attention to differences between species as examples of how genetic mechanisms have been responsible for behavioural adaptation.

Although his views may not be accepted by other behaviour geneticists, he does see behaviour genetics as having a crucial role in integrating research:

Behavioural evolution has been left primarily to the ethologists, who, in spite of their impressive accomplishments, lack genetic sophistication and an appreciation of laboratory techniques, and species-specific analyses have been the province of comparative psychologists, who for the most part are unconcerned with individual differences and ecological adaptation. Each discipline has of course made experimental and theoretical progress, but rarely have their views been sufficiently broad to include the significant principles of all the separate disciplines. It is the field of behaviour genetics, in my opinion, that can forge links between the various disciplines and provide unification within the broad framework of evolutionary principles. (p. 116.)

References

General behaviour-genetics texts

Dixon L. K. and Johnson R. C. (1980) *The Roots of Individuality: A survey of human behavior genetics.* Brooks/Cole, Monterey.
Ehrman L. and Parsons P. A. (1976) *The Genetics of Behavior.* Sinauer, Massachusetts.

Ehrman L. and Parsons P. A. (1981) *Behavior Genetics and Evolution.* McGraw-Hill, New York.

Fuller J. L. and Thompson W. R. (1960) *Behavior Genetics.* Wiley, New York.

Fuller J. L. and Thompson W. R. (1978) *Foundations of Behavior Genetics.* Mosby, St. Louis.

McClearn G. E. and DeFries J. C. (1973) *Introduction to Behavioral Genetics.* Freeman, San Francisco.

Parsons P. A. (1967) *The Genetic Analysis of Behaviour.* Methuen, London.

Plomin R., DeFries J. C. and McClearn G. E. (1980) *Behavioral Genetics: a Primer.* Freeman, San Francisco.

Thiessen D. D. (1972) *Gene Organization and Behavior.* Random House, New York.

Vale J. R. (1980) *Genes, Environment and Behavior: An interactionist approach.* Harper and Row, New York.

Conference proceedings and other edited volumes on behaviour genetics

van Abeelen J. H. F. (ed.) (1974) *The Genetics of Behaviour.* North-Holland, Amsterdam.

Ehrman L., Omenn G. S. and Caspari E. (eds) (1972) *Genetics, Environment and Behavior: Implications for educational policy* Academic Press, New York.

Fuller J. L. and Simmel E. C. (eds) (1983) *Behavior Genetics: Principles and applications.* Erlbaum, Potamac, Md.

Hirsch J. (ed.) (1967) *Behavior-Genetic Analysis.* McGraw-Hill, New York.

Hirsch J. and McGuire T. (eds) (1982) *Behavior-Genetic Analysis.* Vol. 16 in the Benchmark Papers in Behavior Series. Hutchinson Ross, New York.

Kaplan A. R. (ed.) (1976) *Human Behavior Genetics.* Thomas, Springfield, Ill.

Manosevitz M., Lindzey G. and Thiessen D. D. (eds) (1969) *Behavioral Genetics: Method and research.* Appleton-Century-Crofts, New York.

Oliverio A. (ed.) (1977) *Genetics, Environment and Intelligence.* North-Holland, Amsterdam.

Royce J. R. and Mos L. P. (eds) (1979) *Theoretical Advances in Behavior Genetics.* Sijthoff and Noordhoff, Alphen aan den Rijn.

Texts on genetics and intelligence (for those dealing specifically with race and intelligence, see Chapter 8)

Block N. and Dworkin G. (eds) (1976) *The IQ Controversy: Critical readings.* Random House, New York.

Brody E. B. and Brody N. (1976) *Intelligence: Nature, determinants and consequences.* Academic Press, New York.

Cancro R. (ed.) (1971) *Intelligence: Genetic and environmental influences.* Grune and Stratton, New York.

Eysenck H. J. (1979) *The Structure and Measurement of Intelligence.* Springer-Verlag, Berlin.

Kamin L. J. (1974) *The Science and Politics of IQ.* Erlbaum, Potomac, Md.

Taylor H. F. (1980) *The IQ Game: A methodological inquiry into the heredity-environment controversy.* Rutgers University Press, New Brunswick.

Vernon P. E. (1979) *Intelligence: Heredity and environment.* Freeman, San Francisco.

Willerman L. (1979) *The Psychology of Individual and Group Differences.* Freeman, San Francisco.

Additional references

Archer J. (1973) Tests for emotionality in rats and mice: a review. *Animal Behaviour* **21**, 205.

Beer C. G. (1973) Species-typical behavior and ethology. In *Comparative Psychology: a Modern Survey,* Dewsbury D. A. and Rethlingshafer D. A. (eds) p. 21. McGraw-Hill, New York.

Broadhurst P. L. (1960) Experiments in psychogenetics: applications of biometrical genetics to the inheritance of behaviour. In *Experiments in Personality,* vol.1. *Psychogenetics and Psychopharmacology* Eysenck H. J. (ed.) p. 1. Routledge and Kegan Paul, London.

Broadhurst P. L. (1975) The Maudsley reactive and nonreactive strains of rats: a survey. *Behavior Genetics* **5**, 299.

Cronbach L. J. (1975) Five decades of public controversy over mental testing. *American Psychologist* **30**, 1.

Freeman D. (1983) *Margaret Mead and Samoa: The making and unmaking of an anthropological myth.* Australian National University Press, Canberra.

Hirsch J. (1963) Behavior genetics and individuality understood. *Science* **142**, 1436.

Kovach J. K. (1979) Genetic influences and genotype-environment interactions in perceptual imprinting. *Behaviour* **68**, 31.

Lockard R. B. (1971) Reflections on the fall of comparative psychology: is there a message for us all? *American Psychologist* **26**, 168.

Loehlin J. C. (1983) John Locke and behavior genetics. *Behavior Genetics* **13**, 117.

Murphey R. M. and Hall C. F. (1969) Some correlates of negative geotaxis in *Drosophila melanogaster. Animal Behaviour* **17**, 181.

Seligman M. E. P. (1970) On the generality of the laws of learning. *Psychological Review* **77**, 406.

Silverman P. (1978) *Animal Behaviour in the Laboratory: Behavioural tests and their interpretation illustrated mainly by psychopharmacology in the rat.* Chapman and Hall, London.

Skinner B. F. (1966) The phylogeny and ontogeny of behavior. *Science* **153**, 1205.

Thiessen D. D. (1972) A move towards species-specific analyses in behavior genetics. *Behavior Genetics* **2**, 115.

Walsh R. N. and Cummins R. A. (1976) The open-field test: a critical review. *Psychological Bulletin* **83**, 482

Wilcock J. (1972) Comparative psychology lives on under an assumed name — psychogenetics! *American Psychologist* **27**, 531.

2 Mental retardation and the principles of genetics

Topics of this chapter

1 The basic terminology and principles of genetics (new terms are italicized the first time they are used).
2 The three ways (chromosome aberrations, single gene disorders and polygenic effects) by which genetics can contribute to mental retardation.
3 The variability in degree of retardation, in specific cognitive skills, in rate of development and in physical defects which accompany different genetic disorders.
4 The physical, biochemical and behavioural means of diminishing the consequences of mental retardation for the afflicted individuals and for their relatives.
5 The factors which maintain the genetic potential for mental retardation in the population.

This chapter describes the basic principles of genetics relevant to studies of the inheritance of behaviour. We saw in Chapter 1 that surprisingly little of modern genetics is relevant, mainly because the problems and inaccuracies of behavioural assessment are incompatible with sophisticated and detailed genetic analysis — the phenotype we observe in behaviour genetics is often too far removed from the underlying genotype. Therefore the emphasis in this chapter is on mental retardation, a behaviour which can be classified more readily than most and where there are often underlying biochemical disorders more amenable to genetic analysis.

Mental retardation illustrates a behavioural problem that can arise for different reasons. Apart from the three genetic determinants that are the topic of this chapter there are also environmental determinants ranging from such specific events as early meningitis or traumatic injury to the head through to less clearly defined socio-cultural factors (Table 2.1). The very large proportion currently retarded for unknown reasons should decrease with new developments in everything from biochemical analysis to evaluation of the pre- and postnatal environment. One such example is the recent appreciation of the fetal alcohol syndrome and the behavioural problems in children whose mothers drank alcohol while pregnant.

TABLE 2.1   Causes of mental retardation (Derived from L. S. Penrose (1974)
The Biology of Mental Defect (4th edn). Sidgwick and Jackson, London.)

Chromosome aberrations	Down's syndrome	10%	
	Other autosomal syndromes	2%	
	X and Y syndromes	3%	
			15%
Single gene	Autosomal recessive	5%	
	Autosomal dominant	1%	
	X-linked	1%	
			7%
Polygenic			15%
Environmental effects	Specific diseases or injuries	5%	
	Various	15%	20%
Unknown			43%
Total			100%

Definition of mental retardation

Detailed classifications of mental retardation are described in Clarke
and Clarke (1974) but explanation and classification can be confused.
For example, Allen and Allen (1975, p.1) define the category of
'cultural familial mental retardation' as one 'in which there is
evidence of intellectual inferiority in at least one parent and in one or
more siblings. The cultural deprivation presumably stems from
inadequate stimulation provided by parent or parents'. Such a pattern of family resemblance is equally compatible with a genetic explanation and later in this chapter we shall consider the
work of Zigler and others in distinguishing these two possible
determinants.

The very concept of mental retardation is anathema to many
people who claim that there is no benefit in assigning such a label to
people. In particular Mercer (see Cleland 1978) argues against
labelling because (a) it is relative (a person may be called mentally
retarded in one social system but not in another), (b) it is accepted
unthinkingly as an explanation ('he cannot do this task because he is
retarded'), and (c) it puts the blame on the person not on a society
which provides inadequate assistance.

No one would dispute the dangers of misclassifying the child who
is in the category of 'borderline retarded' or 'borderline intelligent' as
it is more positively known. In the 1979 San Francisco law suit of Larry
P versus Wilson Riles, Judge Peckham ruled that standardized IQ
testing to detect such children was unconstitutional since it dis-
criminated against ethnic minorities (see Chapter 8) and led to these
children receiving only a 'limited, dead-end education', not one
which may foster the abilities they did possess.

But many of the examples in this chapter concern people who are
far more severely handicapped, where denying both the label
'mentally retarded' or the more recent alternative 'intellectually
handicapped' and the specialized care that should be available as a
result, would do more harm than good. The levels of retardation asso-

ciated with particular IQ levels may be appreciated better by considering some of their practical consequences as listed in Table 2.2.

Just how much attention should be paid to the exact IQ score is a matter for careful interpretation, since it can be influenced by the sensory and physical handicaps that often accompany more severe retardation. In addition IQ is not everything and social competence must also be considered. Clarke and Clarke (1974) point out that, although 3% of the population have IQs less than 70 and hence the potential to be retarded, less than half of these (at least in the UK)

TABLE 2.2   Levels of retardation and behavioural expectancies at three age levels (modified from *President's Panel on Mental Retardation Chart Book*, Department of Health, Education and Welfare, 1963)

Level	IQ	Preschool: age 0–5 Maturation and development	School: age 6–21 Training and education	Adult: 21 and over Social and vocational adequacy
Profound IV	<25	Gross retardation; minimal capacity for functioning in sensorimotor areas; needs nursing care	Obvious delays in all areas of development; shows basic emotional responses; may respond to skilful training in use of legs, hands and jaws; needs close supervision	May walk, need nursing care, have primitive speech; usually benefits from regular physical activity; incapable of self maintenance
Severe III	25–39	Marked delay in motor development; little or no communication skill; may respond to training in elementary self-help, e.g. self-feeding	Usually walks barring specific disability; has some understanding of speech and some response; can profit from systematic habit training	Can conform to daily routines and repetitive activities; needs continuing direction and supervision in protective environment
Moderate II	40–54	Noticeable delays in motor development, especially in speech; responds to training in various self-help activities	Can learn simple communication, elementary health and safety habits, and simple manual skills; does not progress in functional reading or arithmetic	Can perform simple tasks under sheltered conditions; participates in simple recreation; travels alone in familiar places; usually incapable of self maintenance
Mild I	55–69	Often not noticed as retarded by casual observer, but is slower to walk, feed self and talk than most children	Can acquire practical skills and useful reading and arithmetic to a third to sixth grade level with special education. Can be guided toward social conformity	Can usually achieve social and vocational skills adequate to self maintenance; may need occasional guidance and support when under unusual social or economic stress

require special educational or other assistance because of their inability to cope intellectually. The rest 'function within the limits of the community's tolerance in a welfare state' (p. 24).

Having indicated some of the current issues involved in defining retardation, we can consider the three ways in which genetics can contribute to it.

Chromosome aberrations

The chromosomes

The *chromosomes* are the structures within the cells of the body containing all the genetic material. They can be seen quite easily in a sample of *somatic* cells, the name for any cells other than the *gametes*, the collective term for sperm and eggs. In practice, white blood cells are generally used for somatic chromosome preparations, but skin or other tissues are employed in certain cases. When the cells are suitably cultured, chemically treated and stained, the chromosomes become visible under the microscope as darkly staining structures. The normal human chromosome complement or *karyotype* consists of 46 chromosomes, with other species having different characteristic numbers, for example 42 in the rat and 78 in the dog.

We refer to 46 as the *diploid* chromosome number in humans, to indicate that in females at least, these 46 can be sorted into 23 pairs on the basis of absolute size and relative length of the two arms around the central region, the *centromere*. The arrangement in males is slightly different, as can be seen by comparing the two parts of Fig. 2.1. Forty four (or 22 pairs) of a male's chromosomes are similar to those in females and are called *autosomes*. These have no direct influence on sex determination in contrast to the remaining two chromosomes, the sex chromosomes. In females these consist of two X chromosomes, which are *homologous*, matching in size and shape. In males there is only one X, the other chromosome being a very small, largely non-homologous Y chromosome. Thus the normal male is denoted formally as 46,XY and the normal female as 46,XX where the 46 represents the total chromosome complement and the XX or XY the sex chromosomes. Two extensions to this notation are to call the long arm of each chromosome q, and the short arm p (for 'petite') and to identify numerically the specific region of each arm (see Yunis 1977 for details). For example, a female with *Cri du chat* Syndrome (Fig. 2.8) where part of the short arm of chromosome 5 has been deleted is missing is 46,XX, del(5) (p 13).

Our information on human chromosome structure is surprisingly recent. It was only in 1956 that improved techniques revealed that humans had 46 chromosomes rather than the 48 as previously believed. Major advances occurred in the 1970s with the widespread adoption of several chemical techniques that would differentially stain portions of the chromosome arms. Since specific regions of each

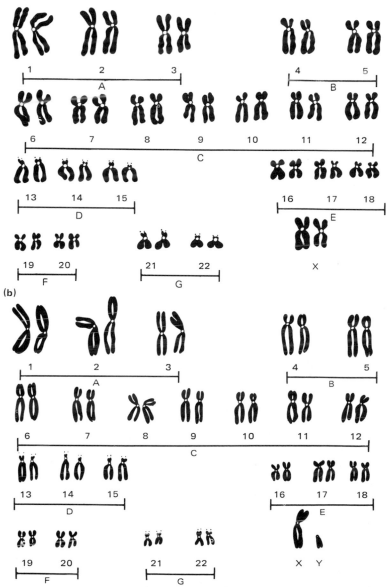

(a)

1 2 3
A

4 5
B

6 7 8 9 10 11 12
C

13 14 15
D

16 17 18
E

19 20
F

21 22
G

X

(b)

1 2 3
A

4 5
B

6 7 8 9 10 11 12
C

13 14 15
D

16 17 18
E

19 20
F

21 22
G

X Y

FIG. 2.1   The karyotypes of (a) a normal female, 46XX and (b) of a normal male, 46XY made with a traditional staining technique. The chromosomes are arranged by size into seven groups, A–G. The X is part of group C and the Y of group G. (Modified from V. A. McKusick (1964) *Human Genetics* (1st edn). Prentice-Hall, Englewood Cliffs, N.J.)

chromosome can be located this way, it is far easier to identify some of the different chromosome aberrations that can occur. For example when there is *trisomy* (three copies rather than two of a particular chromosome) or *monosomy* (one copy rather than two), which chromosome pair is involved? Until staining techniques were available, there was debate as to whether the more common form of Down's Syndrome (mongolism) involved trisomy 21 or trisomy 22. Figure 2.9a shows chromosomes stained by the Giemsa technique of a Down's Syndrome male where the bands clearly indicate 21 is involved.

Similarly Giemsa staining confirmed that the male in Fig. 2.5 with 49 chromosomes had four X chromosomes and not an abnormal number of autosomes. Staining techniques also help to identify the less obvious aberrations including: *deletion* — where part of a chromosome is missing, *duplication* — two copies next to each other of one part of a chromosome, *inversion* — reversal of part of a chromosome and *translocation* — the attachment of part of one chromosome to a member of another chromosome pair. Apart from deletions (Fig. 2.8) and translocations (Fig. 2.9b) we shall not be considering these other aberrations in detail, since individually they are too uncommon for their effects on behaviour to have been adequately documented.

However our knowledge of chromosome aberrations is expanding rapidly, the number of known disorders having increased from 12 in 1971 to at least 41 in 1976 (Yunis 1977). Before we can examine several of the more common aberrations, we must first consider some aspects of how chromosomes function.

Chromosomes and cell division

The existence of chromosome pairing is fundamental to the process of genetics as it permits the integration of the genetic information from the sperm and the egg.

The process involved in *gametogenesis* (production of the sperm and eggs) is called *meiosis* and must be distinguished from *mitosis*, the process involved in the duplication of somatic cells. In mitosis (Fig. 2.2a) the chromosomes divide and then separate to form two identical daughter cells each with the diploid chromosome number. In meiosis the sequence is more complex (Fig. 2.2b) and involves two stages of division to produce *haploid* sperm and eggs with only one copy from each chromosome pair. When these fuse at fertilization, they produce a *zygote* (meaning 'together') with the original diploid number.

The exact details of meiosis involve many stages of little immediate concern here except as they affect sex differences in the timing of meiosis, the errors of cell division during meiosis and the sex ratio. We can examine the sex differences by considering the fate of a single pair of homologous chromosomes. In producing sperm or eggs, these chromosomes go through two divisons in developing from the *spermatocytes* and *oocytes*, diploid cells derived from cells originating shortly after fertilization. Sperm production begins at puberty and continues until death, the entire cycle from spermatocyte to mature sperm taking 64–72 days, with many new cells beginning the process every day. The two divisions in *spermatogenesis* mean that four sperm are produced for every meiosis of a spermatocyte.

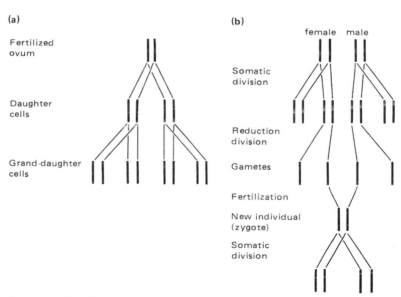

Fertilized
ovum

Daughter
cells

Grand-daughter
cells

(b)

female   male

Somatic
division

Reduction
division

Gametes

Fertilization

New individual
(zygote)

Somatic
division

FIG. 2.2   The difference between (a) mitosis and (b) meiosis in a single pair of chromosomes. (Derived from J. A. Fraser Roberts (1970) *An Introduction to Medical Genetics* (5th edn). Oxford University Press.)

The sequence in females differs in two respects: (a) half the cells called *polar bodies* are discarded at each division, so that in the end there is one mature egg and three polar bodies and (b) the process of division begins before birth and recommences after puberty on a 28-day cycle with the final division only occurring at this stage.

With the whole process of female meiosis taking as much as 40 years, there is a possibility of misdivision and something going wrong with an older egg. To what extent this is due simply to the passage of time or to hormonal changes as menopause approaches is unclear, but several chromosome aberrations, notably the trisomy form of Down's Syndrome and also Klinefelter's Syndrome (see later) are more common in children born to older mothers. The father's age may matter but to a much lesser extent. The other major effect of maternal age is an increased incidence of nonidentical twinning, which (see Chapter 6) results from the independent fertilization of two eggs. So older women are more likely to release more than one egg, as well as to have something go wrong with the eggs they do produce.

The most common misdivision is *non-disjunction* where two chromosomes go to the same pole of the dividing cell rather than to opposite ends. This mistake leads to cells with no copy of that particular chromosome or with an extra copy of that chromosome. Figure 2.3 illustrates what can happen when there is non-disjunction of the X chromosome, leading to two trisomies 47,XXX or 47,XXY and one monosomy 45,XO, where the O indicates a missing chromosome. The other monosomy, 45,YO is inviable since no fetus can survive without at least one X chromosome and the genetic information it contains.

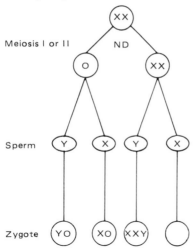

Primary oocyte

Meiosis I or II   ND

Sperm

Zygote

FIG. 2.3   Non-disjunction (ND) of the X chromosomes during meiosis of an oocyte, (O indicates the egg with no Xs). Fertilization by an X or Y bearing sperm produces four possible zygotes of which the 45, YO is inviable. (From H. Zellweger and V. Ionasescu (1978) in *Medical Aspects of Mental Retardation* (2nd edn) Carter C. H. (ed.), p. 123. Thomas, Springfield, Ill.

The final point about cell division concerns the sex ratio. In a normal meiosis, eggs must carry an X chromosome while sperm have a 50:50 chance of having an X or the much smaller Y chromosome. However, the sex ratio at conception (the *primary* sex ratio) is not equal with there being approximately 120 male conceptions for every 100 females. The ratio changes so that at birth the *secondary* sex ratio is 106 males:100 females and by the age of 33 the *tertiary* sex ratio is 100:100, with females predominating thereafter. The change between the primary and secondary sex ratios indicates that males are at greater risk than females before birth. Whether this sex difference in susceptibility to prenatal problems also explains the 30–50% greater incidence of mental retardation in males is considered later in this chapter.

The X and Y chromosomes

Until the new banding techniques became widely available, disorders of the X chromosomes were the best understood because of a phenomenon called *Barr bodies*. To properly karyotype an individual takes time and experience because many of the chromosomes are of similar size. But the Barr body allowed the rapid screening of many individuals for the number of X chromosomes and is still the basis of sex testing of athletes. If somatic cells are taken, usually by scraping inside the mouth, and stained, we get the results shown schematically in Fig. 2.4a and b for normal females and males respectively. The

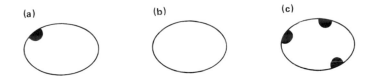

(a)   (b)   (c)

FIG. 2.4   Barr bodies in different karyotypes. (a) The normal 46,XX female
with one Barr body, (b) the normal 46,XY male with no Barr bodies and (c) a
male who is 49, XXXXY with three Barr bodies.

general rule is that the number of Barr bodies is one less than the
number of X chromosomes, so that a 49,XXXXY male with four Xs has
three Barr bodies (Fig. 2.4c) and the karyotype shown in Fig. 2.5a.
Examples of the sort of survey made feasible with rapid Barr body
screening are discussed by Kessler and Moos (1973). Among newborn
males, 2.2 per 1000 are *sex chromatin-positive*, that is, show one or
more Barr bodies. The frequency is 9.4 per 1000 among males in
mental retardation institutions and 5.4 per 1000 in general mental
hospitals, indicating some link with psychosis.

The most widely studied X chromosome disorder from the
behavioural aspect is *Turner's Syndrome* (45,XO) where the females
in most cases lack one of the X chromosomes and hence have no Barr
bodies. The major external characteristic is the folds of skin on the
neck (Fig. 2.6) while internally they usually lack a uterus and have
only rudimentary gonads. The incidence of Turner's Syndrome is
unclear, but is probably one in 4000 female births. What is certain is
the high miscarriage rate with 98% of embryos with Turner's
Syndrome being lost in early pregnancy.

Non-disjunction during meiosis usually in the father is responsible, but Turner's Syndrome can also result from the loss of an X in
the first few cell divisions after fertilization. That is, a normal 46,XX
zygote loses an X so that in subsequent divisions those cells
descended from the one with the missing X will be 45,XO and the
remainder normal 46,XX. The extent of such *mosaicism* depends on
the number of cells when the X chromosome was lost. The fewer cells
at that stage, the greater the proportion that will be derived from the
45,XO cell.

We do not know if mosaicism contributes to the variation in intellectual skills seen in Turner's Syndrome. Two of Turner's original
seven patients were academically outstanding, but the average IQ
tends to be in the 80s. The question we have to constantly consider as
we discuss chromosomal and other disorders is the lack of uniformity
in the effects on IQ. Does the variation occur simply because the
defect is superimposed upon the range of IQs in the normal
population with IQs between 130 and 70? If a defect reduces ability by
20 points, the IQ 130 person still scores 110 and is above average, while
the IQ 70 person now has a score of 50 and would be classed as
moderately retarded. One bias which may creep in is that the sufferer
with above average ability may go unnoticed. In the absence of

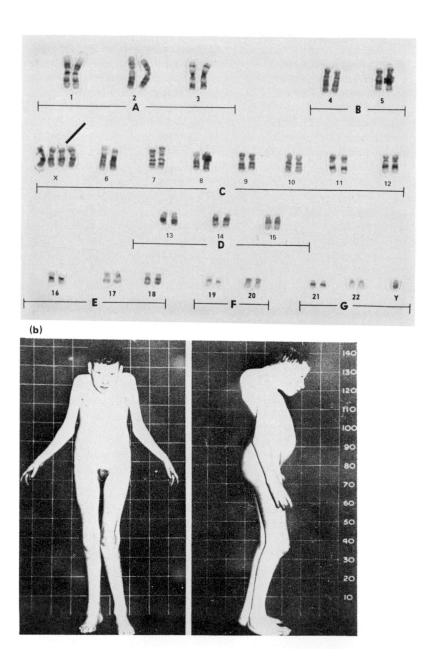

(b)

FIG. 2.5 (a) Karyotype of a 49,XXXXY male with a group of four X chromosomes marked. This karyotype was prepared with Giemsa stain, which gives a distinctive banding pattern to each set of homologous chromosomes — compare with the classical staining technique used for Fig. 2.1. Note that the chromosomes are arranged in six groups labelled A–G, according to overall size and position of the centromere. (Preparation courtesy of Dr Graham Webb, Genetics Department, Royal Children's Hospital, Melbourne.) (b) A 13 year old 49,XXXXY male with severe vertebral deformities, clubfoot and small testes. (From J. Jancar, (1964) in *International Copenhagen Congress on Mental Retardation*, Oster J. (ed.) vol. 1, p. 179.)

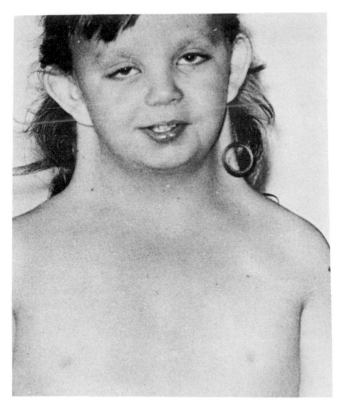

Fig. 2.6   A girl with Turner's Syndrome, showing the typical webbing of the neck. (From R. J. Gorlin and H. Sedano (1971) *Modern Medicine of Australia* June 21, p. 14.)

clearcut physical symptoms only those with marked deficits in ability are likely to be karyotyped and otherwise studied in order to determine the cause of their disability. This bias will obviously enhance the chances of finding an association between the particular defect and mental retardation and emphasizes the need for prospective studies of unselected newborn children. To achieve adequate sample sizes, such studies must obviously be enormous. Six studies (summarized in Mange and Mange 1980) karyotyped a total of 56 952 newborn boys and girls and found 353 abnormal, an incidence of 1 in 161, but only four Turner girls.

With this qualification in mind, we can consider a point about the ability of Turner women which is of more interest than the minor decrement in overall IQ and involves a specific deficit in spatial and temporal ability. Briefly, spatial ability (see Chapter 7) comprises orientation in space and time and performance on those parts of intelligence tests which do not involve solving problems by verbal means. Examples would be tracing a maze or a picture completion task where the person has to find the important missing elements in a drawing. Even when matched with a control sample of women with similar overall IQ scores, Turner women do much worse on such tasks, so that their verbal skills must be superior if they are to be the same in overall performance. Their temporal deficit is evidenced by poor discrimination of rhythm and by the inability to keep tapping at a constant rate, although the significance of these effects is complicated by the hearing impairment found in at least 70% of Turner women. It has recently been argued that both the spatial and temporal deficits reflect an underlying disorder in visual memory and motor co-ordination.

The failure to undergo puberty means that Turner women must also cope with determining their gender identity. The majority develop essentially female behaviour despite the absence of ovaries and can be helped by hormone substitution therapy. In this they are assisted by the observation (Nielson et al 1977) that Turner women are remarkably emotionally stable and have a high tolerance for adversity. This compliant attitude may be more than an adjustment to the problems they have to face. A frequent suggestion is that their pattern of spatial deficit and of emotional stability is consistent with damage to the parietal lobe of the brain on the nondominant side, which in most people is the right-hand side. Oversimplifying, the dominant hemisphere of the brain is usually the left one which controls language skills and is connected to the right hand — hence the preponderance of right-handed people — while the right hemisphere is more concerned with spatial and musical skills. Recent evidence suggests more general right hemisphere dysfunction plus frontal lobe effects on attention span (Pennington and Smith 1983).

If Turner's Syndrome normally results from non-disjunction involving the passage of no X chromosome to one gamete, there should also be people who receive the two X chromosomes from the other half product of the non-disjunction, namely 47,XXY and 47,XXX. 47,XXY is *Klinefelter's Syndrome* where the person is essentially male but with very small testes and often some breast development. There is not the same high spontaneous abortion rate which accompanies Turner's Syndrome and so the incidence is higher — one in 1000 in newborn males, one in 100 in males in mental insti-tutions and as high as one in 20 in men attending infertility clinics. The infertility cannot be treated, but treatment with testosterone can bring on the physical changes associated with puberty in males and alleviate their low self-esteem.

The cognitive deficit, while still not major, is larger in most patients than in Turner's Syndrome. The interesting difference from Turner's Syndrome is that the pattern of deficit is reversed, with Klinefelter patients having adequate nonverbal IQ but a variety of verbal difficulties such as the inappropriate usage of words and poor sentence construction. However, the differences between the syndromes may not be distinct and a poor sense of direction may explain the inability of Turner women to order objects and numbers and of Klinefelter men to order words in sentences.

The other karyotype involving non-disjunction of the X chromo-some is the 47,XXX female. Apart from frequent speech disorders, little is known about any specific cognitive deficits but such women are two to three times more likely to be institutionalized as retarded or psychotic (mainly schizophrenic) than their one in 1000 frequency in the population would suggest. The extra X chromosome may not be the direct cause however, since prematurity, birth complications and a family history of inbreeding and malformations have all been found more frequently among these women than would be expected.

People do survive with even higher numbers of X or Y chromosomes than Fig. 2.5 indicates, although physical handicap usually accompanies such a karyotype (Holmes *et al* 1972). Some degree of retardation is common with these karyotypes and Moor has gone so far as to chart the decline in ability with increasing number of Xs or Ys (Fig. 2.7). This decline is consistent with each extra chromosome but does not imply a direct, causal connection between extra chromosomes and low ability. For example, birthweight is reduced by 300 g for each extra X and by somewhat less for extra Ys, so low birthweight and the complications it brings may also contribute to the reduced intelligence. It must also be remembered that these karyotypes are very rare and are most often searched for among the abnormal population. The chances would be very slim of finding a 49,XXXXY male with no marked physical or mental disabilities even if such a person existed, so it is just possible that Moor's results merely represent the greater chance of finding a very rare karyotype when it is accompanied by marked abnormalities.

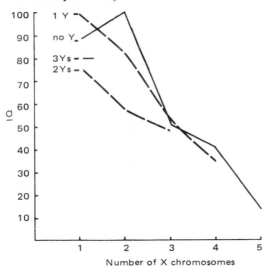

FIG. 2.7   Mean IQ of individuals with abnormal numbers of sex chromosomes. (The entire sample size is only 374, so some of the values are based on very few people.) (From L. Moor (1967) *Revue de Neuropsychiatrie Infantile et d'Hygiene Mentale de l'Enfance* **15**, 325.)

Figure 2.7 summarizes only total IQ scores and not specific cognitive deficits where there may also be consistent effects. In a review of such deficits in sex chromosome anomalies, Pennington and Smith (1983) propose that all disorders with an extra sex chromosome (XXY, XXX and also XYY — see below) are associated with poor speech and language development.

The question of biased sampling raised with Moor's results also applies to the 47,XYY male. This karyotype has received much publicity and the unfortunate epithet of 'supermale' because such men are some 10–15 cm taller than average and are claimed to be more likely to be aggressive or criminal. Or to be precise, compared with an XYY incidence of one in 1080 in newborn males, they are much more likely to be found in prisons (one in 316), in mental institutions (one in 223) or especially in security institutions such as Broadmoor in England which house the 'criminally insane' (one in 51) (Mange and Mange 1980). However the incidence varies greatly between the studies and is complicated by the frequent practice of surveying only males of above average height, since these are much more likely to have the XXY karyotype. This is not to imply that XYY males are more aggressive simply because they are taller and hence can be more confident of winning any fight in which they become involved.

So much is uncertain about the XYY male and is likely to remain so because of the many sampling difficulties. Mange and Mange (1980) review these and discuss the recent survey by Witkin which involved all Danish males born between 1944 and 1947 and subsequently seen by the draft board. Of the 12 XYY males detected among the 4139 males who were more than 184 cm tall, five had criminal records but only for offences involving property not persons. The conclusion was that all XYY males but especially the five 'criminals' were of below average intelligence. Since criminals in their control (46,XY) population also scored lower on the particular intelligence test, they concluded that males of lower intellectual ability were more likely to be convicted of crimes, irrespective of karyotype. Note that this does not mean they are more likely to commit crimes, but only that they are more likely to be caught and successfully prosecuted.

The other dilemma is whether or not it is worth pursuing such research. Of the three problems arising with such research the first is the cost. Remember that this disorder involves an extra Y and not an extra X chromosome, so that it is not detected by Barr body staining and full karyotyping must be used. Even Witkin's massive sample was still not enough. Only five XYY males with criminal records were found which is hardly sufficient for major conclusions. Secondly even if an adequate project could be funded, it would have to be prospective, that is, following the growth of 47,XYY males from birth to see if they demonstrated aggressive tendencies. This close scrutiny could result in a 'self fulfilling prophecy', the reason why such a study in Boston in 1975 was halted. Thirdly even if a link to aggression were demonstrated, what consequences should follow? Should all pregnancies be terminated where a 47,XYY karyotype is detected? Should such males of necessity be treated differently by the courts when and if they commit violent crimes? In 1968 there were two murder trials, one in Melbourne and one in Paris where the 47,XYY karyotype was offered as mitigation.

Whatever the outcome, the importance of the 47,XYY karyotype should not be exaggerated, as Mage observed:

Overwhelming statistical evidence indicates that the XY karyotype is associated with major social problems such as violent crime and war. If we are to provide medical and psychiatric assistance to XYY individuals, let us not neglect the XYs, who in aggregate present a much greater problem to the community. (*New Scientist* (1975) **65**, 270.)

The last point before we turn to the autosomal disorders concerns one way in which the sex chromosomes and the autosomes differ markedly. People survive tolerably well with several additional X or Y chromosomes, whereas even a small addition to the autosomes can have severe consequences. Why is there such a difference? It comes back to the Barr bodies and to a process called *Lyonisation* after Lyon who initially proposed this idea to explain her observations on coat-colour in cats and mice. In essence, the Barr body in the normal 46,XX female represents the condensation and inactivation of one of her X chromosomes. This inactivation begins early in development (by the sixteenth day after fertilization) and means that the female becomes a mosaic with some cells having the X derived from her father active, while others have the maternally derived X. Lyon's example was the calico cat which can only be female and where some patches of hair are orange and others black according to which X chromosome is active. Patterns of biochemical activity in blood and skin cells have confirmed that this hypothesis applies equally well to humans.

Although this inactivation explains why an abnormal number of Xs can be tolerated because only one X is left active, Turner's and Klinefelter's Syndromes show that there are effects on the embryo of the abnormal number of Xs in the weeks before inactivation. To complicate things it now seems likely that only the long arm, q, of the X chromosome is inactivated because symptoms of Turner's Syndrome are also seen in women who are 46,XX, del(X)(p).

Autosomal aberrations

We can illustrate how much more severe are the effects of autosomal aberrations with two examples, one where only a very small piece of one chromosome is absent and one where a small addition is present. The former is *Cri du Chat Syndrome*, 46,XX or XY, del(5)(p13) where part of p, the short arm of chromosome 5 is missing (Fig. 2.8). *Cri du Chat* or cat-cry refers to the high-pitched mewing or wailing due to a combination of several defects in the throat including a very narrow larynx. Profound retardation with IQ less than 10 is common and such people rarely learn to speak, while walking frequently is not achieved until the age of 10. Until recently it was difficult to separate this karyotype from 46,XX or XY, del(4)(p), but banding techniques now make this possible. A separate del(4)(p) syndrome is recognized where there are severe facial anomalies but not the distinctive cat-cry.

A contrasting disorder where a small fragment of extra chromosome is present is the *Cat Eye Syndrome* (47,XX or XY,

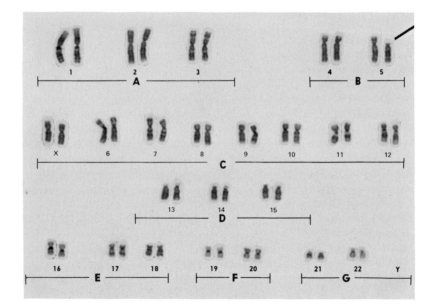

FIG. 2.8   Karyotype of a female with *Cri du Chat* Syndrome — 46, XX, del (5)(p13). Note that only part of the short arm, p, is missing from chromosome 5. The most recent nomenclature terms this part p13, or subdivision 3 of region 1 of p. (Preparation courtesy of Dr Graham Webb, Genetics Department, Royal Children's Hospital, Melbourne.)

+der(22)). The *der* is the notation for an extra small chromosome, probably derived from part of 22 (Hsu and Hirschhorn in Yunis, 1977). The name Cat Eye comes from the vertical pupil of the eye like that of a cat and although motor and perceptual development is retarded, major mental retardation is uncommon.

While many other chromosome aberrations have been described only recently and are not fully documented, some are showing far more consistent phenotypes than others. For example, del(18)(q) is always associated with severe mental retardation, but del(18)(p) is very variable in the degree of retardation with IQs ranging from a borderline 75 down to a severe 35.

Down's syndrome

The best known chromosome disorder is undoubtedly *Down's Syndrome* (DS) originally called Mongolism. At the time he coined the term in 1866, Dr Down implied more than just a similarity to oriental facial features. With the cultural prejudice common to his era he believed that the facial features and the impaired intelligence were a reversion to a lower form of human life, in this case to the Orientals. He also had other categories of retardation corresponding to American Indians, Africans and Malayans but fortunately these have not persisted.

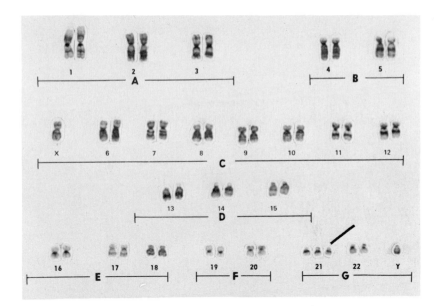

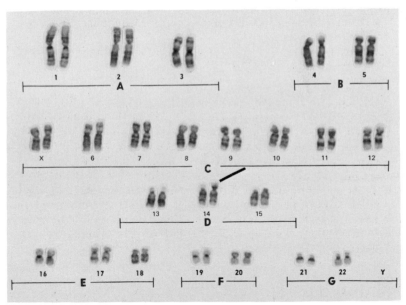

Fig. 2.9   The karyotypes of Down's Syndrome (a) the trisomy 47, XY, +21 and (b) the translocation 46, XX, −14, +t(14q21q). (Preparations courtesy of Dr Graham Webb, Genetics Department, Royal Children's Hospital, Melbourne.)

Two types of DS must be distinguished. Ninety five percent of all DS patients are 47,XX or XY, +21, that is, they have a trisomy of chromosome 21 (Fig. 2.9a). This type of DS occurs more frequently in children born to older mothers (Fig. 2.10), since only 16% of all babies are born to women over 35 but 60% of all trisomy DS babies. The two more recent surveys among those in Fig. 2.10 demonstrate an even greater incidence of DS with advancing age, perhaps because of increased exposure to environmental hazards such as radiation which can enhance the frequency of chromosome aberrations or because of a higher survival rate of DS babies born to older mothers, given that only 30% of DS fetuses survive until birth. Whichever turns out to be the correct explanation, increased risk in older women is assuming greater importance with the trend to delay families until the woman is in her late 20s or early 30s.

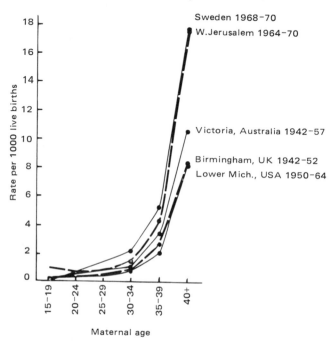

FIG. 2.10   The incidence of Down's Syndrome at different maternal ages. (From Z.A. Stein and M. Susser (1977) in *Research to Practice in Mental Retardation Vol. III: Biomedical aspects* Mittler P. (ed.), p. 45. University Park Press, Baltimore.)

The other 5% of DS individuals result from a different aberration involving a translocation between chromosomes 14 and 21. They are 46,XX or XY,−14,+t(14q21q) which indicates that an extra 21 is positioned at the end of the long arm of chromosome 14 (Fig. 2.9b). This type of DS does not arise through non-disjunction and is no more common in older mothers. Instead it runs in families and can be detected by karyotyping the parents. If one parent is 45,XX or XY, −14,−21,+t (14q21q), they have the normal chromosome complement but not in the usual positions. Some of their children will not survive since they have only one chromosome 21, but depending on the outcome of meiosis (Fuchs 1980) others will survive as normals, as translocation DS, or as *carriers* (like the parent with the translocation and the risk of carrying it to the next generation).

The problems of a DS child start before birth when their growth rate after 24 weeks gestation slows relative to normal. From birth they

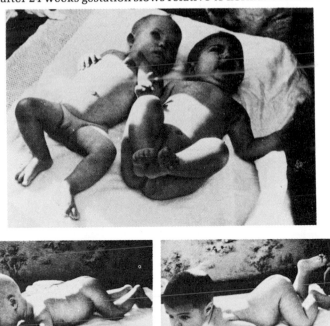

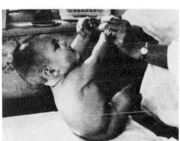

FIG. 2.11   A pair of non-identical twins, where the one on the left has Down's Syndrome. Note how he does not lift his head and limbs in the way his normal brother does. (Modified from V. A. Cowie (1970) *A Study of the Early Development of Mongols*. Pergamon, Oxford.)

differ in such areas a muscle tone, since they cannot support their head or limbs as well (Fig. 2.11). However, mental retardation is not marked in the first 3 years but thereafter gradually increases in severity (Fig. 2.12). Translating these results into means, Morgan reports an average IQ of 72.9 in 1–year old DS infants, dropping to an average 26.8 in DS children above the age of 11. Given that tests of 'intelligence' in very young children are really more tests of sensori-motor coordination and development than of cognitive abilities, this change with age may imply that those popular, remedial programs based on intensive sensori-motor stimulation early in life may not necessarily avert later cognitive deficit.

Figure 2.12 indicates considerable variability among DS children. The range of IQs was 35–107 for children below 1 year of age and 10–55 for children older than 11. It is imperative that we understand why DS children differ so much, not only for the purposes of remedial help, but more for a reassessment of prenatal diagnosis. Detecting and terminating a DS fetus destined to be only borderline or mildly retarded may be regarded very differently from the case of a fetus likely to be profoundly affected. What techniques exist to explain or to predict the variation in DS?

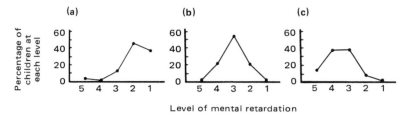

FIG. 2.12   Development and distribution of combined scores on tests of intelligence and of social maturity for Down's Syndrome children of different ages (a) 0–3 years, n = 77, (b) 3–7 years, n = 91, (c) ⩾ 7 years, n = 49. Table 2.2 describes the levels of mental retardation except that level 1 here includes the bottom of the normal IQ range. (From S. B. Morgan (1979) *Mental Retardation* **17**, 247.)

Mosaicism

A mosaic DS individual is one who has 46,XX or XY and also 47,XX or XY, + 21 cells. In the 1960s the first reports appeared of individuals with DS features but with normal or near normal abilities who turned out to be mosaics after karyotyping many cells. Although the proportion of normal cells varies from 5–95% in different mosaic individuals, there is little evidence that ability is related to this proportion. One survey found one girl with only 6% normal cells but an IQ of 93, and another girl with 91% normal cells with an IQ of 60, despite very similar intensive tutoring by her family.

In general mosaics outperform ordinary DS, the discrepancy between the two groups increasing with age. To a large part this reflects the greater verbal skills of the mosaics. Apart from vocabulary and comprehension, their actual speech is better modulated and the intonation improved relative to the droning, unclear speech typical of DS. Since they are able to communicate their problems, mosaics are more likely to receive directed assistance. There is more to the mosaic's advantage over other DS children than language, since they can also perform more adequately on visual–perceptual tasks involving copying shapes and drawing a person, where verbal skills matter less.

Family background

There is conflicting evidence that, among noninstitutionalized DS children, those with higher IQ scores have parents with higher IQs and/or higher educational attainments. It is as if DS reduces performance by a set amount, so that those otherwise likely to be of above-average intelligence will be less retarded than those likely to be of lower ability.

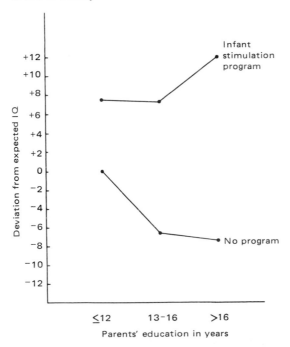

FIG. 2.13   Performance on the Bayley Scales of Infant Development or the Stanford–Binet Intelligence Scale (depending on age) for Down's Syndrome children according to average education level of the parents. Results are expressed as deviations from the average IQ expected in Down's Syndrome children of that age grouping. (Derived from F. C. Bennett, C. J. Sells and C. Brand (1979) *American Journal of Diseases of Children* **133**, 700.)

Figure 2.13 shows a more complicated picture. Children of better educated parents are no higher in IQ when the decrease of ability with age is taken into account. On the other hand, these children benefited the most from an early stimulation program designed to improve their cognitive and social development. Such a difference between groups in their response to a treatment occurs frequently in behaviour genetics and is given the name *genotype-environment interaction*. That is, we explicitly recognize the possibility that genotypes differ in their response to particular environments or conversely, that the effects of a particular environmental treatment depend upon the genotype.

Sex

Although both boy and girl DS patients decline in ability with age, boys are consistently five to seven IQ points lower than girls of the same age. One suggested explanation is that since in the normal population girls usually do better on verbal abilities and boys on nonverbal abilities (see Chapter 7), the specific verbal disability of DS individuals may disadvantage boys more than girls. Another possibility involves the higher mortality of young DS girls, with the severely affected girls dying but boys in a similar situation surviving, albeit with severe retardation.

Behavioural intervention

While mosaicism, family background and sex of the child are factors beyond control, one source of variability in DS children can be manipulated, namely the amount of specific training they receive. Until recently such training came from the family and several cases are reported of exceptional efforts by the parents leading to gratifying results. 'Mr. Bolt' was born in the early 1900s and when tested in the 1950s had an IQ of 83 with good general knowledge. For example he knew the meaning of 'autumnal equinox' and that Goethe wrote 'Faust'. More recently another able DS man wrote his autobiography (*The World of Nigel Hunt* (1967). Garrett Publication, New York).

However, there are two reasons why excessive importance may have been placed on these cases as examples of what intensive training can achieve. Firstly both men lacked the ability to conceptualize or generalize, a point which is obvious if one examines the style of Nigel Hunt's book. Secondly karyotyping of both these men is obscure — there are various reports that Nigel Hunt was a mosaic or a translocation DS — so we remain uncertain as to whether their prowess is likely in the usual trisomy DS.

More recently, there has been a spate of projects concerned with early education of DS children. Many of them are flawed in obvious ways, such as not karyotyping or sampling only from better educated families where there is one parent not in the workforce and hence able to devote more time to the DS child. In a critical review of 105 studies, Rynders *et al* (1978) found only 15, totalling 103 people, with adequate karyotyping, behavioural assessment and consideration of the decrement of ability with age. Their main conclusion is that the considerable variability within each of the three DS karyotypes (Table 2.3) limits the value of the predictions frequently made to parents of new born DS children about subsequent cognitive performance and the advisability of institutional placement.

This variability also limits assessment of the various intensive early intervention programs for DS children. There is the added complication of genotype-environment interaction (Fig. 2.13) where not all children respond similarly to the program and a more fundamental difficulty of uncertainty as to the bases of the programs. Some are concerned with improving muscle tone and early sensori-motor

TABLE 2.3   Average IQ and range of scores in Down's Syndrome children (derived from J. E. Rynders, D. Spiker and J. M. Horrobin (1978) *American Journal of Mental Deficiency* **82**, 440)

DS karyotype	Number of DS	Average IQ	IQ range
Trisomy 21	39	45	18–75
Translocation	14	53	28–85
Mosaic	50	57	14–100

coordination in the belief that improvement in more cognitive skills will follow automatically, while others emphasize improving verbal skills so that the DS child can give and receive more information. Others again emphasize social skills so that the child may be more able to interact with normal children and perhaps even be 'mainstreamed' into the normal school system. Given the variability in the children and in the programs, it is premature to support the frequent claim that all DS children are educable if they receive the correct intervention.

Discussion of chromosome aberrations

This section has shown that disorders of the chromosomes are linked to mental retardation. We do not know if other behaviours are affected, because only intelligence test performance can be assessed adequately in the retarded. Tests used to measure other characteristics such as personality are usually verbal, limiting their usefulness with all but the normal and the mildly retarded, e.g. the work on the personality of Turner girls. Tests are now available to assess social competence in the retarded, but often measure reliably only language development and domestic and vocational activities that relate closely to intelligence. Their measures of conduct disorders such as violent, stereotyped or unacceptable behaviours are not yet adequate. Although preoccupation with the intellectual deficits of the retarded is often criticized, much more development of suitable tests is needed before other behaviours can be reliably related to particular chromosome aberrations.

In the case of disorders involving the X or Y chromosomes, a fairly consistent picture emerges where the greater the deviation from the normal chromosome number, the greater the intellectual retardation. Nothing as obvious emerges with the autosomal aberrations, although the banding techniques have clarified many issues.

Scope for future development is indicated in a final chromosome example, X-linked mental retardation. Gerald (1980) traces the development of this research from the initial observation that there are some 30–50% more men than women in institutions for the mentally retarded. At first the reason was thought to involve undiagnosed single-gene disorders on the X chromosome of the sort discussed in the next section. Then in 1969, Lubs reported an association between mental retardation in males and the presence of an X chromosome with a fragile site. The end part of the long arm was joined to the rest of the chromosome by a thin stalk or constriction which sometimes breaks (Fig. 2.14). Only in 1977 was it appreciated that this phenomenon does not occur in the culture media used subsequent to 1969 to grow the cells. Once this was recognized, the commonness of this disorder became apparent. The fragile X comes usually from the mother and has also been found in other female relatives. Fragile X males are generally infertile, with one of the first fertile X males being found in Melbourne in 1981. X inactivation means that fragile site effects on the abilities of women are less serious and more variable, depending on which cells in which organ have the fragile X functional.

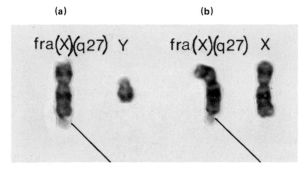

FIG. 2.14   The sex chromosomes of a male with the fragile X syndrome (a) and his sister (b), showing the restriction on the X chromosome. Note the difference between the normal and abnormal X chromosomes of the female. In this case the male is moderately retarded while his sister is of borderline intelligence. (Preparation courtesy of Dr Graham Webb, Genetics Department, Royal Children's Hospital, Melbourne.)

The IQ of fragile X males is usually in the range 50–65 although several males of normal IQ have been reported and conversely very low IQ males have been found in institutions. Fragile X males do much better on tests of verbal than of nonverbal intelligence. This is the opposite of what is found among other males with nonspecific X-linked mental retardation and also among normal males (Chapter 7). Detailed psychological assessment has focussed on language functioning revealing deficiencies in tasks where heard information has to be processed, while information coming through vision or touch is handled more successfully (Hagerman and McBogg 1983). There are problems in speech production and many fragile X boys are so slow in developing language that they are labelled autistic. The mistakes they make in speech are similar to those made by normal children but more frequent, implying that they result from impaired normal development rather than from any specific deficit which would be expected to produce more bizarre effects. However, a recent survey of fragile X males in Melbourne revealed that in figure copying tasks rather like that shown in Fig. 2.17, these men produced unusual patterns quite unlike those seen in normals of similar mental age. This survey also showed that they found it almost impossible to do digit span tasks where numbers have to be repeated back to the tester. They knew the numbers and they could perform adequately on tasks involving memory of pictures, so they do seem to have some specific disabilities in number memory.

Given that almost nothing was known about fragile X until 1977, a quotation from Gerald (1980) indicates just how rapidly our understanding of genetics in mental retardation can change.

It is now clear that the fragile-X syndrome is of major importance. Next to trisomy 21, it is the most common of the causes of mental retardation that can be specifically diagnosed. Because of its familial nature, this genetic defect directly concerns even more persons than trisomy 21 concerns. (p. 697.)

Single-gene disorders

Mendelian genetics

We only have 23 pairs of chromosomes but there are far more than 23 traits where heredity plays a part. Therefore, each chromosome must contain the genetic information relevant to many different characters and we use the term *gene* to refer to the particular chromosome region determining one such character. The techniques of *gene mapping* indicate on which chromosome a particular gene is located. To illustrate what we mean by each chromosome containing many genes, near the tip of the long arm of the X chromosome are the *loci* or physical locations of a group of four genes, one for green colour blindness, one for red colour blindness, one for haemophilia (uncontrolled bleeding due to the lack of a blood-clotting factor) and one for a chemical glucose-6-phosphate dehydrogenase (G6PD). African and Mediterranean populations often lack this chemical and suffer severe anaemia when given certain antimalarial drugs, such as primaquine, or when they eat or inhale the pollen of fava beans.

One implication of this example is that to produce the different phenotypes there must be different *alleles* or distinct forms of the gene, a normal form and an abnormal form leading to colour-blindness, absence of the blood clotting factor, deficiency of G6PD or whatever the gene might control. Some genes have more than two alternative alleles. For example, people may have the blood group A, AB, B or O, depending on the combination of the three A, B and O alleles at the same locus.

The role of alternative alleles in inheritance was first described by the Austrian monk, Gregor Mendel in 1865, although his work remained unrecognized until its independent rediscovery in 1900 by three European botanists. Mendel's contribution to genetics came from the realization that the contemporary 'blending' theory of inheritance could not be correct. If the genetic information possessed by each parent were simply blended together, there would be less and less variation in each successive generation. What was needed instead was a 'particulate' theory of inheritance which would preserve the uniqueness of each parent's contribution.

It was not until 1880 that the movement of chromosomes during meiosis was described and Mendel had to devise his theory of inheritance on the basis of inference from breeding experiments with garden peas. Mendel postulated that each inherited trait was controlled by two factors (what we would now call alleles of the same gene) and that each parent contributed one of the alleles to the new zygote. Figure 2.2b shows that this system follows the process of meiosis. Suppose that each of a pair of homologous chromosomes carries one allele for a particular trait. Then each gamete will have one of the alleles, the pair being restored at fertilization by a corresponding allele from the other parent. Mendel's theory can be summarized in his three principles.

1 Random segregation   When gametes are formed, each receives only one of the two alleles, the choice being made at random. That is, there is a 50:50 chance that a particular gamete will receive a particular allele.

2 Independent assortment   The alleles for each trait are inherited independently of those for other traits.

3 Purity of gametes   The alleles remain unchanged from one generation to the next.

Although these rules still apply, we now know enough to realize that Mendel was fortunate in that the loci for each of the seven traits he studied were on different chromosomes of the pea's seven pairs. If this had not been the case, he would have had to cope with the complication of *linkage*, where, rather than assorting independently, alleles on the same chromosome will tend to be inherited as a group. The situation is even more complex in that when the homologous chromosomes pair during meiosis, there can be *crossing-over* or *recombination* after the first somatic division (Fig. 2.2b), so that part (a) from one chromosome links to part (b) from its homologue and vice versa, breaking up the groups of alleles. The basis of gene mapping is that the closer together two loci are on the same chromosome, the lesser the chance of a crossing-over occurring. Although linkage and recombination are basic concepts in introductory genetics, we need not consider them further here because they have rarely been examined in behavioural studies.

The only other genetic concept we require before we can begin to discuss single genes and mental retardation is that of *dominance* or *recessivity*. Before Mendel began to cross his peas he first demonstrated that the strains bred true and contained no genetic variation — if a strain was self-pollinated, then the progeny would all be like the parent. Nowadays we call such strains pure or *homozygous* from 'homo' meaning 'same'. The two alleles in each member of the strain are identical, so that every gamete and hence every individual is the same. When two such homozygous strains, say for wrinkled and round peas, are crossed, a *heterozygote* ('hetero' meaning 'different') is produced. What Mendel saw was that the heterozygote was not intermediate between the parents and a little wrinkled, but that it was as round as the round parent. The round allele was masking the effect of the wrinkled one and Mendel coined the terms 'dominant' and 'recessive' to describe these. The round allele is dominant over the wrinkled one, or conversely, wrinkled is recessive to round. Mendel therefore appreciated that the phenotype does not necessarily indicate the genotype, since both heterozygotes and round homozygotes are similar in appearance.

The concepts of dominance and recessivity apply equally to humans. For example, people can be blood group B through being homozygous with two B alleles or heterozygous with one B and one O, since B is dominant to O. Similarly A is dominant over O, so that AA and AO are phenotypically the same, but A and B are *codominant* with the AB heterozygote differing from both the AA and BB homozygotes.

Single genes and mental retardation

Although exact estimates vary, humans probably have more than 10 000 genes and in theory it is possible for almost any of these to have an abnormal allele. The number of single-gene disorders being identified as such is increasing by almost 100 every year, so that the 370 disorders known in 1956 had grown to over 3000 by 1980. Table 2.4 is based on 1971 data and indicates only half the present number but reveals one major implication for work with mental retardation. Of the 10% of single-gene disorders with mental retardation as one symptom, the great majority (83%) are autosomal recessive, whereas among the other disorders only 37% are inherited this way. As we work through examples of each mode of inheritance we shall see why autosomal recessive inheritance causes particular problems in predicting who is likely to have affected children. Chapter 4 provides one evolutionary hypothesis as to why mental retardation is more likely to be autosomal recessive.

TABLE 2.4   The inheritance of single gene disorders (derived from G. Lindzey, J. Loehlin, M. Manosevitz and D. Thiessen (1971) *Annual Review of Psychology* **22**, 39)

	Autosomal dominant	Autosomal recessive	X-linked	Total
Without mental retardation as one symptom	786 (55%)	517 (37%)	107 (8%)	1410
With mental retardation	7 (5%)	112 (83%)	16 (12%)	135
Overall	793 (51%)	629 (41%)	123 (8%)	1545

We determine how a single-gene disorder is inherited by means of a *pedigree*, a family history of the disease. To illustrate autosomal dominant, autosomal recessive and X-linked recessive inheritance, we can examine three well-known diseases each with distinctive behavioural effects. X-linked dominant and Y-linked disorders are too rare and often controversial to warrant consideration here.

Autosomal dominant inheritance

Huntington's Chorea is now usually called Huntington's Disease (HD) because not all sufferers show choreic ('jerky') muscle seizures. This disorder attracts attention because of the unique problems it creates for the family of any sufferer. Although it can begin in childhood, the first symptoms of changed personality, diminished facial expression, slurred speech and stiff limbs usually do not appear until after the age of 35. Over approximately the next 10–15 years until death occurs, there is a progressive deterioration of mental and motor performance accompanying the loss of brain cells from various areas of the cortex (Fig. 2.15a). Although relatively rare (one in 16 000 births) there are certain regions especially in the north-eastern USA and in Tasmania, Australia where it is much more common (one in 5500 births in Tasmania). The excess in the USA can be attributed to the migration

(a)

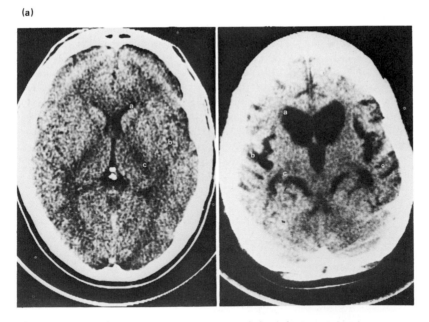

Fig. 2.15   (a) The brain of a normal person (left) and a Huntington's Disease patient (right) scanned by computerized tomography. The changes are that the HD patient has (label a) enlarged ventricles (the fluid-filled hollows of the brain), (b) larger sulci (the fissures which separate portions of the cerebral cortex) and (c) little trace of the caudate nucleus (an arched band of grey matter forming part of the basal ganglia). For a description of the role of these areas see texts such as K. W. Walsh (1978) Neuropsychology: a clinical approach. Churchill Livingstone, Edinburgh. (Photographs courtesy of Dr E. Chiu, Huntington's Disease Clinic, University of Melbourne, Department of Psychiatry.)

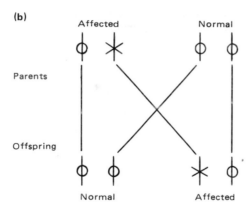

(b)

Affected   Normal

Parents

Offspring

Normal   Affected

FIG. 2.15   (b) The inheritance of an autosomal dominant disorder. The affected individual is virtually always a heterozygote (see text) and half his/her gametes carry the abnormal allele, half the normal form. Half the children on average will be heterozygotes and will suffer the disease, while the other half will be homozygous and completely normal. O, normal chromosome, X, chromosome with abnormal allele.

of three affected individuals from a Suffolk village in 1632 and the Tasmanian figures to the migration in 1848 of an affected woman from Somerset with her 13 children. Figure 2.15(b) shows the fate of children of an HD sufferer, assuming that the affected person is heterozygous rather than homozygous for the HD allele. This is almost always the case as homozygotes can only arise in a proportion of the children born to two heterozygotes, the chances of two such people mating being very small.

Although males and females should be equally likely to have an autosomal dominant disorder, one peculiarity of HD is that this is not always so. There seems to be an excess of males in adult HD cases with the reverse in juvenile HD, while in 75% of the cases of juvenile HD the father is the person who transmits the HD allele. The reason for this is uncertain and is not characteristic of autosomal dominant disorders in general. The other problem is that juvenile onset HD is characterized by muscle rigidity rather than the choreic muscle seizures common in the adult form. However, the existence of a pair of identical twins, one with the choreic and one with the rigid form indicates that the same allele must be involved.

What makes HD so distressing is that the symptoms often do not appear until after the person has had a family. The children are then in the situation of coping with the affected parent for many years, while at the same time knowing that they have a 50:50 chance of suffering the same fate themselves. Attempts have been made to diagnose HD before the symptoms appear, since this would reassure half the population at risk and allow them to have children with no risk of passing on HD. The main methods being tried include brain scans to detect changes in the shape of the ventricles (as seen in

Fig. 2.15a), analyses of muscle movements and changes in specific components of intelligence tests. If particular parts of the brain are degenerating, it would seem feasible that some particular cognitive skills could be more affected than others. However, HD involves mainly the frontal lobes of the brain where damage must be major before performance on formal intelligence tests is impaired.

An alternative ethical viewpoint is whether such diagnostic methods should be developed before there is any means of adequately treating HD. Otherwise early diagnosis puts even greater strain on those detected at this early stage.

Autosomal recessive inheritance

Phenylketonuria (PKU) is the most widely discussed single gene disorder since it illustrates three features, a well understood biochemical disorder, a screening program for early detection and an adequate treatment. However, its apparently clearcut nature is now being questioned (Murphey 1983). In 1934 Følling first described mentally retarded patients who excreted phenylpyruvic acid. By the 1950s it was known that the homozygotes did not synthesise an enzyme, phenylalanine hydroxylase. Enzymes are chemicals which control the rate of other chemical reactions, in this case, the metabolism of one essential *amino acid*, phenylalanine to another amino acid, tyrosine. (The 20 or more amino acids are the constituents of all the varied proteins in the body.) The result is that phenylalanine builds up to excessive levels in the liver and spreads through the blood stream, affecting the brain in particular. The brain weight is reduced, there is a deficiency of myelin (the sheath around nerve fibres), of dendrites (the multiple fine processes linking nerve fibres) and of synapses (the junctions with other nerve fibres) as seen in Fig. 2.16a. The outcome of all this is that untreated people with PKU usually have an IQ less than 20, with only one in 50 exceeding an IQ of 60. Only one third of them learn to talk and many have abnormal reflexes and posturing (Fig. 2.16b).

The mode of inheritance is shown in Fig. 2.16c. The feature of autosomal recessive inheritance is that an affected individual must be homozygous for the deleterious allele. This can only happen if both parents are either actual PKU sufferers or much more commonly, *carriers* (heterozygotes, who have the allele but appear phenotypically normal). In the latter case, remembering Mendel's law of random segregation, half the gametes of each carrier will have the PKU allele and half the normal allele. If we call the normal allele P and the PKU allele p, we have the situation

		sperm	
		½P	½p
	½P	¼PP	¼Pp
eggs			
	½p	¼Pp	¼pp

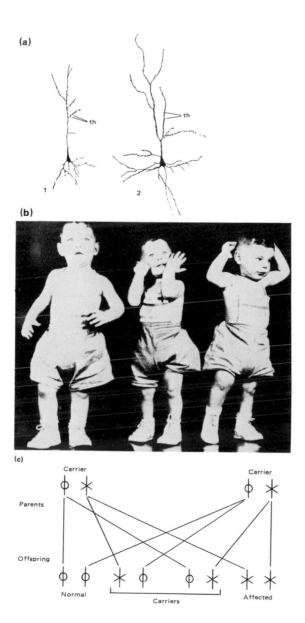

FIG. 2.16   (a) Comparison of a nerve fibre from the cerebral cortex of (1) a 60 year old man with untreated PKU and (2) a normal 2 year old. Note that the PKU man has fewer dendrites and fewer 'thorns' (labelled th), the sites of synapses. (b) A set of triplets where the two on the right are identical and have PKU. The posturing of their arms is characteristic of more severely affected PKU individuals. (c) The inheritance of an autosomal recessive disorder. Both parents are phenotypically normal but are carriers. Half their gametes carry the PKU allele and hence one quarter of their offspring are homozygous for this allele and are affected while half are carriers and one quarter are homozygous normal. O, normal chromosome, X, chromosome with abnormal allele. (a and b from L. B. Holmes et al., (1972) Mental Retardation: An atlas of diseases with associated physical abnormalities. Macmillan, New York.)

By the laws of probability only one quarter of their progeny will have both PKU alleles (pp) and hence have PKU, while one half will be carriers (Pp) and one quarter normal homozygotes (PP).

The incidence of PKU is about one in every 11–17 000 births (or one in 5000 in the Irish and Scots) so that it is by no means the most common autosomal recessive disorder. For example, cystic fibrosis (abnormal functioning of mucous-secreting glands so that the lungs and pancreas become blocked and cease to function) has an incidence in Caucasians of between one in 1500 and one in 2400 live births. What makes early detection of PKU so important is that it only starts to affect abilities after birth. Until that time the gene in the fetus controlling phenylalanine hydroxylase is inactive and all genotypes are similar. Thereafter progressive mental retardation is experienced as phenylalanine levels increase, unless a special treatment is introduced. The treatment consists of limiting the amount of phenyl-alanine through a diet low in protein which usually contains phenylalanine and a special milk substitute which again has no phenylalanine. Since every child differs biochemically, frequent blood samples are needed to monitor phenylalanine levels — too high a level impairs brain development, too low a level affects weight gain and general health.

It is important that the diet should begin as soon after birth as possible, since IQ can drop as much as two points for every week that treatment is delayed (Table 2.5). For this reason many countries have

TABLE 2.5   Age at which dietary treatment began and most recent IQ scores for groups of PKU children (derived from C. S. Shear, N. S. Wellman and W. L. Nyhan (1974) *Heritable Disorders of Amino Acid Metabolism: Patterns of clinical expression and genetic variation*, Nyhan, W. L. (ed.), ch. 9. Wiley, New York)

Age when diet introduced	Mean IQ	Range
0–2 months	99	72–132
3–18 months	69	58–89
1.5–6 years	49	24–81
> 6 years*	38	20–75

*Six of eight patients were untreated.

introduced PKU screening for all newborns. But Stine (1977) among others has suggested that these programs were premature. Any system that requires up to 20 000 children to be processed in order to detect one case of PKU can only be marginally efficient on a cost/benefit basis. Secondly, there are at least five other reasons apart from PKU for elevated phenylalanine levels (Nyhan 1974). Children with the so-called Type II hyperphenylalanaemia have levels of phenylalanine which are initially as high as in untreated PKU children, yet their intelligence is normal. A more practical problem once a PKU child has been detected is the nature of the diet. Apart from its expense, it can be so monotonous that parents sometimes find it necessary to introduce other foods for variety despite the detrimental effects on intelligence (Sarason and Doris 1969). It now seems that the brain is sufficiently developed by the age of 6 or 8 for the diet to be stopped without any adverse effects on IQ, although irritability and declining school performance may require its reintroduction.

While the screening programs now mean that almost all cases of PKU are detected (but not necessarily treated, as this is a decision for the parents), many adult cases are encountered among the relatives of known PKU sufferers. It is estimated that 10% of these are either normal or only mildly retarded even though they received no special diet. Among these adults who are more severely affected, dietary therapy cannot improve their IQs but can reduce irritability, seizures and other behavioural problems.

The adult group causing the most concern at present is treated PKU females. Although the treatment in childhood limited the effects of phenylalanine on their own brains, the levels in their bodies are still excessive and can damage their unborn children. Out of 94 births to 28 treated PKU women, Hsia found eight PKU children and 79 without PKU (seven were not classified). Of the 79 without PKU, 61 were retarded and seven died, while the data on the other 11 were incomplete. The reintroduction of the low phenylalanine diet during pregnancy has been tried but with limited success, leading to the recommendation (discussed in Nyhan 1974) that sterilization or at least therapeutic abortion should be routinely available for such women.

These problems indicate that PKU detection and treatment is no longer the simple success story that it appeared to be a few years ago. This point becomes even clearer if we go beyond the overall IQ scores of treated children. Such PKU children tend to be more clumsy, to score lower on performance than on verbal IQ and to be delayed by up to 3 years on tests which measure visual perceptual handicap of the sort indicated in a classroom situation by confusion between '6' and '9' or between 'b', 'p' and 'd' and by the inability to copy patterns (Fig. 2.17). Pennington and Smith (1983) note poor performance on mathematics and on certain spatial tests and suggest that there is right hemisphere brain damage, possibly connected with the skills that are developing at the time the diet is discontinued.

The other problem area is the variability between PKU children. In 1946 Penrose first commented on the resemblance between retarded brothers and sisters in the degree of defect — if one was more severely retarded than was usual for a particular disorder, then so too were the other affected siblings. Figure 2.18 illustrates this point for PKU by contrasting two families. In Family A, the normal child, the treated PKU child and the untreated PKU child all have considerably higher IQs than their counterparts in Family B. The discrepancy is so great that the untreated PKU child from family A does as well as the treated PKU children from Family B.

The Austrian Screening Program has revealed one complication in the relatives of PKU children. The parents (who must be heterozygous) and the heterozygous siblings were six to eight points lower on verbal IQ than expected, suggesting some effect of having just one PKU allele. Bessman considers that such results may help

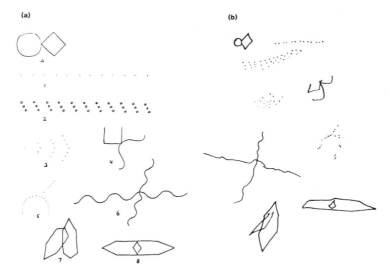

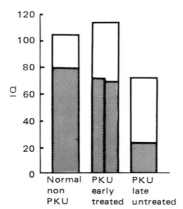

FIG. 2.17   (a) The Bender visual-motor gestalt test. (b) The performance of a 12.5 year old PKU boy, treated since the age of 2. His overall IQ is 76, but his copying of the Bender test pattern is only equivalent to that of a normal 5.5 year old. (From R. Koch, M. Blaskovics, E. Wenz, K. Fishler and G. Schaeffler (1974) in *Heritable Disorders of Amino Acid Metabolism.* Nyhan W. L. (ed.), p. 109. Wiley, New York.)

FIG. 2.18   IQs of children in two PKU families. One PKU child from family a (□) and two PKU children from family b (▧) had been started on dietary treatment by 2 weeks of age. One other PKU child from family a was not diagnosed until 8 years and received no treatment while the other family b PKU child began treatment at 2 years but was still severely retarded. (From C. S. Shear, N. S. Wellman and W. L. Nyhan (1974) in *Heritable Disorders of Amino Acid Metabolism* Nyhan W. L. (ed.), p. 141. Wiley, New York.)

explain the large group of the mentally retarded where the cause of their retardation is unknown (Table 2.1). PKU heterozygous females are partially deficient in the synthesis of tyrosine with only minor effects on their own ability. But the effects on their children exposed *in utero* to abnormal tyrosine levels may be much more severe. So without the parent or the child actually being homozygous for the disorder, mild mental retardation may still result. Given the myriad of biochemical disorders, this heterozygote biochemical imbalance could be a frequent course of retardation. Although this theory is attractive, the IQ deficits in PKU or any other heterozygotes studied to date are not sufficiently large to generally lead to retardation. However, it may still help to explain the greater number of retarded children born to female than to male retardates (see Table 2.7 and accompanying discussion).

X-linked recessive

The *Lesch–Nyhan Syndrome* is one of the rarer genetic disorders, but it serves as an example of X-linked inheritance because of its unique features. The most striking symptom is compulsive self-mutilation where the child chews away his lips (Fig. 2.19a) and fingers unless he is permanently restrained or has his teeth extracted. The children are not insensitive to pain and scream in agony when they bite themselves.

All the patients to date have been males, which is to be expected with an X-linked recessive disorder. To express the disorder males have only to get the X chromosome with the abnormal allele from one parent, since the Y chromosome cannot mask its effects. On the other hand females need to get an abnormal X from both parents, an unlikely event in the case of a rare allele. The characteristics of an X-linked recessive disorder are affected males and carrier females (Fig. 2.19b).

The disorder results through deficient activity of the enzyme hypoxanthine guanine phosphoribosyl transferase (HGPRT). People with intermediate levels of HGPRT show gout rather than the self-mutilation and it is gout and related kidney disorders that usually lead to death of Lesch–Nyhan males before puberty. The time-course of the disorder and accompanying mental retardation differ from some of our earlier examples such as Down's Syndrome. Lesch–Nyhan boys show normal mental and motor development for the first 6 months, but thereafter muscle spasms and the self-mutilation begin and both physical and mental development slow down. In the end the average IQ is below 50 and height and weight are below the third percentile, that is, such boys are smaller and lighter than 97% of normal children.

Although the abnormal enzyme levels can be altered by chemical means, the self-mutilation does not change. One possibility is to try to alter the self-mutilation by behavioural means, but this creates problems with such commonly-used aversive stimuli as electric shocks. These boys are so used to pain that they do not learn from painful cues (Fig. 2.20). However, they do learn quickly from a combination of rewarding other behaviour and ignoring finger-chewing, so that their moderate mental retardation is not inhibiting their ability to learn, given the right training situation.

(a)

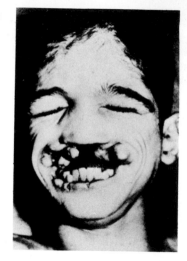

(b)

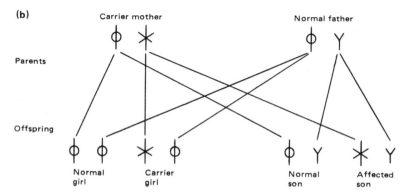

FIG. 2.19 (a) A 14 year old boy with Lesch–Nyhan syndrome who has almost totally destroyed his upper lip. (From W. L. Nyhan (1968) *Federation Proceedings* **27**, 1027.) (b) The pattern of X-linked inheritance. If the father is normal and the mother a carrier, half the sons will be affected and half the daughters will be normal but carriers. O, normal chromosome X, chromosome with abnormal allele. Y, normal Y chromosome.

Inborn errors of metabolism

Phenylketonuria and the Lesch–Nyhan Syndrome exemplify the 'one gene–one enzyme' hypothesis. Both these disorders involve and illustrate a point first made by Garrod in 1902 when he proposed that 'inborn errors of metabolism' resulted from inactive or missing enzymes. Because enzymes are involved in every stage of metabolic sequences, an abnormal enzyme at one step affects every subsequent reaction in the sequence. Consider the sequence

$$A \xrightarrow{E1} B \xrightarrow{E2} C \xrightarrow{E3} D$$

where A is metabolised to D through three steps each with a unique

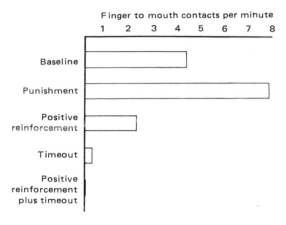

Finger to mouth contacts per minute

FIG. 2.20   The average response of four Lesch–Nyhan boys to treatment for finger-chewing. Punishment was electric shock to the fingers, positive reinforcement involved stopping self-injury, reassuring the child and encouraging other behaviours and time-out was pointedly ignoring the child for 5 seconds. (Derived from L. Anderson, J. Davies, M. Alport and L. Herrman (1977) *Nature* **265**, 461.)

enzyme indicated as E1, E2 and E3. A mistake with any one enzyme disrupts the entire sequence, so that either not enough of the final product is available (as with HGPRT in Lesch–Nyhan Syndrome) or there is an excess of one of the earlier products which cannot be metabolised further (as with phenylalanine in PKU).

The detailed evidence that has accumulated since 1902 for the 'one gene–one enzyme' hypothesis is adequately covered elsewhere (Mange and Mange 1980; Stine 1977) along with our rapidly growing body of knowledge of molecular genetics. Molecular genetics focusses on *deoxyribonucleic acid* (DNA) the basic chemical structure in which all the genetic information is coded and on how this information is used to specify the amino acids which constitute the enzymes. Increased information has modified the 'one gene–one enzyme' hypothesis so that it is now more correctly called the 'one gene–one polypeptide' hypothesis, a *polypeptide* being a chain of amino acids with each enzyme consisting of one or more such polypeptide chains. That is, the function of one enzyme may be affected by several genes each determining a different constituent polypeptide. Not all disorders involve enzymes and there are many autosomal dominant disorders involving structural proteins such as those which make up bone and muscle. Enzyme disorders are generally autosomal recessive and, as Table 2.4 showed, predominate in single-gene causes of mental retardation.

The sheer complexity and detail of genetic transmission at a molecular level explains why a bewildering variety of disorders can result but are of little help in understanding the behavioural repercussions of the disorders, a point adequately illustrated by the *mucopolysaccharidoses*. This group of disorders is characterized by the lack of particular enzymes leading to an excess of mucopolysaccharides (combinations of sugar molecules) in many tissues and especially in the urine. But knowledge of the biochemical defect in itself is not yet sufficient to enable prediction of the behavioural outcome. Omenn (1976) gives the example of two syndromes, Hurler and Scheie, that have opposite effects on ability despite involving the same enzyme deficiency and having the same mucopolysaccharides in the urine. The former results in severe retardation and the latter in normal intelligence. Similarly Hurler and another syndrome, Hunter, have the same excess urinary mucopolysaccharide and severe retardation, but Hurler children are apathetic and placid, while Hunter children are hyperactive.

The mucopolysaccharidoses are examples of what Omenn (1976) classes as metabolic disorders 'intrinsic to the brain' since they involve abnormal levels of enzymes normally present in the brain. Another example is Lesch–Nyhan Syndrome where the boys lack the HGPRT normally present in the brain, particularly in the basal ganglia, the control area for involuntary movements. The effects on this brain region would explain their muscle spasms but not their self-mutilation. Omenn distinguishes these intrinsic disorders from metabolic disorders 'extrinsic to the brain', such as PKU where the brain is damaged by abnormal levels of normal metabolites produced by other organs.

Other single gene disorders

Although the inborn errors of metabolism are the biochemically best understood single gene disorders, they do not provide a complete picture of the ways in which single gene disorders can give rise to retardation. Apart from metabolic disorders and such progressive disorders as Huntington's Disease, Holmes et al (1972) recognize three other classes of mental retardation where single genes may be involved.

1   Central nervous system malformations. There are at least nine genetic syndromes involving hydrocephaly, the excessive accumulation of cerebrospinal fluid within the ventricles or hollows of the brain. The result is brain damage, mental retardation and often death unless a drainage shunt is provided as soon as the disorder is detected.

2   Syndromes of multiple deformities. Many disorders can be classified by a pattern of deformities, each of which is found in other syndromes but which occur together only in the one particular syndrome. An example is the Laurence–Moon Syndrome where there is obesity, retinal degeneration, polydactyly (extra fingers and toes), very small genitalia, sterility, and moderate retardation. Such multiple effects resulting from a single genetic defect are a phenomenon called pleiotropy, but it is worth emphasizing that just because a disorder leads to multiple deformities it is not necessarily genetic. One in 500 of the severely retarded have the Rubinstein–Taybi Syndrome with facial abnormalities including a beaked nose and cataracts in the eyes, excessive body hair, unusual angles of the

thumbs and big toes and greatly reduced stature. Yet there is no evidence that this is a single-gene disorder and its cause remains unknown.

3 Neurocutaneous syndromes. Neurocutaneous syndromes involve both the skin and the nervous system, the most common of these (one in 2500–3000 births) being the autosomal dominant *neurofibromatosis*. The usual symptoms are patches of increased pigmentation (*cafe-au-lait* spots) and multiple skin tumours. Mental retardation occurs in over 25% of the cases and may be related to the degree of pigment and tumour formation — those who are more afflicted physically are also more likely to be affected mentally.

Discussion of single gene disorders

Single gene disorders have a bewildering range of effects on intellectual and physical growth. Omenn (1976) considers this an advantage for research in behaviour genetics since it provides an opportunity to observe the connection between genes and behaviour in far more detail than is possible in the normal population, where we only deal with genes in an abstract sense (Chapters 3 and 6) and not through knowledge of their biochemical effects. But have any general trends emerged from single gene research to date?

One is the concept of critical periods, times during development when biochemical levels must be within fine limits if later abilities are not to be impaired. We have seen this with phenylketonuria where early introduction of the correct diet is important and with hydrocephaly where the shunt needs to be implanted as soon as possible. Lenneberg (1968) extends this concept to include a wider range of genetic disorders including thyroid gland disfunction and blood group incompatibility. *Eurythroblastosis fetalis* (or Rh incompatibility) results from the mother and the baby having different phenotypes at the Rhesus blood group system. In certain circumstances (Mange and Mange 1980; Stine 1977) the different blood types can react causing severe jaundice in the child and damage to several regions of the brain including the brain stem, cerebellum and optic nerves. Children who survive have several behavioural deficits including cortical deafness, where they hear artificial sounds but rarely attend to language and are slow in learning to speak. In contrast if adults or older children contract this type of jaundice, irreversible damage is far less likely.

But general conclusions and prognoses for individual cases are limited both by deficits in specific components of ability and by inter-individual variation. Phenylketonuria (Table 2.5) is not the only disorder in this situation. In *galactosaemia*, exclusion from the diet of milk sugar which contains galactose, is sufficient to bring IQ within the normal range, but there remains some social maladjustment, behavioural problems and visual–motor deficits which impede progress at school. In *homocystinuria*, the level of the enzyme cystathionine is reduced and there are consistent skeletal deformities. In terms of individual variation, half the patients are of normal IQ and the rest range from mild to severe retardation.

One reason for such variability may be the genetic background on which the disorder is imposed. Therefore we now consider the range of variation in the population as a whole.

Polygenic inheritance

Polygenic inheritance is such a fundamental part of behaviour genetics that Chapter 3 is devoted to explaining its principles. All that is being included here is sufficient to complete the account of mental retardation.

As its name implies, polygenic inheritance involves the action of many genes on a particular trait. When we consider genes this way, we can no longer describe each individual gene in terms of its bio-chemical and other properties, but think only of each allele as adding or subtracting a given amount from the phenotype. Our approach is therefore quantitative rather than qualitative. The example of height will illustrate this distinction. People may be very short in stature because of such single gene effects as the autosomal dominant *achondroplasia* which reduces height by 40–50 cm. But the presence of such qualitative single gene effects is not enough to describe the range of heights in the normal population. Instead we must think of each person's height being determined by many genes acting quantitatively with some alleles increasing and some decreasing height. In addition, environmental effects such as nutrition will influence the final height.

If intelligence and other behaviours are viewed in the same way, there is the possibility that some people may have an IQ of less than 70 and be classed as mentally retarded because of polygenic effects. They have no single-gene or chromosomal abnormality but score so low because of a combination of their environment and of many alleles for low intelligence. Figure 2.21 illustrates this hypothesis.

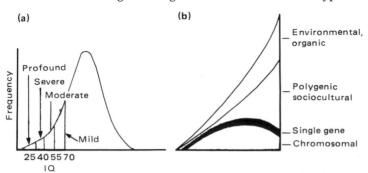

FIG. 2.21   (a) The distribution of intelligence in the population as a whole and (b) the relative contributions of different causes of retardation in the group with IQs less than 70. (Modified from H. Zellweger and V. Ionasescu (1978) in *Medical Aspects of Mental Retardation* (2nd edn) Carter C. H. (ed.), p.123. C. C. Thomas, Springfield, Ill.)

There are more people with IQ below 70 than there are above 130, due to the addition of a group affected by single gene and chromosome aberrations and by major environmental trauma to those who constitute the bottom end, genetically and environmentally, of the normal population. Zigler (1967) labelled the former group as 'organic' and the latter as 'cultural-familial' retardation. Others refer to the latter as 'sociocultural', although this tends to obscure the possibility that genes may play some part.

How do we set about distinguishing these two groups? Firstly as Zigler did, by the severity of the retardation. Many of the single gene and chromosome disorders we have discussed lead to severe or profound retardation with IQ scores less than 40. If people simply have a poor environment or genetic endowment, they might be expected to be only a little below normal and hence only mildly or moderately retarded. But we have already seen that the variability within most single gene or chromosome disorders is too great for every mentally retarded person to be classified with certainty this way and we have to turn to a second approach.

If cultural-familial retardation results from very poor environmental and genetic conditions, one would expect their relatives also to score below average since they will have some genetic similarity and possibly also environmental similarity. On the other hand, single genes segregate and chromosome aberrations generally affect only the one individual in the family (except for the fragile-X syndrome and the translocations). Therefore their relatives should be of normal ability except for the few severely retarded with the same disorder.

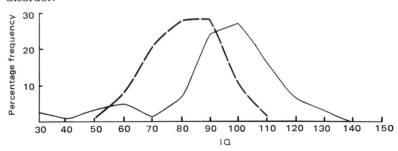

FIG. 2.22   Distribution of IQ in the siblings of two groups of mentally-retarded children. ---, siblings of children with IQ 50. — siblings of children with IQ 30-50. (From J. A. Fraser Roberts (1952), *Eugenics Review* **44**, 71.)

Roberts divided a sample of 271 mentally retarded children into two groups, those with IQ scores above 50 and those with IQ scores between 30 and 50. (In the terminology of that time, these were referred to as 'feeble-minded' and 'imbeciles' respectively.) He then tested 562 of their siblings with the results shown in Fig. 2.22. The more severely retarded had siblings with an average IQ about 100 and the mildly retarded had siblings with an average close to 80. This is the difference we would expect if the former group were retarded because of specific genetic or environmental disorders, while the latter represents families at the lower end of the normal population with many alleles that decrease IQ in every family member. Roberts has been criticized for using the siblings' scores in allocating the retarded children to his two groups thereby exaggerating the differences between the groups. But Fig. 2.22 shows that although the more severely retarded had siblings with an average IQ of 100, they also had many more siblings with IQs less than 50, compared with the mildly retarded. This result fits with the existence of the two groups, since there is a finite probability of siblings of the severely retarded having the same single gene disorder or chromosome translocation.

Other data sets are free of this possible bias and include the major pedigree study carried out by Reed and Reed. Starting with 289 people classified in Minnesota between 1911 and 1918 as mentally retarded, they traced some 82 000 descendants, spanning a total of seven generations. If we examine first the 242 out of the 289 who had both IQ assessment and siblings, we find major differences among these siblings according to the degree of retardation of the initial 242 (Table 2.6). We can refer to the 242 as the *proband* group (that through which the other individuals were traced). The more severely retarded probands are less likely than the mildly retarded probands to have retarded siblings, but if they do have retarded siblings then they will probably be quite severely retarded, exactly the same pattern as that just described in Robert's data. In contrast, the less retarded were those with a family history of low ability.

TABLE 2.6   Siblings of Reed and Reed probands (derived from C. A. Johnson, F. M. Ahern and R. C. Johnson (1976) *Behavior Genetics* **6**, 473)

IQ levels of probands	Percentage of probands with retarded siblings	Average IQ of retarded siblings
0–19	36.2	28.2
20–29	43.8	34.2
30–39	54.8	35.8
40–49	64.9	47.6
50–59	59.5	58.1
60–79*	80.5	60.5

*Only one proband had an IQ above 69.

One could argue that the siblings of the retarded are an abnormal group, since they grow up with the stress of a retarded child in their family. The Reed and Reed data indicate that there is much more to the determinants of retardation than this possible environmental effect. People with a retarded sibling were five times more likely to have a grandchild with IQ less than 80. That is, even a retarded uncle, aunt or grand-uncle or -aunt is associated with below-average IQ.

Many other analyses of the Reed and Reed data could be discussed but we shall confine ourselves to one topic, the excess of males among the retarded. In the Reed and Reed data, 2.2% of all males and 1.6% of all females were retarded. Pauls explained this result by a 'differential genetic loading', a situation where males are more susceptible to the genetic and environmental influences causing retardation. If polygenic inheritance were involved in this excess of males, one would expect differences between the relatives of male and female retarded probands — females would have to have more alleles or more environmental influences for low ability before they exhibited retardation and therefore would be more likely to have retarded relatives. Pauls showed (Table 2.7) that this sex effect was consistent across first-degree relatives (children and parents), second-degree relatives (cousins, uncles, aunts etc.) and third-degree relatives (second cousins etc.).

TABLE 2.7   Percentage retarded among relatives of male and female retardates in the Reed and Reed study* (derived from D. L. Pauls (1974) *Behavior Genetics* **9**, 289)

Degree of relationship	Relatives of male probands	Relatives of female probands
First	10.9	16.8
Second	5.4	7.7
Third	3.3	4.6

*Data are averaged over male and female relatives. In each case more males were affected.

'Differential loading' remains a rather abstract explanation, since it fails to explain why males are more susceptible. Sex linkage is an alternative hypothesis, but estimates indicate that this would only explain some 5% of the cases. The fragile X syndrome has also been suggested but it is still too early to say if this will be sufficient to explain all the excess of the males. It is worth noting there are other disorders more common in males. Boys are three times more likely to stutter and up to five times more likely to be reading disabled, even though the latter diagnosis is restricted to children of normal IQ. That is, an excess of retarded males would not explain this sex difference.

Polygenic disorders

We have only discussed the polygenic explanations of mental retardation in terms of alleles for high and low ability. But there is one major group of physical disorders, the *neural tube defects*, which are polygenically determined and frequently can lead to mental retardation. In the developing embryo, a tube-like structure called the neural tube forms the basis of the brain and spinal cord. If this tube fails to fuse at the head end, the brain does not form and the fetus is *anencephalic* ('without a brain') and often stillborn. If it fails to fuse at the lower end, *spina bifida* ('divided spine') develops, often with a hernia of the membranes surrounding the spine. The pressure of spinal fluid then leads to the large protrusion seen low on the back of spina bifida babies (Andrews and Elkins, 1981).

The incidence of the various neural tube defects (NTD) is very high in the British Isles (two to three per 1000 live and stillbirths) and particularly high in Belfast (10.3 per 1000 births). The incidence is higher in families with a history of NTD and the risks of parents having another child with NTD are five, 10 or 21% depending on whether or not they have already had one, two or three children with NTD, a pattern consistent with polygenic inheritance. The more alleles or environmental determinants the parents have for NTD, the more likely it is that any child they have will cross the threshold from normal to abnormal neural tube development. Although NTD is a polygenic disorder, prenatal diagnosis is possible because of a substance, alphafetoprotein, present in excess in the amniotic fluid of NTD fetuses. It is thought that it leaks out from the open neural tube into the amniotic cavity.

The extent of the genetic determination has recently been thrown open to question by the demonstration that folic acid taken before conception by women at risk for NTD can greatly reduce the incidence of defects. A major controversy exists over the evaluation of the benefits of folic acid. In the course of clinical trials, can one deny it to those women who would most likely benefit from it?

Over 80% of spina bifida children develop normally, unless their disorder has been accompanied by hydrocephaly, requiring a shunt to reduce pressure of the excess cerebro-spinal fluid. Only one third of those with shunts have normal ability, while the others show normal syntax (sentence and phrase construction) but do poorly on other measures of verbal ability, such as comprehension, and on all measures of performance ability. Despite their poor verbal skills, 40% of these children show the 'cocktail party' or 'hyperverbal' syndrome where they frequently repeat phrases or chatter constantly but unproductively. Anderson and Spain (1977) give a very full account of the behavioural problems in spina bifida and of the general issues faced by a family with a handicapped child.

Discussion of polygenic inheritance

The familial or polygenic determinant of mental retardation has had a long and bitter history. As we saw in Chapter 1, it was one basis in the earlier part of this century for restricted migration into the USA of certain ethnic groups considered to be 'intellectually inferior'. It was also a cornerstone of the eugenics movement in the period until 1930, where the aim was to 'improve the quality of the genes', by such methods as sterilization of the retarded. Sarason and Doris (1969) review this unfortunate period in the development of behaviour genetics in more detail than we are able to here.

The data we have considered are very different from the earlier subjective reports such as that on the Kallikak family (Fig. 1.7) where the only thing they got right was the greater importance of the mother's phenotype as a predictor of mental retardation! However the controversy has not ended, although it now takes a very different form, centering around the treatment of 'cultural-familial' retarded children.

Such children cannot yet be effectively treated by dietary or other biochemical methods similar to those used with some single-gene disorders. One possibility is to change their home, cultural and school environment by 'compensatory education' programs. In the USA, many of the children needing this help come from disadvantaged ethnic minorities, especially the Blacks and it is impossible to separate any discussion of the success of these programs from the factors making minorities more likely to be involved in the programs in the first place. These issues are considered in Chapter 8 but our discussion should lead us to anticipate two features of these programs (a) that they must begin early in the same way as dietary treatments for clearcut genetic disorders such as PKU (Table 2.5) and (b) that they must cope with genotype–environment interaction (Fig. 2.13) where children may respond differently to the same program.

Human population genetics

We have only discussed how genes cause disorders, not how the abnormal genes are there in the first place. Often disorders 'just happen', as in the case of non-disjunction causing chromosome aberrations such as Turner's Syndrome. Single gene defects may also occur in the same way in the absence of any family history, in which case they are called *mutations*. The rate of such spontaneous mutations is about four per 100 000 gametes, although this is not constant and is much higher for some disorders. For example, the mutation rate for neurofibromatosis is 13–14 per 100 000 gametes.

Spontaneous mutations are just chance errors in copying the enormous amount of information in our genes. However, mutations and also chromosome aberrations may be *induced* by a specific *mutagen* or factor in the environment. The major mutagens are ionizing radiation (such as X-rays), ultraviolet radiation and chemicals. Mutagenicity testing of chemicals is now a major field of research, trying to identify which of the thousands of chemicals to which we are routinely exposed are mutagenic.

Of more relevance to behaviour genetics at the present time are abnormal alleles which do not arise by mutation in the parental generation, but which run in families and also in ethnic groups. We have seen many of these already — G6PD deficiency in Mediterranean populations, Huntington's Disease in southern England and now in Tasmania, phenylketonuria in the Irish and Scots, and spina bifida in the British Isles especially in Ireland.

One of the disorders most closely confined to one ethnic group is the autosomal recessive *Tay–Sachs* disease (also called infantile amaurotic idiocy). Although damage to the central nervous system begins prenatally, the first symptoms are not obvious until about 6 months and thereafter the child's skills degenerate until death occurs by the age of 4. One person in every 25 among the Ashkenazi Jews (those from Central Europe) is a carrier, compared with one in every 400 for other Jewish and non-Jewish groups. We can estimate the incidence of such children as $1/25 \times 1/25$ (the chance of two carriers mating) $\times \frac{1}{4}$ (since it is autosomal recessive) or one in 2500 in the Ashkenazi Jews and $1/400 \times 1/400 \times 1/4 = 1$ in 160 000 for others.

Stine (1977) discusses such ethnic group differences and their implications at length. Here we consider the principles of Darwinian
evolution to understand why these differences originated and how
they are maintained in the population.

Charles Darwin was a Victorian gentleman who, almost by chance,
travelled as naturalist on a survey trip to the Galapagos Islands. Struck
by the variation among the 14 species of finches in one area, he noted
that the major differences were in the beaks which were highly
appropriate for the eating habits of each species. He marshalled more
and more evidence of such purposeful variation and its likely
derivation from an original ancestor, publishing his results in 1859 in
The Origin of Species. Darwin's central thesis was the principle of
natural selection. Any population will show variation for a character.
If individuals with a particular level of this character are more likely
to survive and to pass their genes on to the next generation, then more
of the progeny are likely to exhibit this desirable level. Others will be
selected against in what he called 'survival of the fittest' but which
can be more accurately termed 'reproduction of the fittest'. A fundamental requirement for Darwinian evolution is that the variation be
genetic, since an advantageous trait will not pass from generation to
generation if it is solely environmental.

Darwin had the major problem that his ideas preceded the
understanding of Mendelian genetics. The ideas of Mendel and
Darwin in appreciating the factors that maintain and create diversity
were first brought together in 1908 by Hardy and by Weinberg in what
has come to be known as the *Hardy–Weinberg law*. The law states that
the frequency of particular alleles will remain constant from one
generation to the next if five conditions are fulfilled.

One consequence of the law is that the genotypic frequencies can
be predicted. If there are two alleles A and a with frequencies p and q
where $p + q = 1$, then the frequencies of A and a in the next generation
will also be p and q and the genotypes AA, Aa and aa will be present in
the proportions p^2, $2pq$ and q^2 respectively, the so-called *Hardy–
Weinberg equilibrium*, since A mates p^2 times with A to give AA, etc. If
the progeny are not present in these proportions, one or more of the
following five conditions has been violated.

1 Large population size   If a population is small, then it is unlikely
that it will contain exactly p A alleles and exactly q a alleles. In such a
case there are two related factors that can disturb the
Hardy–Weinberg equilibrium. The first of these is the *founder* effect.
We have seen this already with the incidence of Huntington's Disease
in the USA and in Tasmania. A few individuals contributing a rare
allele to a population when it is still very small ensure that this allele
will be much more common than in the large population from whence
they came.

The best-documented founder effect (Stine 1977) is the autosomal
dominant disorder *porphyria variegata* whose symptoms include
bouts of emotional disturbance. Its high incidence in white South

Africans can be attributed to one Dutch free burgher Gerrit Jansz or his wife, who settled there in the late 1600s.

The second influence is *random genetic drift*. Some families may reproduce prolifically, while others may be childless. In a small population these fluctuations may not average out across genotypes and will thus change the allele frequency. At the present time, populations in industrial nations tend not to be sufficiently restricted geographically for drift to matter except for traditional religious sects such as the Old Order Amish in Indiana, Ohio and Pennsylvania. One of the original families of this sect, Mr and Mrs Samuel King, introduced the allele for Ellis–van-Creveld Syndrome, a form of dwarfism which often leads to early death. Despite the deleterious effects of this allele, the Kings and their descendants had larger families than the other community members causing the frequency of the allele to 'drift' higher.

2 Random mating If the genotypes do not mate at random, then the $p^2 : 2pq : q^2$ ratio will not be observed in their offspring. Rather than being random, mating may be assortative where like phenotypes mate. In the normal population there is considerable *positive assortative mating* particularly for IQ, so that people of similar abilities are more likely to mate with consequence for any polygenic model of inheritance (Chapter 6).

Assortative mating involves like phenotypes, and should be distinguished from *inbreeding* which involves like genotypes from within the same family grouping, usually first or second cousins but also incestuous relationships. Autosomal recessive disorders will become more frequent because inbreeding raises the chances of matings between people heterozygous or homozygous for the same recessive allele. Five percent of PKU cases result from such consanguineous marriages, 27% of Laurence–Moon cases and 40% of all Tay–Sachs cases among white Gentiles. Apart from their recognizable defects, inbreeding also leads to an increase in malformations and mortality (Table 2.8) as well as a decrease in intelligence (see Chapter 6).

TABLE 2.8   Effects of inbreeding on the incidence of defects (derived from L. L. Cavalli-Sforza and W. F. Bodmer (1971) *The Genetics of Human Populations*. W. H. Freeman, San Francisco)

Country	Type of defect	Ratio of abnormals among progeny of first-cousin matings to that among non-inbreds
France	Conspicuous abnormalities	3.76
Italy	Severe defect	1.87
Japan	Major morbid conditions	1.38
	Minor defects	1.26
Sweden	Morbidity	2.26
USA	Abnormality	2.26

However, it may be dangerous to take the data in Table 2.8 as an indication of the perils of inbreeding. Consanguineous marriages are more common in poorer communities and poverty and poor antenatal care may contribute to the higher mortality. One major survey of nineteenth century statistics found that the increased mortality in consanguineous marriages was attributable to tuberculosis, a consequence of their environment not their genetics. It is also worth noting that human populations survive where inbreeding is common (Bittles 1980). Some inbreed through choice, others through geographic or religion isolation, such as the Samaritans in Israel and the Old Order Amish and Hutterites in the USA.

3 Mutations New mutations are the source of genetic variability and the basis for change. Mutations can be the means of maintaining an allele at a sufficient frequency in the population, despite deleterious effects that would otherwise lead to its elimination. One form of dwarfism, achondroplasia is an example where high and early mortality is compensated for by a high mutation rate.

4 Migration If a considerable proportion of one genotype leaves, then the Hardy–Weinberg equilibrium will be disturbed. Although human examples of such a situation are rare, there are several examples from rodent populations (Chapter 5) where this occurs. The only human equivalent on a sufficient scale is Tay–Sachs disease and the other disorders which distinguish Ashkenazi Jews from the other Jewish communities. A significant proportion of the Jews enslaved in 70 AD when the Romans captured Jerusalem (the event which marks the start of the Diaspora) followed the Romans back to western and central Europe as the forebears of the Ashkenazi Jews.

5 Selection If one or more genotypes is favoured by selection, that is, if it is more likely to survive and pass its genes on to the next generation, then gene frequencies will not remain stable from generation to generation. Selection may be *natural*, through biological effects on mortality and fertility or *artificial*, where certain genotypes are deliberately bred together to achieve desired characteristics in their progeny. The latter is one of the methods of studying polygenic variation in animals described in Chapter 3.

The important thing about natural selection is that it can help maintain *polymorphism*, a situation where several alleles exist at appreciable frequencies in a population. The classic example is sickle-cell anaemia, an autosomal recessive disorder which is particularly common amongst Blacks. The sickle-cell allele (Hb^s) leads to red blood cells developing a characteristic sickle shape and has such a range of deleterious effects (Stine 1977) that it is difficult to see why natural selection does not lead to its elimination, leaving only the normal (Hb^a) allele. Hb^s is maintained because of a balance. While Hb^{s}Hb^s individuals have a high mortality from sickle-cell anaemia, the Hb^{a}Hb^s heterozygotes are less susceptible to malaria. So one disadvantage for Hb^s (anaemia) is balanced by one advantage (resistance to malaria), ensuring that the allele remains in the population, at least in areas where malaria is endemic. The frequency of Hb^s is now declining in US Blacks, both because malaria is under control and because of intermarriage with whites where Hb^a is the predominant allele.

The possibility of such a balanced polymorphism can be considered for any single gene disorder, where the deleterious allele is present at too high a frequency to be explained by mutation. One approach to schizophrenia indicates that it may be determined by an autosomal dominant allele (Chapter 7). If this were the case, what biological advantages could maintain the allele at a sufficient frequency to explain the incidence of schizophrenia? Suggestions not yet substantiated are along the lines that schizophrenics are sufficiently cushioned from reality as to be less susceptible to shock following injury or surgery.

Discussion of human population genetics

The maintenance of alleles from one generation to another provides a different perspective on the gene-behaviour relationship than our earlier discussions in this chapter. It is also particularly relevant to animal behaviour and the growing trend (Chapter 1) of concentrating on behaviours relevant to the animal in its natural habitat and not to artefacts of the laboratory setting, unrelated to the evolutionary pressures that shaped the animal's behaviour.

We shall refer to these principles of population genetics throughout the remainder of this book and particularly in Chapter 8 when we examine the differences between Whites and Blacks in IQ. Apart from the question of whether or not these differences are partially genetic, the other issue is why should genetic differences in intelligence exist between populations? Founder effects, drift, assortative mating, differences in mutation frequencies, selective migration and different intensities of selection for ability have all been proposed, but without any clear solution. Perhaps if we paid more attention to the mechanisms by which these differences could have originated in the past, we may be better equipped to explain their existence at the present time.

The treatment of genetic disorders

Although treatment of single gene and chromosome disorders is largely the province of the medical specialist, the potential contributions of behaviour geneticists and psychologists are receiving increasing recognition. We need to be able to explain why people with a particular disorder vary in their degree of retardation and to indicate which skills are most impaired and which can be built upon to provide some degree of normal functioning. Berg (in Clarke and Clarke 1974) presents a particularly devastating discussion of the implications of individual differences for the traditional 'text book' accounts of many of the disorders we have discussed.

Apart from postnatal screening and both biochemical and behavioural treatment, there can be earlier intervention by counselling before pregnancy and by prenatal diagnosis. Genetic counselling can only help those already known to be at risk, for such reasons as a family history of a disorder or because of their ethnic group. For example, detecting Tay–Sachs heterozygotes among Ashkenazi Jews

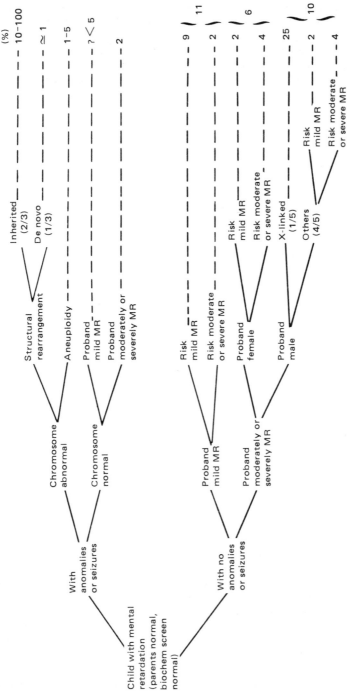

or sickle-cell heterozygotes among Negroes are effective strategies, given a sufficient level of community involvement (Stine 1977).

However, genetic counselling is not always so easy when a bio-chemical defect cannot be detected. In such cases Fig. 2.23 indicates the involved process through which a genetic counsellor can arrive at the risk of families having a second retarded child. The rationale behind each decision point and all the terms should be familiar by now except for *structural rearrangement* which is an inclusive term for translocations, inversions and similar chromosome aberrations and *aneuploidy* which is a general term for variations in chromosome number. The recurrence risks in Fig. 2.23 are empirical, that is, based on data and not just genetic theory and so for reasons of sampling may vary from some of the figures we have discussed here.

Although one can talk of the importance of genetic counselling, this is of little use unless the advice can be put into practice. The USA has one genetic counselling centre for 800 000 people, so that many people who need counselling never receive it. Among the families that do receive counselling, many of the members may be of low IQ because of the very disorder about which they are being advised. Hence, almost half the families counselled misunderstand the information presented to them (Stine 1977).

Genetic counselling can only advise about the risks in general and cannot say what is going to happen in an individual case. Two Tay–Sachs heterozygotes can be told they have a 25% chance of having an affected child or a woman over 40 can be advised of the chances of a Down's Syndrome child but *prenatal diagnosis* is needed to determine if a particular fetus is affected. The last few years have seen major advances in *amniocentesis*, the process of drawing off some amniotic fluid from around the fetus. Fetal cells present in the fluid can then be cultured for biochemical or chromosome analysis and, if the parents wish, the pregnancy terminated if abnormalities are found. Over 100 disorders are detectable this way and the number will grow rapidly with the recent introduction of fetoscopy where the fetus is actually observed and a small blood sample taken for analysis of disorders not detectable from the amniotic fluid. Apart from medical difficulties with amniocentesis, one other problem indicated earlier is that the severity of the disorder cannot be predicted, so that a Down's Syndrome fetus who is only mildly retarded may be aborted.

One of the most important advances in molecular genetics has been recombinant DNA technology where the complete chemical code for the gene determining a particular disorder can be isolated and transferred to other organisms such as bacteria for cloning larger quantities. At present the main use of such DNA probes is to determine whether an individual has the allele for a particular disorder which otherwise is difficult to detect, e.g. prenatal detection of anaemias where the disorder is of the gene controlling postnatal haemoglobin levels. One such probe has recently been developed for Huntington's Disease, possibly rendering redundant the searches for phenotypic indicators discussed earlier for people at risk.

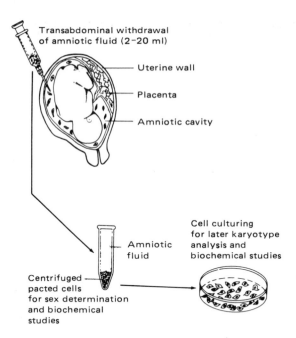

Transabdominal withdrawal
of amniotic fluid (2-20 ml)

Uterine wall

Placenta

Amniotic cavity

Cell culturing
for later karyotype
Amniotic   analysis and
fluid   biochemical studies

Centrifuged
pacted cells
for sex determination
and biochemical
studies

FIG. 2.24   The process of amniocentesis to obtain cells initially for Barr body staining and some biochemical tests, or after culturing, for chromosome analysis or further biochemical screening. (From G. J. Stine (1977) *Biosocial Genetics: Human heredity and social issues.* Macmillan, New York.)

The alternative approach receiving more publicity is that of genetic engineering, which involves changing the abnormal allele or introducing into the genotype an allele that will produce the missing enzyme. This procedure is successful in bacteria and in human cells grown in culture and has for example been used to restore HGPRT function in Lesch–Nyhan cells. But its application to actual treatment of an affected individual has not yet been proven. Nor, if we refer back to Table 2.1, is it likely to substantially reduce the number of cases of mental retardation. Such genetic engineering is only relevant to the 7% of cases with a single gene disorder, since only in these cases can an allele causing a biochemical imbalance be manipulated. Even then, such a variety of disorders exist that it is unlikely that genetic engineering will be available for all or even a majority of the defective alleles.

Conclusion

Chromosome aberrations and single gene disorders provide unequivocal evidence that genes can influence every aspect of intelligence at many different stages of development. It is important to realize that the genes are not directly affecting intelligence as such, but disturbing metabolism in ways which produce mental retardation as one of the symptoms. The discussion of obvious genetic defects and intelligence prepares us for a more difficult question: must genes also be involved in the determinants of intelligence in the rest of the population?

Another aspect of the question of whether genetic disorders say anything about normal behaviour is to regard such disorders as 'natural experiments' where the individual has been manipulated in ways that would otherwise be ethically impossible. For example, males in general do better on nonverbal IQ tests and females better on verbal tests, a sex bias which could be due to the sex chromosomes or to sex hormones or to sex-typed socialization. In the normal individual these factors cannot be separated, but may, using single gene disorders such as 'testicular feminization' where 46,XY males develop as apparently normal females and show the verbal and nonverbal pattern of performance characteristic of their phenotypic, not their genotypic sex. However as noted earlier fragile X males also show a superiority on verbal tests and they are clearly male, albeit often sterile. A frequent physical symptom of fragile X is macro-orchidism (enlarged testes).

A more general point that should be emphasized is that there is no such thing as 'mental retardation', which a person either has or has not. Instead there is a group of people with deficits of differing degrees in different intellectual skills, occurring at different stages of development and with differing associated physical and biochemical disorders. For convenience, we call this group the 'mentally retarded'. But there has been very little systematic attempt to characterize these different patterns of disability. Zigler made a beginning with his distinction between the 'organic' retardates who often have specific deficits in ability and the 'cultural-familial' group who simply develop more slowly than normal, but who have generally similar types of skills to normals, taking into account the different non-intellectual factors they are likely to encounter in the course of growing up. This development-difference controversy remains a major issue in mental retardation (Zigler and Balla 1982).

If we are to learn more about how the intellectual skills of the retarded are organized and selectively impaired we need to (a) differentiate children by the exact cause of their defect, (b) utilize an appropriate series of tests, taking into consideration their physical handicaps and span of concentration (Allen and Allen 1975) and (c) consider the rate of development of their abilities in the way mosaic and trisomy Down's Syndrome children have been contrasted.

Such a study would bring together many of the fragmented results presented in this chapter where differences between disorders are confused by differences in the tests and biases in selecting the children. The adoption of standardized assessment programs of this sort on a large scale is perhaps the only means of ever collecting adequate information on the behavioural effects of many of the rarer single gene (Nyhan 1974) and chromosome (Yunis 1977) disorders. Information is currently collated internationally on some of the physical manifestations of chromosome disorders and there is no reason why the same could not be done for behaviour, if a satisfactory and uniform set of tests could be agreed upon.

However, in the same way that the Behaviourists of the 1920s completely rejected the role of genetics in behaviour (Chapter 1), their successors who believe in treating mental retardation solely by behaviour modification would reject these studies and indeed the whole theme of this chapter. It does not matter to them what caused a particular child to be retarded, but only which components of his or her behaviour need to be altered.

In their discussion of language development in the severely retarded, Rohr and Burr spell out the contrasting position presented in this chapter.

Behaviourists generally regard knowledge of mentally retarded persons' etiology as irrelevant in planning their training program. Although this may be true for some areas affected by mental deficiency, this study suggests that language abilities of retarded persons are affected differently in various etiological classes of mental retardation. Knowledge of distinct patterns of language deficiency in different etiological classes may be important in planning language-training programs that meet the specific needs of the individual. (Rohr and Burr (1978) *American Journal of Mental Deficiency* **82**, 553.)

Discussion topics

1 J. Beckwith and J. King's paper 'The XYY syndrome : a dangerous myth' (1974), *New Scientist* **64**, 474, presents an extreme view of the risks associated with prospective studies of XYY boys. Do the potential benefits of such research outweigh the risks and are their comments relevant to other screening programs apart from the XYY ones? (There is extensive relevant correspondence in *Science* in late 1974–early 1975 when the Boston XYY program was under discussion.)

2 In Chapter 5 we shall discuss Wilcock's claim that much of the single-gene research on rodents is 'trivial' in that it demonstrates no effects of genes on behaviour apart from those that could have been predicted from the physical effects. For example, if a single gene disorder impairs vision, then performance on any learning task involving visual cues will be impaired. Can the same criticism be made of human single gene research?

3 Comment on the following extract from Stine (1977, p. 508): "With respect to the cost of screening, the uncertainty of preventing mental retardation by dietary control, and the fact that PKU is a relatively rare disease, it is questionable whether legislation should have occurred for this particular genetic defect ... Perhaps the money spent on this program could have been more effectively spent for better diagnostic and treatment facilities for the 99 out of 100 retarded children who do not have PKU."

4 In Mittler (1977, vol. 1) and B. Z. Friedlander, G. M. Sterritt and G. E. Kirk (eds) (1979) *Exceptional Infant* vol. 3, Bruner/Mazel, New York (especially the article by J. McV. Hunt, p. 545) research on the cultural-familial and on the organic retardates are all included together and evidence from one group often quoted in the context of the other. Should the two groups be regarded as interchangeable in this way?

References

Annotated bibliography

Carter C. H. (ed.) (1978) *Medical Aspects of Mental Retardation* (2nd edn). Thomas, Springfield, Ill. (An extensive, detailed discussion of the medical aspects of many different forms of retardation. Chapters 1, 2 and 9 are most helpful.)

Clarke A. M. and Clarke A. D. B. (eds) (1974) *Mental Deficiency: The changing outlook* (3rd edn). Methuen, London. (A major text on mental retardation and behaviour. The first six chapters cover similar ground to the present chapter but with different emphases — Clarke and Clarke are less convinced about specific cognitive deficits in retardation.)

Holmes L. B., Moser H. W., Halldorrsson S., Mack C., Pant S. S. and Matzilevich B. (eds.) (1972) *Mental Retardation: An atlas of diseases with associated physical abnormalities.* Macmillan, New York. (Not for the faint-hearted! Descriptions with detailed pictures of a wide range of chromosomal, single gene and other disorders.)

Mange A. P. and Mange E. J. (1980) *Genetics: Human aspects.* Saunders College, Philadelphia. (One of the good modern texts on human genetics. Chapters 7, 8 and 9 on chromosome aberrations are very helpful and Chapter 6 on sex differences presents another side to genetic disorders apart from mental retardation.)

Mittler P. (1977) *Research to Practice in Mental Retardation* (3 vols). University Park Press, Baltimore. (The sections on early intervention in Vol. 1 , on assessment and on cognition and learning in Vol. 2 and the four genetics sections in Vol. 3 are important summaries of current research in these areas.)

Nyhan W. L. (ed.) (1974) *Heritable Disorders of Amino Acid Metabolism.* Wiley, New York. (A crucial reference on inborn errors of metabolism with much information on behaviour hidden among the biochemistry.)

Stine G. (1977) *Biosocial Genetics: human heredity and social issues.* Macmillan, New York. (Less detailed than Mange and Mange but with far more emphasis on the social problems arising from genetic disorders. The discussions of ethnic group differences are recommended.)

Additional references

Allen R. M. and Allen S. P. (1975) *Intellectual Evaluation of the Mentally Retarded Child: A handbook* (rev.). Western Psychological Services, Los Angeles.

Anderson E. M. and Spain B. (1977) *The Child with Spina Bifida.* Methuen, London.

Andrews R. J. and Elkins J. (1981) *The Management and Education of Children with Spina Bifida and Hydrocephalus.* Education Research and Development Committee, Canberra. Report No. 32.

Bittles A. H. (1980) Inbreeding in human populations. *Journal of Scientific and Industrial Research* **39**, 768.

Cleland C. C. (1978) *Mental Retardation: A developmental approach.* Prentice Hall, Englewood Cliffs, N.J.

Fuchs F. (1980) Genetic amniocentesis. *Scientific American* 242(6), 37.

Gerald P. S. (1980) Editorial : X-linked mental retardation and an X-chromosome marker. *The New England Journal of Medicine* **303**, 696.

Gibson D. (1978) *Down's Syndrome : The psychology of mongolism.* Cambridge University Press.

Hagerman R. J. and McBogg P. McK. (1983) *The Fragile X Syndrome: diagnosis, biochemistry, intervention.* Spectra, Denver.

Kessler S. and Moos R. H. (1973) Behavioral aspects of chromosomal disorders. *Annual Review of Medicine* **24**, 89.

Lenneberg E. H. (1968) The effect of age on the outcome of central nervous system disease in children. In *The Neuropsychology of Development: A symposium,* Isaacson R. L. (ed.) p. 147. Wiley, New York.

Murphey R. M. (1983) Phenylketonuria (PKU) and the single gene : an old story retold. *Behavior Genetics* **13**, 141.

Nielsen J., Nyborg H. and Dahl G. (1977) *Turner's Syndrome.* Acta Jutlandica, Aarhus.

Omenn G. S. (1976) Inborn errors of metabolism : clues to understanding human behavioral disorders. *Behavior Genetics* **6**, 263.

Pennington B. F. and Smith S. D. (1983) Genetic influences on learning disabilities and speech and language disorders. *Child Development* **54**, 369.

Rynders J. E., Spiker D. and Horrobin J. M. (1978) Underestimating the educability of Down's Syndrome children; examination of methodological problems in recent literature. *American Journal of Mental Deficiency* **82**, 440.

Sarason S. B. and Doris J. (1969) *Psychological Problems in Mental Deficiency* (4th edn). Harper and Row, New York.

Yunis J. J. (1977) *New Chromosomal Syndromes* Academic Press, New York.

Zigler E. (1967) Familial mental retardation : a continuing dilemma. *Science* **155**, 292.

Zigler E. and Balla D. (eds) (1982) *Mental Retardation: The developmental-difference controversy.* Laurence Erlbaum, Hillsdale, N.J.

3 Polygenic variation and the analysis of normal individuals

Topics of this chapter

1 The principles of biometrical genetics, by which variation in a normal population can be explained by several genes acting in conjunction with environmental effects.

2 The methods of analysing inbred animal strains and their crosses and the importance of inbred strains in efficient experimental design.

3 Ways of analysing non-inbred animal stocks by familial relationships or by artificial selection and the scope for behavioural analysis provided by selected lines.

Introduction

In Chapter 2 we introduced the topic of polygenic variation, the situation where several loci determine a particular characteristic. When this happens we can no longer rely on the direct evidence of genetic involvement provided by chromosome aberrations or even on the clearcut patterns of inheritance seen in the pedigrees associated with single gene disorders. Instead there is a wider range of possible genotypes, depending on the combination of alleles at the different loci. In addition, environmental influences on the phenotype must be recognized more explicitly.

Since most psychological research is concerned with behaviour in this category, an understanding of polygenic variation is fundamental to the appreciation of individual differences in behaviour. Two things should be stressed from the outset. Firstly, the methods are not biased towards the detection of genetic effects and can equally well show that a trait is determined solely by the environment as by genetics. That is, a genetic analysis is just as much an environmental analysis and can often tell a great deal about the environmental factors of importance.

Secondly, while the principles of polygenic analysis are equally applicable to all organisms, plant as well as animal, the methods will differ according to the species. Plants can be self-pollinated and hence can be inbred and made completely homozygous very easily. Animals can be inbred more slowly by mating relatives or can be bred selectively for high or low performance. The human mating system cannot be interfered with and we must turn instead to such groups as twins or adopted individuals for our genetic analysis.

One implication of these different methods that can create considerable confusion is that different nomenclatures have developed for polygenic analysis, even though they all originate from the pioneering statistical work in about 1920 of Sir Ronald Fisher in the UK and Sewell Wright in the USA. One nomenclature calls the field *biometrical genetics* (Mather and Jinks 1977). It was largely developed at Birmingham from work on plants and on animals such as the vinegar fly, *Drosophila melanogaster*, which have a very short generation time and where inbred strains are readily available. The other nomenclature uses the name *quantitative genetics* (Falconer 1981). It developed more in Edinburgh and in the USA and owes a lot to scientists interested in domestic animals where inbreeding is impractical. This distinction of research emphases is an oversimplification as biometrical genetics are now applied extensively to human behavioural data where inbreeding is most definitely not the norm. However, even though biometrical and quantitative genetics use the same principles to answer the same questions, literature in this area can often involve their contradictory systems of notation. For example Mather and Jinks use the symbols 'd' and 'h' where Falconer uses 'a' and 'd' to mean the same thing. In this book we use the notation of Mather and Jinks, where appropriate.

Statistics in biometrical genetics

The term biometrics implies a knowledge of statistics and an acquaintance with basic statistical concepts such as mean, variance, correlation, regression and analysis of variance is helpful in understanding the material in this chapter. These concepts will be introduced gradually throughout the chapter but are covered in much more detail in any introductory statistics text for biologists or psychologists e.g. Sokal and Rohlf (1973), Mendenhall *et al* (1978).

We need statistics to describe differences and similarities in patterns of behaviour within or between groups of individuals. With polygenic inheritance we know nothing about the underlying genotype and we try to infer the relative genetic and environmental determinants of variation from the phenotype. Any one individual's phenotype only provides information about such determination when comparisons are made between the phenotypes of individuals of different degrees of genetic relationship. If this information is not available, no conclusive evidence of the degree of genetic involvement is possible.

This point can be appreciated from Fig. 3.1 which shows the IQ score differences typically found between US Whites and Blacks. Some of the arguments about whether or not the racial differences are partly genetic or are solely environmental are presented in Chapter 8. The reason for these arguments is that there are few interracial adoptions, particularly of White children into Black families and few families have both White and Black offspring with one parent in common, far less the identical twins, one White one Black, which some people claim we need for any conclusive results! Therefore we simply do not have the genetic relationships needed to provide unequivocal evidence about the determinants of the racial differences. While Fig. 3.1 is very controversial, it also clearly illustrates two statistical concepts.

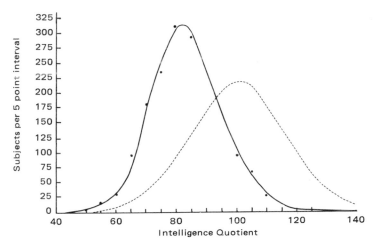

FIG. 3.1   Distribution of IQ scores on the Stanford–Binet test for Black ($\overline{X}$ = 80.7, V = 154) and for White ($\overline{X}$ = 101.8, V = 269) children, tested in 1960. (—— Black, ---- White) These Black children come from the southeastern USA and Black children from the northern states have a higher mean. (From W. A. Kennedy, V. van de Reit and J. C. White, (1963) *Monographs of the Society for Research on Child Development* **28**, no. 6)

Means and variances

Most of the discussion centres around the average or *mean* difference between Whites and Blacks. The mean, for which we shall use the symbol $\overline{X}$, is the sum of the scores of all people tested divided by the number of people. The formula for this is $\overline{X} = \Sigma X/N$, meaning the sum ($\Sigma$) over the score (X) of each of N people divided by N.

But Fig. 3.1 shows another racial difference, in that the Black scores are clustered more around their mean with fewer people scoring far above or far below this level. We say that the Black scores have a smaller *variance* (V) or spread of scores. The formula for a variance is $V = \Sigma (X - \overline{X})^2/N$. The deviation from the mean $(X - \overline{X})$ is squared partly to take into account deviations above and below the mean which would otherwise cancel each other out and also to give extra weight to values far from the mean. To get the average deviation, this is then divided by N or more usually $N - 1$ which gives a better estimate of the variance in a population if we are only measuring a sample from that population. Because it can be tedious to calculate each deviation and then square it, the formula for the variance is expressed more conveniently as

$$V = \frac{\Sigma X^2 - (\Sigma X)^2/N}{N - 1}$$

The Black variance is 154 compared with 269 for the White sample. This result can be explained equally well genetically (US Blacks derive from small groups enslaved in limited areas of West Africa, so that the genetic variability would be expected to be less than in the White population which originates from the migration of larger groups from many different areas of Europe) or environmentally (Blacks may have access to a far more restricted range of environments and opportunities). Another possibility is that the difference in variance is an artefact of the *scaling*, the way in which the tests are scored. This is a frequent problem with polygenic traits, arising because measures of behaviour are designed to differentiate most adequately between those close to the average. Groups towards either extreme of the range will show a smaller variance, merely because the tests cannot discriminate so well between members of such groups. Falconer (1981) and Mather and Jinks (1977) describe the techniques of *transformation* used to rescale data to eliminate this problem. If we were to transform the data in Fig. 3.1 by taking the logarithm of each score (logarithms spread out lower scores more than the high scores) then the variances of Blacks and Whites become similar, although the difference in means remains.

Intelligence test items are usually selected and scored to give a mean of 100 and a variance of 225 for the White population (Fig. 3.1) with a distribution of total scores which is 'normal', that is forming a bell-shaped curve with most in the middle and fewer the more extreme one goes. In fact the distribution of intelligence is not exactly normal but follows what is technically called a Pearson Type IV distribution, which means that there are slightly more individuals at each end of the distribution than the normal curve would predict.

With a normal distribution the *standard deviation* (SD), the square root of the variance, predicts the proportion of the population with a particular score. For example, 68% will have a score within one standard deviation above and below the mean and 96% will have a score within two standard deviations of this mean.

An example of such a prediction given by Jensen (1980) is the proportion of Blacks expected in classes for the educable mentally retarded (EMR) discussed in the Larry P. versus Wilson Riles court case in Chapter 2. Given that (a) 28.5% of the San Francisco school population is Black, (b) the IQ cutoff for EMR classes is 75 and (c) the average IQ for Blacks in California is 85 compared with a White mean of 100, then on the basis of the normal distributions, EMR classes would be expected to contain 68% Blacks and 32% Whites (Fig. 3.2). The actual Black percentage is 66%, slightly less than expected. Similarly, if the cutoff for placement in gifted classes is IQ 135, 2% Blacks and 98% Whites would be expected. The actual figures are 5.5% Blacks, 62.2% White, the remaining 32.3% being Orientals who make up a smaller proportion of the entire school population than the Blacks but a far larger proportion of the gifted children. Chapter 8 examines the consequences of such racial differences in educational opportunities.

The variance and standard deviation predict the scatter of individual scores around their mean. If several samples were taken

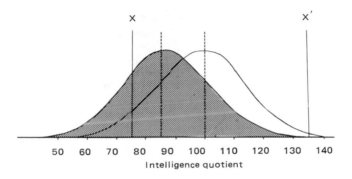

FIG. 3.2   Two normal distributions of IQ with means of 85 and 100. Two cutoffs of 75(X) and 135(X') illustrate the effects of the mean difference of 15 on the proportions outside the cutoff scores.

from the one population, one would expect the sample means to be distributed around the population mean with a certain amount of variation, although not as much as with individual scores. The variation among sample means is inversely proportional to N, the size of the sample — the larger the sample, the more likely it is to be close to the population mean and the smaller the variation between sample means. Dividing the variance by N gives the *variance of the mean* (V_m) and either the square root of this or the standard deviation divided by $\sqrt{N}$ (which are equivalent procedures) give the standard error (SE). To assess the range of sample means, the SE can be used in the same manner as the SD is used with individual values.

What determines the normal distribution?

Following the rediscovery of Mendel's work there was controversy over the inheritance of quantitative traits. How could Mendel's 'discrete factors' (genes) give rise to a continuous range of phenotypes? Was the variation genetic or due solely to environmental factors? The answers came in 1908 from work carried out by Nilsson–Ehle in Sweden and Johannsen in Denmark. Nilsson–Ehle examined wheat grains that vary in colour from red through to white. With a series of crosses he demonstrated that three genes were involved, the degree of redness depending on the total number of colour alleles present.

Johannsen examined seed weight in 19 lines of dwarf bean, obtained by selfing individuals picked from a population whose seed weight ranged from very light to very heavy. The 19 lines differed in seed weight and these differences persisted from generation to generation. There was some variation in seed weight within each line but the progeny of heavy and light seeds from the same line were similar in weight.

From these results, Johannsen concluded that variation within each pure line was environmental while that between his lines was genetic, reflecting the genotypes of the 19 seeds which were selfed.

Thus the continuous variation shown for seed weight by this species involved both genetic and environmental differences.

From such work and the subsequent pioneering efforts of Fisher and Wright, the following biometrical principles emerged: (a) there are several genes involved, each obeying Mendel's laws and (b) each allele has quantitative effects, adding or subtracting from the total phenotypic score. There is no need to recognize specific alleles each with biochemical or other distinguishing effects. (c) the effects of each allele are small relative to the total variation but their effects are cumulative and (d) environmental effects obscure any discontinuities between the genotypes.

It is possible to explain the normal distribution from these points. By chance alone most individuals would be expected to have a balance of alleles and environmental effects, some increasing and some decreasing performance. Therefore they would cluster in the middle of the distribution with fewer and fewer individuals having an excess of alleles or environmental effects for extremely high or low performance.

The next step is to ask precise questions about the determinants of performance for a particular behaviour, such as: what is the relative importance of genetic and environmental influences? Are the genes mainly autosomal or X-linked? Are the alleles for increased or decreased performance dominant? For convenience, such information is sometimes given the general name *genetic architecture*.

We shall introduce the methods of answering such questions about the genetic architecture through the simplest case of inbred animal strains and then move to the more complex situation of non-inbred animal stocks. Methods specific to humans are discussed in Chapter 6.

Inbred strains and the principles of biometrical genetics

An inbred strain is one made homozygous at most, if not all, loci through several generations of mating between close relatives. The result of inbreeding can be illustrated theoretically with a single locus with alleles A and a and equal gene frequencies ($p = q = \frac{1}{2}$). (Falconer (1981) and Mather and Jinks (1977) extend the method to situations where the alleles are present at different frequencies.) According to the Hardy–Weinberg law which assumes random mating, the genotypes from a cross between two Aa heterozygotes are present in the proportions:

$$\frac{1}{4}AA \quad \frac{1}{2}Aa \quad \frac{1}{4}aa$$

If mating occurs only between genetically identical individuals

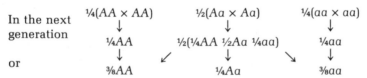

The frequency of heterozygotes has been reduced by 50%. After further generations (20 generations of brother–sister mating is the usual criterion (Festing 1975)) we have two inbred strains, one where all animals are homozygous *AA* and one homozygous *aa*. Thereafter, except for the rare case of a mutation, these must breed 'true' with every individual within that strain being genetically identical in every generation.

The possible number of inbred lines increases with the number of loci. With one locus, only two lines *AA* and *aa* are possible, but if a second locus with alleles *B* and *b* is added, then there are four possible lines *AABB*, *AAbb*, *aaBB* and aabb. One constraint on the number of lines is the effect of alleles which cause death or sterility when homozygous, representing the usual deleterious effects of inbreeding (Table 2.8). In one particular animal, the house mouse *Mus musculus* this is less of a problem because of some unique features of mouse population structure which we shall discuss in Chapter 5; some 300 inbred mouse strains have been established this century.

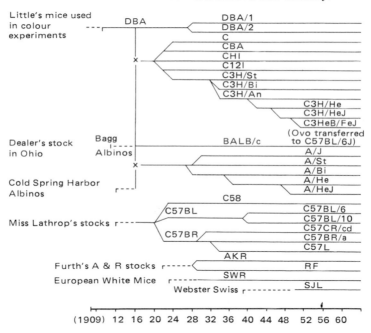

FIG. 3.3   Origins of some standard inbred strains of mice. (From J. Staats. (1966) in *Biology of the Laboratory Mouse* (2nd edn), Green E. L. (ed.) p. 109, McGraw-Hill, New York.)

There are also practical reasons for so many widely used inbred mouse strains. Rats have a slightly longer generation interval and hence are less likely to be used for inbreeding, whereas *Drosophila* have such a short generation time that laboratories often inbreed their own strains rather than bringing in established lines. Figure 3.3 lists the origins of the most common mouse strains. Notice that several of the strains such as A, C3H and C57Bl have distinct *substrains*, genetically different groups within the strains arising through mutation or contamination by accidental interbreeding with other mice.

Why do we need inbred strains?

There are four practical reasons why inbred strains are important in all behavioural research, not just in behaviour genetics.

1 *Reduced individual variation* If the variation between individuals can be reduced, then a more definitive conclusion can be drawn about the effects of a particular treatment on behaviour, as illustrated in Fig. 3.4. Here three inbred mouse strains are compared with a heterogeneous stock on a shuttle box task, where the animal learns to avoid an electric shock by shuttling into the adjacent compartment when a warning signal is presented. The performance of the heterogeneous stock is inconsistent, some mice learning quickly and some hardly learning at all. In contrast, all the mice from a particular inbred strain learn at similar rates. The genetic variation is still present in the inbred strains, but it is solely *between* the strains, the differences *within* the strains being environmental.

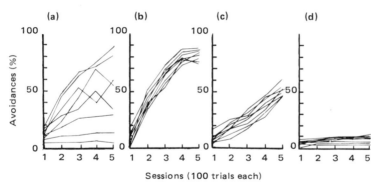

FIG. 3.4   The performance over five avoidance learning sessions each of 100 trials of individual mice from a heterogeneous stock (a) and from three inbred strains (b,c,d) on a shuttle-box avoidance learning task. (From D. Bovet, F. Bovet-Nitti and A. Oliverio (1969) *Science* **163**, 139.)

The advantage of the inbred strains is obvious. With the heterogeneous stock alone, one would doubt the value of a learning task where the performance varies so much between animals. Festing (1975) has quantified the advantage in experimental design which results from using inbred strains. The most widely used statistical test for comparing two experimental groups is the *t test* where the difference between the means is considered relative to their standard errors (the larger the SEs, the less conviction one has that the difference between the means is statistically significant and due to anything other than chance). If the SE within each group is reduced by using an inbred strain, that is by eliminating the genetic variation, then the chance of finding a significant result is increased.

Figure 3.8 demonstrates the role of genetic variance in increasing individual differences within a group. There is a much greater spread of scores in the backcross generation where (as explained later) there is genetic segregation and heterozygosity at some of the loci, than in the other generations within which there is genetic homogeneity. If one were testing the effects say of a drug or environmental treatment, it would be better to use one of the genetically homogeneous groups where one can be more certain that any differences result from the treatment than from random individual variation.

2 *Consistency of behaviour* The genotype of an inbred strain remains consistent from one generation to the next in contrast to a heterogeneous stock where the breeding population varies in genetic constitution. Figure 3.5 indicates just how consistent inbred strains can be. Eight inbred strains were measured in 1953 and in 1966 in laboratories on opposite sides of the Atlantic, the task being the open-field test where activity is measured by the number of squares an animal enters in a novel environment. Despite the difference in time and in location the ordering of the strains is very consistent.

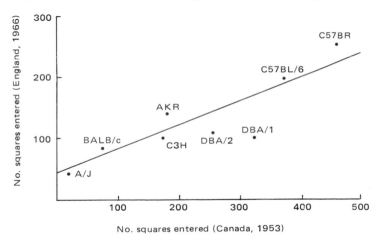

FIG. 3.5   The correspondence between the open-field activity of eight mouse strains tested in Canada in 1953 and England in 1966. The correlation (see later) between the two tests is 0.86, indicating a very high degree of similarity between performances on the two occasions. (From M. F. W. Festing (1975) *Food and Cosmetics Toxicology* **13**, 369.)

3 *Experimental design* The other aspect of consistency is that the same animal is not required for different measurements as long as another of the same genotype is available. Hence animals can be compared on different tasks without the problem of experience on one task confounding the subsequent ones, or animals can be sacrificed at different stages of the experiment for biochemical or other assays. An example where the use of strains avoids confounding is the question of whether an animal which learns one task quickly will necessarily perform well on a different learning task, that is whether or not there is a general learning ability or only task-specific

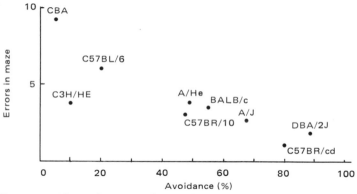

FIG. 3.6   The performance of nine inbred strains at the end of training on an avoidance learning task and a maze-learning task. (Derived from Figs. 4 and 5 in D. Bovet, F. Bovet-Nitti and A. Oliverio (1963) *Science* **163**, 139).

effects. Strains which make the most avoidances in the shuttle box tend to make the fewest errors on a maze-learning task (Fig. 3.6).

4 *Knowledge of the animal* With inbred strains one is aware of idiosyncratic factors which may complicate experimental results. For example, apart from cannibalism of their young which often happens when they are disturbed and which necessitates particular care in routine handling, some C3H substrains also carry an allele for retinal degeneration. This means that changes in performance as mice get older may not be due to maturation but to mice becoming blind and unable to see the cues.

Other differences are even more subtle. Some substrains of BALB/c often lack a corpus callosum, the bridge of fibres connecting the right and left hemispheres of the brain with potential consequences for many aspects of behaviour (Chapter 5). Strains also differ in the precise location of the different brain structures so that the stereotaxic atlases, the sets of coordinates for positioning electrodes or cannulae in precise locations in the brain have to be developed for each individual strain and cannot necessarily be generalized from one to the other.

What determines strain difference?

The mere observation that strains differ tells us nothing about the causes of these differences until we carry out breeding experiments crossing the strains in different ways. We can approach the analysis of such breeding experiments through examining the genetic and environmental influences on firstly the means of the different generations and secondly the variances within and between the generations.

Components of means

Take the simplest case of two inbred strains differing at a single locus, with alleles A and a. The genotypes of the two strains which we can call P_1 and P_2 ('P' for 'parent') are AA and aa respectively. They can be crossed to produce an F_1 (for 'first filial generation') with genotype Aa and the quantitative differences between the genotypes defined by three parameters:

$m = \frac{1}{2}(AA + aa)$, the average of the parental genotypes.

$d = \frac{1}{2}(AA - aa)$, the *additive* effect of the locus, the extent to which each of the homozygotes deviates from m.

$h = Aa - \frac{1}{2}(AA + aa)$, *dominance*, the difference of the heterozygote from the average of the two homozygotes. If there is no dominance, the heterozygote lies exactly in between them.

The three genotypes can thus be defined as

$$P_1 (AA) = m + d$$
$$P_2(aa) = m - d$$
$$F_1(Aa) = m + h$$

and arranged along a single scale as shown below

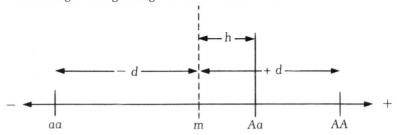

By convention AA is always taken as the homozygote with the higher score, so that d is positive. However A need not be the dominant allele, in that h will be positive if A is dominant and negative if a is dominant. The differences between the generations are now defined in terms of how genes work. The additive effect is the 'gene dosage' effect, the way in which aa is changed by substituting one 'dose' of A to give Aa or two 'doses' to give AA. Dominance recognizes that we must consider genotypes and not just genes, in that the value of Aa is more than merely the substitution of A for a.

One value of this approach is in predicting the scores of other generations. For example if F_1 females were crossed what would be the expected mean score of the F_2 progeny? Given equal frequencies of A and a the F_2 progeny would segregate: $\frac{1}{4}AA$ $\frac{1}{2}Aa$ $\frac{1}{4}aa$ with means $m + d$, $m + h$, $m - d$ respectively. The overall mean is simply the sum of these taking the frequencies into account

$$\frac{1}{4}(m + d) + \frac{1}{2}(m + h) + \frac{1}{4}(m - d) = m + \frac{1}{2}h$$

For example, if the scores of the aa, Aa and AA genotypes respectively were -2, $+1$ and $+2$, then $m = 0$, $d = 2$ and $h = 1$, with the expected mean of the $Aa \times Aa$ cross being $0 + \frac{1}{2}(1) = \frac{1}{2}$. Any other generation can be estimated in this fashion. You can check that the backcross B_1 of Aa with AA gives two genotypes $\frac{1}{2}AA$ and $\frac{1}{2}Aa$ and mean of $m + \frac{1}{2}d + \frac{1}{2}h$ with an expected value of $+1\frac{1}{2}$. Similarly the backcross B_2 to the recessive parent, aa is $m - \frac{1}{2}d + \frac{1}{2}h$ with expected value $-\frac{1}{2}$.

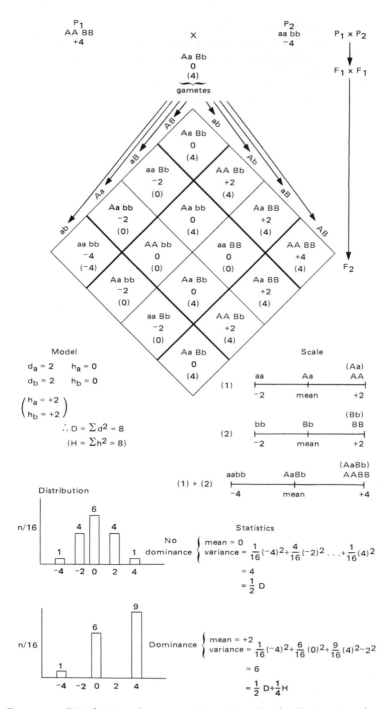

FIG. 3.7   Distribution of scores and statistics for the F_2 in a two locus situation, assuming either no dominance or (in brackets) complete dominance. The heavy lines on the chequerboard indicate genotypes with the same phenotype in the no-dominance situation. (See Mather and Jinks (1977) for a fuller description.)

We can extend this model to any number of genes as long as they act independently. If we have one locus with alleles A and a and a second locus with alleles B and b, the assumption of independence means that the difference between AA and aa is the same in $AABB$ compared with $aaBB$ as in $AAbb$ compared with $aabb$, that is differences between alleles at one locus are independent of differences in alleles at other loci. If this is not the case, we have the phenomenon of *non-allelic interaction* or *epistasis*.

Figure 3.7 shows what happens to the F_2 when there are two loci with additive effects d_a and d_b both equal to 2 and dominance effects h_a and h_b, under the situation where there is either no dominance ($h_a = h_b = 0$) or complete dominance at both loci ($h_a = h_b = 2$). Sixteen genetic combinations are produced, one of which has decreasing alleles only, one increasing alleles only, and the rest with some combinations of both. In the absence of dominance, the genotypes fall into five phenotypic classes forming the elementary normal distribution. When dominance is present there are only three phenotypic classes and the distribution is skewed. The presence of skewness can be used to detect dominance but is often not large enough to produce clearcut results. As the number of loci increases so too does the number of phenotypic classes and eventually when there are enough classes the distribution appears continuous.

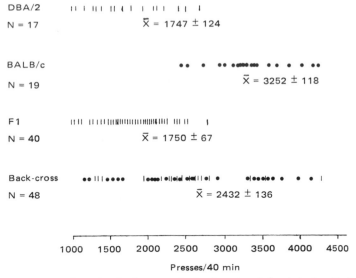

FIG. 3.8   The individual scores of mice from two inbred strains, their F_1 and backcross of the F_1 to the higher scoring parental strain (BALB/c). The measure is the number of barpresses for intracranial self-stimulation in a 40 minute period. | , Pigmented mice. ●, albino mice. The means (shown ± SE) are used to calculate the backcross mean of $m + \frac{1}{2}[d] + \frac{1}{2}[h]$ as follows:

$$m = \frac{1}{2}(1747 + 3252) = 2499.5$$
$$[d] = \frac{1}{2}(3252 - 1747) = 752.5$$
$$[h] = 1750 - \frac{1}{2}(1747 + 3252) = -749.5$$
$$\text{Expected mean} = 2499.5 + \frac{1}{2}(752.5) + \frac{1}{2}(-749.5) = 2501$$

(Data from P. Cazala and J. L. Guenet (1976) *Behavioural Processes* 1, 93.)

It is rarely possible to determine the number of loci in a polygenic system affecting behaviour, far less the additive and dominance effects at each locus but this does not matter. If we use the terms $[d] = d_a + d_b + \ldots$ and $[h] = h_a + h_b + \ldots$ to mean the cumulative effects over all loci, then as long as the overall additive and dominance effects can be calculated from the parental and F_1 means, the means of the other generations can be predicted.

The worked example in Fig. 3.8 involves the prediction of the mean of the backcross B_1 for the frequency of intracranial self-stimulation, a task where the animal presses a lever at a high frequency in order to receive an electrical current to a 'pleasure centre' in the lateral hypothalamus of its brain. The observed B_1 mean is 69 less than that predicted from the m, $[d]$, $[h]$ values, but this deviation from the expected of less than 3% can be attributed to chance alone, since it is much less than the SE.

There are two further points that follow from estimating these genetic parameters:

1 The parameters may vary with the strains involved. Figure 3.9 shows the activity of male mice of three strains and their F_1s when placed in the open-field. The F_1s between BALB/c and C3H showed negligible $[h]$ values, while $[h]$ values were large for the crosses to C57Bl. Figure 3.3 suggests that the history of the strains may explain this difference. BALB/c and C3H are related and hence may differ at fewer loci, so there will be fewer heterozygous loci and less dominance.

2 It is possible to test the validity of the genetic model. When we calculate m, $[d]$ and $[h]$ from two parental lines and their F_1 and use these values to predict the F_2 or another generation, we are testing the genetic model. If our prediction is inaccurate and the observed mean differs significantly from the expected mean, then something is wrong with the model and it must be modified.

A prediction is inaccurate because it fails to take into account one or more factors contributing to differences between the genotypes. Apart from non-allelic interactions, common factors include sex-linkage and effects of the maternal genotype. The next step is to modify the models specifying each genotype to include parameters for these additional factors. The methods of calculations are beyond the scope of this text but are detailed in Mather and Jinks (1977) with behavioural examples in Hay (1972).

We consider only one example, the role of maternal effects in the open-field activity of rats from the Maudsley Reactive (MR) and Non-reactive (MNR) strains bred selectively for emotionality in the open-field. The two reciprocal F_1s differ from each other, especially in the stressed group, where the mothers of the rats had been required to learn a shuttle box task prior to mating (Table 3.1). A model involving only m, $[d]$ and $[h]$ would not predict this difference between the F_1s and hence an additional parameter $[d_m]$ was introduced to specify the effects of the maternal genotype. In the control condition its effect is fairly small, but it is far larger in the condition where the mothers were stressed. The negative $[d]$ and positive $[d_m]$ counteract to give a

TABLE 3.1   Analysis of the maternal effect on the open-field ambulation of rats whose mothers were reared in control conditions or were stresses prior to mating (abbreviations described in text) (derived from J. L. Jinks and P. L. Broadhurst (1974) in *The Genetics of Behaviour* van Abeelen J. H. F. (ed.), p. 1. North-Holland, Amsterdam)

Generation	Expectation	Means Control	Stressed
P_1 (MNR)	$m + [d] + [d_m]$	44.97	35.33
P_2 (MR)	$m - [d] - [d_m]$	22.25	36.53
F_1 { MNR♀ × MR♂	$m + [d_m] + [h]$	33.45	37.40
F_1 { MNR♂ × MR♀	$m - [d_m] + [h]$	28.77	24.95

Parameter	Estimates	
m	36.11	35.93
$[d]$	6.53	-6.83
$[d_m]$	2.34	6.23
$[h]$	-5.00	-4.75

situation of 'maternal buffering' in the stressed groups where the MNR and MR rats have an intermediate phenotype irrespective of their genotype.

This demonstrates the main value of biometrical models in providing a greater understanding of the determinants of behaviour. Just knowing that there are additive and dominance effects may be of little consequence, but knowing that there are major maternal effects may be of great significance in understanding the behaviour. Table 3.1 shows that the means of each generation differ considerably between the control and the stressed situation but the biometrical model clarifies the situation — m and [h] are unchanged and only [d] and [d$_m$] alter, so that one can be more precise about how the stress is affecting the genotypes.

If enough generations and their reciprocal crosses are available, then many different models can be fitted. The model chosen is that which minimizes the difference between observed and expected across all genotypes as measured by a statistical test called the *Chi-squared* (χ^2) *test*. The smaller the χ^2 value the better the 'fit' between the observed and expected. It may be that with a particular set of data a model with m alone is sufficient and the χ^2 is not reduced significantly by adding any of the genetic parameters. In this case we would conclude that genetic factors are not involved, the differences between the genotypes being due to random environmental variation. Table 3 in Hay (1972) illustrates the use of this criterion in determining which of ten parameters were needed to explain differences between the means of 14 genotypes — some behaviours required few parameters to satisfactorily explain differences between the genotypes while others required as many as six parameters specifying additive, dominance, epistatic and X-linkage effects.

We return to the model-fitting approach in Chapter 6 because of the significance it has acquired in human behaviour genetics. If a model specifying simple genetic and enviromental effects adequately fits the data, then there is no need to introduce some of the more complex effects that have been hypothesized to explain human behavioural variation.

The limitation of analysing generation means

One major problem with the analysis of generation means may already be apparent. The assumption is that the alleles are *associated*. One parent has all the alleles for increased performance, the other all the alleles for decreased performance. This need not be the case and the increasing alleles could equally well be *dispersed* between the two parental genotypes. One could be ++.−−, homozygous for the increasing alleles at the first locus, homozygous for the decreasing alleles at the second locus, while the other parent is the opposite, −−.++. While these differ genotypically, their phenotypes are identical and [d] = 0. With more loci and with unequal effects at each locus, the parents may differ to some extent but the possibility remains that the calculated value of [d] may grossly underestimate the magnitude of additive effects.

A similar situation can apply with dominance effects that are assumed to be *directional*, that is at every locus dominance is in the same direction whether it be for high or low performance. But dominance could be *ambidirectional*, for increased performance at some loci, for decreased performance at others. The value of [h] may then be very small despite there being considerable dominance. For this reason, [h] is strictly called *potence* rather than dominance to indicate that it is not really an estimate of all the dominance effects.

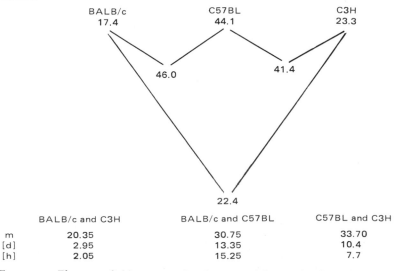

	BALB/c and C3H	BALB/c and C57BL	C57BL and C3H
m	20.35	30.75	33.70
[d]	2.95	13.35	10.4
[h]	2.05	15.25	7.7

FIG. 3.9   The open-field activity of male mice of three inbred strains and their hybrids, measured as the number of squares in the arena entered in 2 minutes. m,[d] and [h] are calculated as in Fig. 3.8. (Data from A. Rose and P. A. Parsons (1970), *Genetica* **41**, 65).

Ambidirectional dominance and dispersed additive effects imply that our analysis of the means alone may tell us little about the genetics of the situation. Together they can explain a phenomenon known as *heterosis* or *overdominance* and seen in the BALB/c x C57Bl cross in Fig. 3.9 where the F_1 scored higher than either parent. If the F_1 mean is elevated by directional dominance and the parental means limited by dispersion, it is possible for the F_1 to be outside the range of parental scores. The same situation under the name *hybrid vigour* is widely used in commercial breeding and Fig. 3.10 is a graphic illustration of its effects on plant growth. The effect is largest on the F_1 with a mean of $m + [h]$, less on the F_2 with mean $m + \frac{1}{2}[h]$ and is reduced correspondingly in subsequent generations.

FIG. 3.10   Hybrid vigour in maize. On the left are two parent strains, followed by typical plants from the F_1, F_2 ... F_8 generations of inbreeding towards the right. (From E. W. Sinnott, L. C. Dunn and Th. Dobzhansky (1958) *Principles of Genetics* (5th ed). McGraw-Hill, New York.)

While comparison of the means may provide limited information about genetics, in Chapters 4 and 5 we examine its usefulness in behavioural analysis and in understanding evolutionary pressures. If we want to know more about genetics, we must turn to the variances.

Genetic components of variances

Analysis of the variances rather than the means has two advantages. Firstly the larger variances of the segregating generation can reveal genetic effects even when the mean phenotypes do not. To take a two locus example, the two genotypes $++.--$ and $--.++$ may have the same phenotype and appear genetically identical but their F_2 will show segregation into different genotypes ranging from $++.++$ to $--.--$. Secondly, the question can be addressed of how much of the phenotypic variation is genetic and how much is environmental. Figure 3.8 showed that each generation has a characteristic variance as well as a mean level of self-stimulation. This variance can be predicted by the principles of biometrical genetics. In the same way that additive and dominance effects contribute to the mean, so also do they contribute to the variance. Since individuals in a generation may differ from each other because of additive or dominance effects, the variance must be divided into additive and dominance components, called D and H respectively.

The theory behind the D and H components of variance is shown in Fig. 3.7. If we have two alleles with additive effects $d_a = d_b = 2$, then $[d] = 4$ and $D = \Sigma d^2 = 4 + 4 = 8$ (the squaring is analogous to the variance formula as introduced earlier). Similarly, if there is complete directional dominance $h_a = h_b = 2$, $[h] = 4$, and $H = \Sigma h^2 = 8$. A simple single locus example is sufficient to calculate the extent to which D and H contribute to the F_2. If there are three genotypes in the proportion $\frac{1}{4}AA$ $\frac{1}{2}Aa$ $\frac{1}{4}aa$, their respective means are $m + d$, $m + h$, $m - d$ and the overall mean $m + \frac{1}{2}h$. m is constant and can be eliminated since it does not contribute to the variation. The variance within the F_2 using the usual formula for a variance is

$$\frac{1}{4}(d)^2 + \frac{1}{2}(h)^2 + \frac{1}{4}(-d)^2 - (\frac{1}{2}h)^2 = \frac{1}{2}(d)^2 + \frac{1}{4}(h)^2$$

Analysis of the crosses in Fig. 3.7 confirms this result. The variance is $\frac{1}{2}D$ if there is no dominance, $\frac{1}{2}D + \frac{1}{4}H$ if there is dominance. The genetic contribution to the variance in any other segregating generation such as the backcross in Fig. 3.8 can be calculated similarly.

Environmental effects can also contribute to the variance and can be estimated from the variances within the parental and F_1 generations. Although the means of these generations differ, every individual within that generation has the same genotype, so that the only variance between the individuals must be environmental. We recognize two sources of environmental variance, that between individuals within a family (E_w) and that between family means, or in the case of animals between litter means, (E_b). With families of size N, variance between family means must contain at least the component E_w/N, due to the contribution of between individual variance over and above any real environmental differences between families.

The values of D, H, E_b and E_w can be calculated from the statistics of different generations just as $[d]$ and $[h]$ were, if sufficient genotypes are available. More segregating generations will be needed since the parental and F_1 variances involve only environmental effects. There are few behavioural examples of biometrical analyses based on variances among inbred strains and their crosses, since the techniques of variance analysis are used far more with non-inbred groups where analyses of means are not possible. The most widespread function with inbred strains is in calculating the *heritability* which is easier to appreciate in the alternative notation of Falconer (1981).

The different notations

In the Birmingham notation (Mather and Jinks 1977) and the Edinburgh notation (Falconer 1981) there are similar terms for additive and dominance contributions to the means — except that they are called respectively 'd' and 'h' in the former system and 'a' and 'd' in the latter. The distinction in the variances is more complex. As we have just seen, the Birmingham notation uses the additive and dominance components to derive the variance in an F_2 and indeed in any random-mating population as $\frac{1}{2}D + \frac{1}{4}H + E_b + E_w$. The Edinburgh model starts with the approach that the total phenotypic variation (V_P) in an F_2 or a random-mating population comprises genetic (V_G) and environmental (V_E) components with the genetic part being divisible into additive (V_A) and dominance (V_D) components:

$$V_P = V_A + V_D + V_E$$

Proportions of V_A and V_D can be derived for other generations and are interchangeable with D and H, as long as one remembers $V_A = \frac{1}{2}D$ and $V_D = \frac{1}{4}H$.

Heritability

Heritability is the statistic most widely used and abused in summarizing the results of genetic analysis. It is the proportion of the total variation which is genetic and is usually written as h^2. Thus

$$h^2 = \frac{V_G}{V_P} = \frac{V_A + V_D}{V_A + V_D + V_E} = \frac{\frac{1}{2}D + \frac{1}{4}H}{\frac{1}{2}D + \frac{1}{4}H + E_b + E_w}$$

are all alternative ways of expressing the heritability. Figure 3.11 is a pictorial representation of heritability. Given a particular level of heritability (80% in the case of Fig. 3.11) the phenotypic variation would be reduced considerably if genetic variation were eliminated, that is if every one were of identical genotype, but would be reduced to a far lesser extent if environmental differences were eliminated. However, Fig. 3.11 indicates that there are still environmental differences despite the high heritability.

The use of heritability estimates is frequently criticized because they are limited in two major ways:
1   Heritability estimates are only valid for that population at that point in time and are sensitive to changes in environmental and genetic variation. If we take a human example, educational attainment, it is clear that as more people get access to adequate educational resources, variation due to the environment may diminish and hence the heritability may rise. Similarly if the sample is restricted in other ways, for example by involving only the families of college academics, the genetic as well as the environmental variation may be limited and a heritability value obtained that is unrelated to that obtained in a more representative sample of the population.

(a)

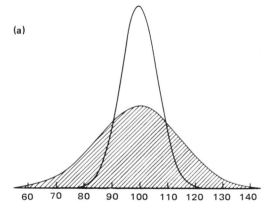

| 60 | 70 | 80 | 90 | 100 | 110 | 120 | 130 | 140 |

(b)

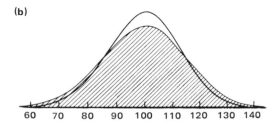

| 60 | 70 | 80 | 90 | 100 | 110 | 120 | 130 | 140 |

FIG. 3.11   The implications of a heritability of 80%. Compared with the total phenotypic variation (shaded curve), the distribution in (a) shows how the variation would be reduced if all genotypes were identical and (b) the converse, if all environmental causes of variation were eliminated leaving only genetic variation. (From H. J. Eysenck (1971) *Race, Intelligence and Education.* Temple Smith, London.)

2 A high heritability does not mean that the behaviour cannot be changed. The heritability only indicates the range of variation due to the genetic and environmental factors currently in operation. If some new environmental influence or some new training program is introduced, there may be major changes in performance irrespective of the original heritability.

'Narrow' and 'broad' heritability

There are two types of genetic variation, additive and dominance and therefore there can be two heritabilities. One involves all genetic effects as discussed above. It is called the *broad* heritability (h^2_b) to distinguish it from the *narrow* heritability (h^2_n) which includes only additive effects. Thus

$$h^2_n = \frac{V_A}{V_A + V_D + V_E} = \frac{\frac{1}{2}D}{\frac{1}{2}D + \frac{1}{4}H + E_b + E_w}$$

The narrow heritability is obtainable directly from inbred strains, which are all homozygous and where dominance cannot be involved. If data on several strains are available h^2_n is easy to calculate using the statistical technique of *analysis of variance*, where the variation within strains due only to environmental effects is compared with the variation between strains where variation is partly additive genetic, partly environmental. As well, there may be strain differences in maternal effects which are both environmental and part of the genotype. The larger the narrow heritability, the greater the between strain variation relative to that within strains.

Alternative breeding programs

The only breeding program discussed so far has involved two inbred strains and the crosses such as the F_2 which can be derived from them. This technique has two limitations in that it involves just laboratory inbred animals and relies only upon two parental genotypes. Many arguments surround the use of laboratory rodents which may be a very selective sample of the available *gene pool* or range of variation in the wild and breeding programs have been developed that can involve both wild and laboratory mice (Chapter 5). The problem of relying on only two strains was shown in Fig. 3.9, where the genetic architecture depended on the history of the pair of strains involved.

It would be better to have more genotypes involved and in the course of biometrical genetics (Mather and Jinks 1977) many techniques of this sort have been developed. One of the most relevant is the *diallel cross* where several strains are crossed in all possible combinations to give a wider sampling of genotypes. In the full-diallel each strain is used as both the male and female parent, permitting analysis of reciprocal differences due to X-linkage or often to maternal effects. If reciprocals are ignored, there is the *half-diallel* situation such as that shown for nine *Drosophila* strains in Fig. 3.12. In this case the task was an avoidance learning situation discussed in more detail in Chapter 4. This particular diallel shows marked potence for high learning ability. More detailed analyses of the diallel are available to measure additive, dominance, maternal, X-linked and epistatic effects (Mather and Jinks 1977).

The best analysis to use depends upon the objective. For a broad overview of genetic variation the diallel is ideal because it involves such a variety of strains. The combination of crosses between two strains is more suitable for a detailed analysis of phenomena such as maternal influences because quite unusual crosses can be created to examine particular effects. Only rarely have both types of analyses been compared with the same behaviours and the same strains. While the results usually correspond (Hay 1972) there is some divergence depending on the features of the strains chosen for the detailed analysis.

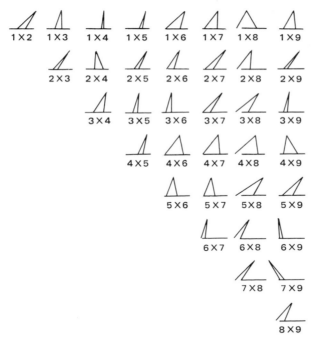

FIG. 3.12   A 9 × 9 diallel cross for avoidance learning in *Drosophila*. For each of the 36 F_1s the base of the triangle represents the learning score on a scale from 0.2 to 0.4 for their parental strains numbered 1-9, while the apex is the score of the F_1. (See Fig. 4.15 for more details of the test and scoring procedure.) Most triangles lean to the right as there is directional dominance for a high rate of learning — 31 F_1s score higher than the average of their parents. (From D. W. Fulker (1979) in *Theoretical Advances in Behavior Genetics* Royce J. R. and Mos L. P. (eds). p. 337. Sijthoff and Noordhoff, Alphen aan den Rijn.

Genotype–environment interaction

The concept of genotype–environment interaction was introduced in Chapter 2, but it is particularly easy to observe in inbred strains where the genotypes are so clearly defined. Figure 3.13 demonstrates that while strains differ in performance, and environmental enrichment reduces the time it takes mice to reach food at the end of a maze, there is also genotype–environment interaction. Some strains such as A/J, RF and DBA improve only marginally if at all as a result of the enrichment, whereas BALB/c, C3H and C57Bl improve markedly.

Genotype–environment interaction is a situation that has been found in many aspects of animal behaviour. Erlenmeyer-Kimling (1972) listed 33 behaviours on which two or more strains were measured after exposure to different treatments and in 14 of these the treatment had opposite effects on some of the strains. The term 'treatment' covers almost any environmental variable including drug administration. Chapter 5 introduces the area of *psychopharmaco-*

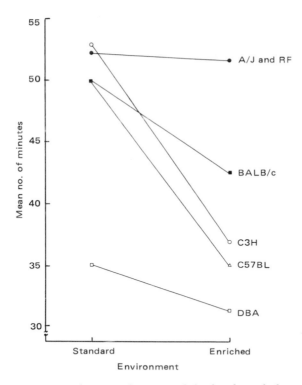

FIG. 3.13   The time taken to reach food at the end of a complex maze for six inbred strains of mice reared in standard and in enriched environments. Two strains A/J and RF performed identically. (From L. Erlenmeyer-Kimling (1972) in *Genetics, Environment and Behavior: Implications for Educational Policy* Ehrman L., Omenn G., S. and Caspari E. (eds), p. 181. Academic Press, New York.)

genetics, or genetic differences in drug effects on behaviour and Fig. 3.14 provides an initial example of why genotype–environment interaction makes the use of inbred strains so important in psychopharmacology. With the non-inbred mice used in Fig. 3.14 there is no consistency of response, **some** becoming more active, some less active in response to the drugs. The overall conclusion is that there is no uniform drug effect, whereas it can be shown that the opposite result emerges if less variable inbred strains are used.

One of the few consistent forms of genotype–environmental interaction is that on average hybrids tend to be less affected by environmental trauma. Figure 3.15 is the summary of what happened to mice of 16 different genotypes (four strains, 12 hybrids) in a 4 x 4 diallel cross when either left undisturbed until testing at 10 weeks (U), or handled and exposed to a maze (H) or trained on a shock avoidance task (S). Apart from the variety of response patterns, the thing to note is the difference between the inbred and hybrid averages. The inbreds change considerably when exposed to the intermediate level of stimulation, whereas the hybrids are affected by only the most extreme situation. This phenomenon has been given the name

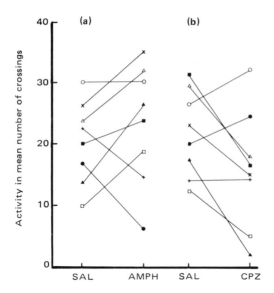

FIG. 3.14   The activity in a tilt-floor box of individual random-bred mice when given saline and 1mg/kg of either (a) amphetamine (AMPH) or chlorpromazine (CPZ). (From A. Oliverio (1974) in *The Genetics of Behaviour*, van Abeelen J. H. F. (ed.) p. 375. North Holland, Amsterdam.)

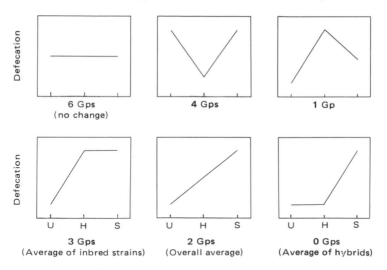

FIG. 3.15   Open-field defecation of 16 genotypes (Gps) of mice in a 4 x 4 diallel cross. The treatments are undisturbed (U), handled (H) and trained to avoid shock (S). (From N. D. Henderson (1969) *Annals of the New York Academy of Sciences* **159**, 860.)

heterozygous buffering, implying that hybrids are less affected by environmental stimulation in general.

There have been few attempts to find if environmental responsiveness is in itself under genetic control. The mouse strain C57Bl tends to respond to more treatments than do other strains especially BALB/c but it is difficult to be more precise. Erlenmeyer-Kimling (1972) suggests that C57Bl may be affected more by specific and traumatic effects such as shock and noise than by background variables such as cage illumination or degree of environmental enrichment, with the reverse pattern applying to BALB/c.

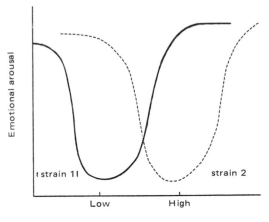

Intensity of stimulation

FIG. 3.16   Hypothetical example of how with a consistent U-shaped relationship between stimulation and emotional arousal, two strains with different means can show opposite arousal responses at two levels of stimulation. (Derived from N. D. Henderson (1968) *Developmental Psychobiology* **1**, 146.)

The importance of finding a basis to environmental responsiveness is that it may create some consistency out of what appears to be chaos, with the nature of genotype–environment interactions being totally unpredictable. Figure 3.16 shows an attempt by Henderson to define one factor that may underlie differences between genotypes. If two strains each show the same U-shaped relation between prior stimulation and emotional arousal (meaning that both over- and under-stimulation have deleterious effects) but differ in the optimal level of stimulation, then testing at only two points on the stimulation continuum will produce contradictory results. The treatment increases the arousal of strain A but reduces the arousal of strain B although the strains really respond in the same way to arousal. The same situation could apply to a wide range of behavioural and bio-chemical factors that change in a U-shaped fashion with environmental stimulation.

Henderson warned that his example meant that the simple situation of two genotypes tested in two environments was sufficient only to reveal genotype–environmental interaction, not to contribute to any understanding of the nature of this interaction. Although genotype–environment interaction studies in behaviour genetics have not advanced much further, it has been possible in experimental plants to breed for high or low environmental responsiveness quite independently of breeding for overall performance. Therefore genotype–environment interaction is part of the phenotype and much more than just an artefact of scaling or the outcome of unreliable measurement, two criticisms that have been levelled at this concept.

This result has important implications for genetic analyses of the variances. As well as interactions with the major environmental variables, there may also be interactions at the micro-environmental level to do with the random uncontrollable differences in food, temperature etc. within and between litters. For this reason the variance within the F_1 may not be the same as those of the parental strains. If there is heterozygous buffering the F_1 variance will be less. On the other hand there are several cases where the F_1 variance is greater and it has been suggested by Caspari that this responsiveness may be advantageous for the animal.

TABLE 3.2   The variances within six generations of mice on a measure of total activity during initial 10 seconds of electric shock (derived from T. G. Newell (1970) *Journal of Comparative and Physiological Psychology* **70**, 37)

Generation	Male	Female
P_1 (BALB/c)	17.3	12.2
P_2 (DBA/8)	32.7	11.1
F_1	10.0	12.6
F_2	26.2	20.0
B_1	12.2	16.6
B_2	22.3	13.5

At the same time, the variances of the segregating generations may also be affected producing results of the sort shown in Table 3.2. In the case of the males all three segregating generations were less variable than the DBA/8 parent. In contrast the female variances showed a completely consistent pattern with both parents and the F_1 having similar variances, all of which were less than those of the segregating generations. The narrow heritability was calculated as 0.63 for the females, but no heritability estimate could be obtained for the males. The paper from which Table 3.2 is taken gives several other examples of the effects of genotype–environment interaction on variances and the distortions such as negative additive and dominance variances which can result (variances must be positive because they are calculated on deviations squared and any number squared is positive).

Discussion of inbred strains

The fact that inbred strains have been used to illustrate the principles of biometrical genetics should not obscure their relevance for the behavioural scientist who has no interest in genetic variation. Apart from their role in experimental design discussed earlier, an additional concern is that differences between inbred strains can limit the generality of behavioural models. Two examples of this are:

1   When food-deprived DBA/2J mice are reinforced on a schedule where food is delivered intermittently, they exhibit an excessive

water intake or polydipsia which does not occur in C57Bl/6J mice.

2 Much work has been done on the protein synthesis inhibitor, cycloheximide which affects memory storage, such that good initial retention is followed by complete loss within 6 hours of training. While DBA/2J mice show this phenomenon, C57Bl/6J respond very differently and demonstrate complete memory impairment immediately after training.

Even though inbred strains are often criticized as being 'non-animals' — they do not exist in the wild and may be unable to survive if placed there — they do have a major part to play in laboratory research. They may not be representative of the species as a whole, but they can be used to test just how representative a behavioural result may be. Does it apply to all genotypes or is it confined to only a few?

Non-inbred experimental animals

When inbred strains are not available two different approaches exist for the genetic analysis of behaviour. One is based upon the resemblance between animals of different degrees of relationship. This we shall consider only briefly, partly because it has not been widely used and also because its value is limited. Once the genetic architecture has been determined by this method there is not a great deal one can do in terms of subsequent behavioural analysis. This is in contrast to the other technique, selective breeding, where the interest is often less in the genetic aspects of selection and more in the range of behaviours altered as a result of the selection process.

Resemblance between relatives

In order to compare the degree of resemblance between relatives we need a statistical measure of the similarity of their scores. Figure 3.17 shows what happens if we have two measures, X and Y plotted against

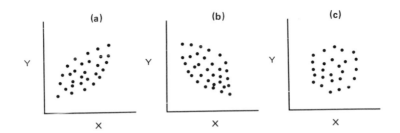

FIG. 3.17   Possible relationships between two variables, X and Y where X and Y are (a) positively related — as X increases, Y increases (b) negatively related — as X increases, Y decreases and (c) are independent so that high values of X are equally likely to be associated with high or low values of Y and vice versa.

each other. X and Y can either be measures of different things on the same individual, e.g. height and weight, or they can be measures of the same thing in different individuals, e.g. X could be a brother's height, Y his sister's height. In the former case there are N individuals, in the latter N pairs of individuals. If higher values of Y are associated with higher values of X, the result in Fig. 3.17a is obtained, if the reverse, Fig. 3.17b and if there is no relationship, the random pattern in Fig. 3.17c is obtained.

To calculate a statistic which measures the extent of any association, consider the distribution divided into four quadrants around the means $\overline{X}$ and $\overline{Y}$.

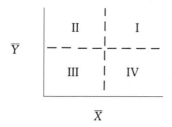

If the X and Y scores for a given individual are both above their respective means, the product $(X-\overline{X})(Y-\overline{Y})$ is positive. In quadrants II and IV, one of the $X-\overline{X}$ or $Y-\overline{Y}$ values is negative, the other positive, so that $(X-\overline{X})(Y-\overline{Y})$ is negative. In quadrant III both are negative, so their product is positive.

Hence $\Sigma\,(X-\overline{X})(Y-\overline{Y})$, the *sum of products* for all points in the distribution serves as a measure of association between X and Y. If the association is positive, most points in the distribution lie in quadrants I and III and the sum of products will be positive. In a negative association, most points lie in II and IV and the sum of products will be negative. If there is negligible relation between X and Y, the points will be equally scattered over the four quadrants and the sum of products will tend to be zero or very small.

Since the sum of products increases with the number of observations, association is usually expressed by the *covariance* (COV_{xy})

$$\frac{\Sigma(X-\overline{X})(Y-\overline{Y})}{N-1}$$

or more conveniently

$$\frac{\Sigma XY-(\Sigma X\Sigma Y)/N}{N-1}$$

where the alternative formula and the division by $N-1$ rather than N are analogous to the formulae for the variance discussed earlier.

The only other thing needed to make the covariance a useful statistic is to take into account the units in which X and Y are measured, by dividing by SD_x and SD_y the standard deviations of X

and Y. The end result is the Pearson product moment *correlation* coefficient (r_{xy})

of $COV_{xy}/(SD_x.SD_y)$ or in more detail

$$r_{xy} = \frac{\Sigma XY - (\Sigma X \Sigma Y)/N}{\sqrt{(\Sigma X^2 - (\Sigma X)^2/N)(\Sigma Y^2 - (\Sigma Y)^2/N)}}$$

(The divisor N-1 is omitted as it is common to both the numerator and denominator.)

Table 3.3 is an example of correlation, showing that even though boys are taller than their sisters, their heights are correlated. The maximum value of a correlation is +1 if there is a perfect positive correlation between two variables, ranging through 0 if no relationship exists, to -1 if there is a complete negative relationship. Errors of measurement and unpredictable random variation mean that a perfect correlation is rarely observed.

TABLE 3.3   Example of correlation and regression

The heights of 11 pairs of brothers and sisters were measured in centimetres, giving

Family	1	2	3	4	5	6	7	8	9	10	11	
Brother (X)	180	173	168	170	178	180	178	186	183	165	168	$\bar{X} = 175.36$
Sister (Y)	175	162	165	160	165	157	165	163	168	150	157	$\bar{Y} = 162.45$

$\Sigma(X - \bar{X})^2 = 478.54;\ \Sigma(Y - \bar{Y})^2 = 428.73;\ \Sigma(X - \bar{X})(Y - \bar{Y}) = 259.18$

$$r = \frac{259.18}{\sqrt{478.54 \times 428.73}} = \frac{259.18}{452.95} = +0.5722$$

While brothers tend to be taller than their sisters, there is a positive correlation between their heights, i.e. in families where the boy is tall, so is his sister.

If the heights given for brothers had in fact come from the fathers, then it would be possible instead to calculate b the slope of the regression line (the average increase in Y per unit increase in X)

as $$b = \frac{259.28}{478.54} = 0.54$$

The advantage of regression over correlation is that given any value X_i of the independent variable we can predict the corresponding Y_i value of the dependent variable using the formula

$$Y_i = \bar{Y} + b(X_i - \bar{X})$$

If the father were 180 cm tall, the predicted height of his daughter would be

$$Y_i = 162.45 + 0.54\ (180 - 175.36) = 162.45 + 2.51 = 164.96$$

The correlation coefficient is used when two variables are inter-related. If we can go further and say that one variable depends upon the other, then a different statistic, the *regression coefficient* (b) is available. In the example in Table 3.3 one cannot say that the height of a brother depends upon that of his sister or *vice versa*. But suppose fathers had been measured rather than brothers. Then the height of the girls would partly depend upon their father's height and certainly not the reverse. We call the father's and the girl's heights the *independent* and the *dependent* variables respectively. The regression formula is very similar to that of the correlation coefficient except that the denominator is the variance of the independent variable so $b = COV_{xy}/V_x$ if X is the *independent variable* or $b = COV_{xy}/V_y$ if Y is the independent variable. This makes sense in that the correlation coefficient where neither X nor Y is the independent variable has the harmonic mean of their variances $\sqrt{V_x.V_y}$ or $SD_x.SD_y$ as the denominator. b is the slope of the regression line predicting how the dependent variable alters with changes in the independent one.

The genetics of family resemblance

The next step is to explain covariances, correlations and regressions in genetic terms and Table 3.4 lists the genetic expectations for many of the more commonly-observed relationships. Falconer (1981) presents detailed derivations of these formulae but the basic principle is straightforward. An offspring gets half its genes from each parent, so that the genetic covariance between parent and offspring will equal half the additive variance. An alternative way of thinking about it is that for every locus the chance is half that the offspring will inherit one particular allele a parent has rather than the other. In the case of two full-siblings (abbreviated to full-sibs), they will have half their additive genes in common, plus an additional resemblance due to dominance since they have genotypes as well as genes in common. For one quarter of the loci both sibs will receive the same combination of alleles from their parents and their resemblance will be increased by one quarter of the dominance variation. That is, while an offspring may get an A allele from one parent and so partly resemble him or her, two offspring of the same parents may both be Aa, sharing genotypes and not just genes.

TABLE 3.4   Resemblance between relatives

Relationship	Covariance (in both notations)	Correlation	Regression
Parent, child	$\frac{1}{4}D$ or $\frac{1}{2}V_A$	$\frac{1}{2}h^2_n$	$\frac{1}{2}h^2_n$
Midparent, child	$\frac{1}{4}D$ or $\frac{1}{2}V_A$	$h^2_n\sqrt{2}$	h^2_n
Full-siblings	$\frac{1}{4}D + \frac{1}{16}H$ or $\frac{1}{2}V_A + \frac{1}{4}V_D$	$> \frac{1}{2}h^2_n$	—
Half-siblings	$\frac{1}{8}D$ or $\frac{1}{4}V_A$	$\frac{1}{4}h^2_n$	—
Uncle, nephew			$\frac{1}{4}h^2_n$
Aunt, niece			
First cousin	$\frac{1}{16}D$ or $\frac{1}{8}V_A$	$\frac{1}{8}h^2_n$	—

This raises an interesting point. If full-sibs have $\frac{1}{4}D + \frac{1}{16}H$ (or $\frac{1}{2}V_A + \frac{1}{4}V_D$) in common, and the total genetic variance is $\frac{1}{2}D + \frac{1}{4}H$ (or $V_A + V_D$), then they must differ by $\frac{1}{4}D + \frac{3}{16}H$ (or $\frac{1}{2}V_A + \frac{3}{4}V_D$). In other words, half their genes and the majority of the dominance effects will be different, so that there are actually more genetic effects making full-sibs *different* from each other than making them *similar*. This result has many implications explored in Chapter 7 and 8 in the context of human society where genetics is often condemned as a means of maintaining the status quo, of retaining resources within those families that have them already.

Family members will also be more similar as a result of environmental influences. As noted earlier, we recognize E_b as the environmental effects that are common to members of the same family, making them more alike and that therefore contribute to environmental differences between families. The remaining environmental effect is E_w, the within family differences due to birth order, competition between siblings or any of the multitude of environmental factors that may differentiate between offspring of the same parents. Our covariance should include some contribution of E_b to take into account the fact that parents pass on environmental cues as well as genes to their offspring and that full-sibs share the same environment as well as the same genes and genotypes. One major difficulty is that although each covariance can be specified in terms of the same additive and dominance genetic parameters, any environmental effect may be unique to that covariance. For example, the nature of the environmental similarity between parents and children differs from that between siblings. If estimates of heritability derived from different relationships differ considerably, then such environmental factors specific to each relationship may be involved.

The general formulae for both correlation and regression involve dividing the covariance by some combination of one or more variances. Since heritability comprises the genetic divided by the total variance and the covariance between relatives is an estimate of some proportion of the genetic variance, we have only to find the correct denominator to be able to use correlations and regressions as estimates of heritability. If we assume that the total phenotypic variation (P) is the same in each generation and that there are no biases such as assortative mating then $P = \frac{1}{2}D + \frac{1}{4}H + E_b + E_w$ and the correlation between offspring and parent (r_{op}) is

$$r_{op} = \frac{1}{4}D / \sqrt{P.P} = \frac{1}{2}h_n^2$$

since the variance of both offspring and parents is P.

A more complicated example is that of the correlation between offspring and midparent, $(r_{o\bar{p}})$, the midparent being the mean of the scores of both the father and mother. Since the variance of a mean is $1/N$th the individual variance, the variance of the midparent values must be $\frac{1}{2}P$. Hence

$$r_{o\bar{p}} = \frac{1}{4}D / \sqrt{\frac{1}{2}P.P} = h_n^2 / \sqrt{2}$$

Regression can be calculated in a similar fashion but with a different

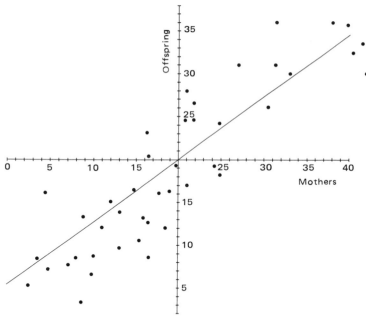

FIG. 3.18   The regression of offspring on mothers for avoidance learning in mice. The axes intersect at the means of the performance of all mothers and offspring. (From A. Oliverio (1971) *Journal of Comparative and Physiological Psychology* **74**, 390).

denominator. While the regression of offspring on parent is

$$b_{op} = \tfrac{1}{4}D/P = \tfrac{1}{2}h_n^2 = r_{op}$$

the regression of offspring on midparent $(b_{o\bar{p}})$ has a different denominator from r_{op} and is a different estimate of heritability

$$b_{o\bar{p}} = \tfrac{1}{4}D/\tfrac{1}{2}P = h_n^2$$

That the regression of offspring on midparent is a direct estimate of the narrow heritability makes it a convenient and widely used measure in genetics, especially as it can be shown mathematically that this estimate is free of bias due to assortative mating, something which does not apply to the regression of offspring on one parent.

Regression in this context means the same as regression in a non-statistical context. The regression coefficient *b* predicts how far the mean of the offspring will regress from the parental mean back towards the population mean. For example, Fig. 3.18 shows that avoidance learning has a considerable genetic component ($h^2 = 0.52$) with the offspring of high-scoring mothers learning quickly, while the offspring of low-scoring mothers learn slowly. But because the heritability is less than 1, the regression line is not as steep as it would be if there were complete genetic determination. That is the offspring are regressing to the mean, the progeny of low-scoring mothers doing relatively better than their mothers and the progeny of high-scoring mothers relatively worse.

'Of course there aren't any women. How can a woman contribute to a sperm bank?'

FIG. 3.19 Regression to the mean is only one potential problem facing a recently developed sperm bank where all the donors are Nobel Prize winners. (From *New Scientist* (1980) **88**, 140.)

But certainly in the human case and perhaps too with animal examples, the cause of regression to the mean may not be genetic. In Terman's famous study of 'gifted children', a group with mean IQ of 152, their children had a mean IQ of 133, still well above the population mean but not as high as their gifted parent. This result need not imply that IQ is largely genetic for the reason mentioned earlier that the covariances include an undetermined environmental component. One could argue that the 'gifted children' were so able not because of their genes but because of especially favourable environmental conditions. Their children in turn would be expected to receive a good environment for the nurture of ability but probably not as good as that of the parents, so that their IQs would be lower. It is impossible on the basis of this set of data alone to decide between the conflicting genetic and environmental explanations. Fulker and Eysenck (1979) provide other examples of regression including one based on the Reed and Reed data discussed in Chapter 2.

In experimental animals there is little need to use the methods of regression and correlation. The uncertainties introduced by dominance and by common environmental effects mean that it is usually far more efficient to use inbred strains or some alternative breeding program.

Artificial selection

Artificial selection provides the most obvious evidence for genetic determination. It involves starting with a heterogeneous population

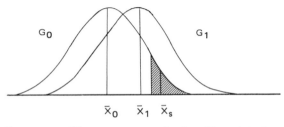

FIG. 3.20   The response to selection. The original population has mean $\bar{X}_0$ and distribution G_0 (for generation 0), out of which a fraction with mean $\bar{X}s$ are selected. The offspring of the selected group have a mean $\bar{X}_1$ with distribution G_1.

containing much genetic variability and over a period of several generations selectively breeding high-scoring males and females together and low scoring males and females together. By this stage if the trait is at least partly under genetic control there should be two distinct lines, one where all individuals score high and one where all score low.

The genetic basis for this technique depends on the polygenic model with many alleles increasing or decreasing performance. High scoring individuals would be expected to have more increasing than decreasing alleles and low scorers the reverse. Artificial selection concentrates all the increasing alleles in one line, all the decreasing alleles in the other. Each generation of selection should move the mean of each selection line further away from the mean of the original population as shown in Fig. 3.20. The response to selection will depend upon two factors:

1 S, the selection pressure   When a trait is polygenically determined, we expect the highest scoring individuals to have more of the increasing alleles. So the more extreme the individuals chosen — say the top 5% instead of the top 20% — the more increasing alleles the group should have and the higher should be the mean of their progeny.

2 h_n^2, the narrow heritability   Selection is only going to work if the character is genetically determined to a reasonable extent. If $h_n^2 = 0$, that is if the high scoring individuals result only from transient environmental factors, then their offspring would be expected to have the same mean as the original population. They would demonstrate the phenomenon of regression to the mean which we discussed earlier.

The Tryon 'maze-bright' and 'maze-dull' strains

The Tryon strains are the best known example of artificial selection. In the 1920s when much work on animal learning involved complex mazes, Tolman initiated selection in rats for maze performance and produced considerable divergence between the lines after only two generations. However, the problem was that selection was based on a combination of maze errors, running time and number of completely

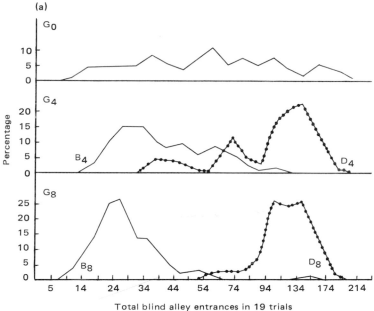

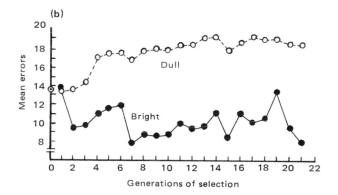

FIG. 3.21   Response to selection of the Tryon maze-bright and maze- dull rat strains. (a) The distribution of scores in the original population G_0, and in the bright (B) and dull (D) lines at generations 4 and 8 of selection. (Derived from R. C. Tryon (1940) *National Society for the Study of Education*, 39th Yearbook, part 1, p. 111.) (b) The changes over all 21 generations of selection — these are rescaled results and are not on the same scale of measurement as (a). (From G. E. McClearn (1963) *Psychology in the Making* Postman L. J. (ed.), p. 144. Knoff, New York.)

correct trials and the unreliability of this complex scoring system even within the same rat tested on different occasions led him to cease the experiment. Tryon began a more elaborate selection experiment using the number of errors made by the rats on a 17-choice maze known to have high reliability. Because selected individuals may be of like genotype as well as phenotype, he deliberately tried to reduce inbreeding by limiting the number of matings between full sibs and by picking only fertile, healthy animals.

Figure 3.21a shows how the lines became increasingly distinct during the course of eight generations. In the original unselected population there was wide variation in the number of errors. After four generations of selection, the two lines were starting to emerge although still showing considerable overlap, but after eight generations the lines were completely distinct. Selection continued for a total of 21 generations but little additional response to selection was observed (Fig. 3.21b). The lines have since been maintained by random mating within each line and are known as the Tryon 'maze bright' (TMB) and 'maze dull' (TMD) strains.

FIG. 3.22   The breeding of 'maze bright' (TMB) strains.

The heritability measured as the response to selection in this situation is 0.21. But apart from the time and effort involved in such a breeding program, there are several factors confounding the response which means that artificial selection is not the optimal nor necessarily an accurate means of determining the genetic architecture.

1 *Environmental effects*   During the several years which a selection experiment may occupy there can be many changes in the environment. For example there may be new personnel with different skills in handling the animals, the techniques of behavioural assessment may improve with the constant repetition or may be changed for some reason. It is therefore advisable to incorporate a control line, a group of unselected animals maintained and tested in the same manner as the selection lines. Any change in their behaviour between generations reflects changes in the environment not in genetics and can be taken into account in the analysis of the selection lines. Perhaps the most

significant drawback of selection is that one cannot readily change the environment or testing schedule once the experiment has begun or it will be impossible to measure the response to selection. Experimental design usually evolves through experience but in selection one is locked into the same design, long after flaws have become apparent.

2 *Inbreeding and directional dominance* Apart from the general effects of inbreeding such as poor viability and low fertility which may impede the response to selection, there can also be specific genetic problems if a trait showing directional dominance is being selected. Suppose there are two loci with A and B the increasing alleles and a and b the decreasing alleles. The aim of selection is to create a high line $AABB$ and a low line, $aabb$. If A and B are both dominant, the genotypes $AABb$, $AaBB$ and $AaBb$ will all perform equally as well as the desired $AABB$ and are just as likely to be chosen for breeding. Thus the presence of directional dominance implies that the response to selection is likely to be asymmetrical with the line involving the dominant alleles responding more slowly and more erratically since it carries decreasing alleles masked by the dominant increasing alleles. In Chapter 5 we examine the evidence for directional dominance for high learning ability but Fig. 3.21b illustrates its consequences for selection. The TMB strain fluctuates far more than the TMD strain which exhibits a smooth, consistent response. As in Fig. 3.1, these differences between selection lines can be seen as a problem of scaling when one gets far away from the mean and Falconer (1981) reviews this topic.

3 *Maternal effects* In selection as in other breeding programs one must recognize that the mother passes on pre- and postnatal environmental cues as well as her genes. Therefore selection should involve some system of cross-fostering to isolate the postnatal variables and reciprocal crosses between the two lines to detect prenatal effects. Table 3.1 illustrated one case where the Maudsley reactive and non-reactive rat strains selected for open-field behaviour were influenced differentially by maternal stress.

4 *Changes in other behaviours* One cannot alter a single feature of an animal's behaviour by selection without expecting other behaviours to be altered, a point illustrated by Ewing's selection experiment in *Drosophila* for short wings. He found that as well as a change in wing length there were changes in courtship depending on how the flies were maintained. No change occurred with single-pair matings (one male, one female in a culture vial), whereas in mass-matings (10 or 50 of each sex in the vial) there was an increased frequency of wing vibration, one of the components of male courtship. In the single-pair situation there could be no competition between males for the females, in contrast to the mass-mating where the males with the alleles for fast wing vibration could compensate for the reduced stimulation provided by their short wings. In this way they were favoured by the females and the alleles for fast vibration were more likely to be passed on to the next generation along with those for short wings.

5 *Genetic drift at other loci not directly relevant to the trait under selection* Since only a few individuals are selected for mating, the problems of genetic drift and founder effects arise (Chapter 2). By chance one may pick individuals with unusual genotypes and these may come to predominate in future generations of selection. The selected variable will not be affected, but there can be major consequences for any study of correlated changes in other variables (see below).

6 *The limits imposed by consequences of the behaviour* To take a hypothetical but plausible example, selection for aggression in female rodents could produce animals that were so aggressive towards males that mating was unlikely. If they did produce young, the mothers may limit the survival of the next generation by refusing to feed them and even eating them.

Correlated responses

The extent to which other behaviours change during selection can be both the downfall and the major justification for a behavioural selection experiment. Given the problems listed above, selection is likely to be at best a very cumbersome method of understanding anything about the genetics of behaviour and the major reason for selection is what it tells us about behaviour. Chapter 5 examines the behavioural implications of the TMB and TMD lines — do they differ on other learning tasks, can the poorer performance of the TMD lines be improved by environmental manipulation, are there differences between the lines in brain chemistry which may help us understand the basis of learning? Therefore the establishment of selection lines is very much the beginning of research and not an end in itself.

But the fact that selection is based on a single trait may mean that selection is not necessarily for that trait but for some underlying variable which the experimenter may never have considered. Although not fully substantiated by later work, Searle suggested that the difference between the TMB and TMD strains may not be in learning so much as in motivation and emotionality. The TMDs learn poorly because they are less motivated by the food reinforcement and because they react differently to the features of the apparatus, and not because they are 'dull'. (See Ch. 5 for more details.)

Merely to observe that some other feature is altered in selection lines does not imply a causal connection. An example comes from mouse lines selected for high and low blood pressure (Fig. 3.23). In humans there is considerable controversy over the effects of alcohol on blood pressure but controlled experimentation is almost impossible. The selection lines initially suggested that alcohol and hypertension (high blood pressure) may be linked since the hypertensive line showed a much higher voluntary consumption of ethanol that the hypotensive line. However when the F_2 between the two lines was studied, the mice with high blood pressure turned out to have a slightly lower alcohol consumption than the low blood pressure mice. The explanation is that purely by chance the line selected for high blood pressure also happened to have the alleles for alcohol preference. In the F_2 where segregation was possible, this fortuitous connection broke down.

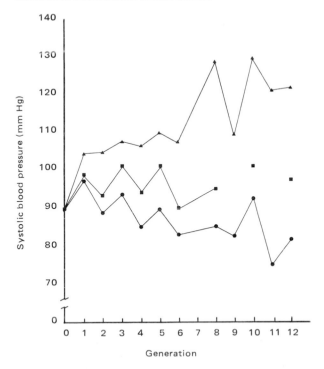

FIG. 3.23   Response to selection for systolic blood pressure in male mice. Note (a) fluctuations between generations in the randomly bred control line (■) and the two selection lines (▲ high; ● low), indicating environmental changes and (b) the more extreme fluctuations in the selection lines following the move to another laboratory between generations 6 and 7. (From G. Schlager (1974) *Genetics* **76**, 537.)

At the end of selection, one has distinct lines that can be subjected to the same biometrical analyses described earlier for inbred strains. The paper from which Fig. 3.23 comes has a very detailed example of how selection and more comprehensive genetic analysis can be combined in this way.

Discussion of non-inbred animals

The statistical details of analysis of familial relationships and of selection lines can be found in Mather and Jinks (1977) and in Falconer (1981) and have not been discussed here. Although selection has a vital part to play in behavioural if not genetic analysis, the paucity of studies of familial resemblance in experimental animals suggests that most scientists find this approach of limited use, compared with what can be done so much more easily with inbred strains.

However, the situation may change as the range of species under investigation increases. Once one moves from standard laboratory species of mouse and rat to some of the more exotic rodents, the use of inbred strains or any method such as selection demanding a prolonged period of controlled maintenance may have to be replaced with familial analysis which determines the genetic architecture with only one or two generations of controlled breeding. This possibility may arise with the increasing interest in species of native rodent, whose adaptations to the environment may well differ from the laboratory species. A limited amount of research has been done on such species as gerbils, spiny mice and voles, but much remains unexplored, especially in Australia where the range of marsupial mice presents a unique situation for the study of maternal behaviour.

Conclusion

This chapter introduces the main methods used to assess the extent and nature of genetic variation in animal behaviour. These methods allow us to do much more than merely make some estimate of 'heritability'. Instead, techniques have been mentioned which assess additive, dominance, epistastic, X-linkage, maternal and environmental effects, as well as genotype–environment interaction. The contributions of these multitude of potential influences on behaviour can be separated through the approach of model-fitting which seeks the most parsimonious explanation of behaviour.

The methods introduced in this chapter may seem remote from the single gene and chromosome techniques of Chapter 2. Early this century there was considerable friction between the 'Biometricians' and the 'Mendelians' over the role of polygenic versus single gene determinants. Although there is still controversy over whether behaviour genetics would advance faster if it stuck to single gene effects with defined biochemical pathways or concentrated on variation in the normal population with its more nebulous polygenic determinants, there is now quite an overlap of the two approaches. For example at the end of selective breeding, chromosome substitution lines can map the loci involved in that behaviour (Chapter 4), inbred strains can be genetically recombined to isolate single gene influences (Chapter 5) and single gene polymorphisms such as the blood group systems can be linked to psychiatric disorders (Chapter 7) and to intelligence (Chapter 8). However, as was pointed out in Chapter 1, it is of little use merely to acquire information on genetics. It is what one does with it that counts. The next two chapters concern two such questions in animal research:

1 What does the genetic architecture tell us about the behaviour? Already we have seen that a consideration of genetics aids efficient experimental design in many areas of behavioural research, but can we go further than this and use the genetic information to analyse behaviour in more detail?

2 Why does a particular behaviour have the genetic architecture it does? A different perspective on the genetics of behaviour is to consider why certain behaviours show features such as directional dominance for high performance, or maternal buffering, or g–e interaction, while other behaviours do not. These differences can tell us something about the ways in which behaviours have evolved and how natural selection has acted upon them.

Discussion topics

1 To many of us who have worked for a long time in the field of comparative psychology, it is a matter of shame and regret that only an amateurish beginning has yet been made by psychologists in the utilization of pure lines of animals in fundamental research in the nature–nurture area (Stone, 1947, quoted by P. L. Broadhurst (1979) in *Theoretical Advances in Behavior Genetics*, Royce J. R. and Mos L. P. (eds), p. 58. Sijthoff and Noordhoff, Alphen aan den Rijn. Using any recent issue of journals such as *Animal Behaviour*, *Developmental Psychobiology* or *Journal of Comparative and Physiological Psychology* as a source of examples, consider if most behavioural researchers of today are much more aware of genetic variation than their counterparts of 1947.
2 P. L. Broadhurst and J. L. Jinks (1961, *Psychological Bulletin* **58**, 337) presented what remains one of the most detailed critiques of biometrical genetic analysis of animal behaviour. But what exactly does a biometrical analysis tell us and why do we need this information?
3 In his review of genotype–environment interaction, Henderson states: "My brief presentation of examples of such interactions makes it appear that these interactive effects are largely chaotic and provide comparative psychologists with an extremely difficult if not impossible task of trying to adequately define behavioral patterns parsimoniously. In fact this is not the case at all, despite the seemingly diverse treatment effects found in various genotypes." (*Developmental Psychobiology* (1968) **1**, p.150.)
Do you feel such interactions are a hindrance or do you agree with Henderson that they are a help to understanding behaviour?

References

Annotated bibliography

Festing M. F. W. (1975) A case for using inbred strains of laboratory animals in evaluating the safety of drugs. *Food and Cosmetics Toxicology* **13**, 369. (A convincing account of the importance of inbred strains for adequate experimental design.)

Henderson N. D. (1968) The confounding effects of genetic variables in early experience research: can we ignore them? *Developmental Psychobiology* **1**, 146. (A discussion of the importance of genotype–environment interactions in behavioural research. Although based around rodents and the effects of early experience, the results have general implications.)

Jinks J. L. and Broadhurst P. L. (1974) How to analyse the inheritance of behaviour in animals — the biometrical approach. In *The Genetics of Behaviour*, van Abeelen J. H. F. (ed.), p. 1. North Holland, Amsterdam. (Many examples of how biometrical models of differing complexity can be applied to animal behaviour.)

Additional references

Erlenmeyer-Kimling L. (1972) Gene-environment interaction and the variability of behavior. In *Genetics, Environment and Behavior: Implications for educational policy* Ehrman L., Omenn G. S. and Caspari E. (eds), p. 181. Academic Press, New York.

Falconer D. S. (1981) *Introduction to Quantitative Genetics* (2nd edn). Longmans, London.

Fulker D. W. and Eysenck H. J. (1979) Nature and Nurture: Heredity. Ch. 5 in *The Structure and Measurement of Intelligence*, Eysenck H. J. (ed.) p. 102. Springer-Verlag, Berlin.

Hay D. A. (1972) Genetical and maternal determinants of the activity and preening behaviour of *Drosophila melanogaster* reared in different environments. *Heredity* **28**, 311.

Jensen A. R. (1980) *Bias in Mental Testing*. Free Press, New York.

Mendenhall W., McClave J. T. and Ramey M. (1978) *Statistics for Psychology* (2nd edn). Duxbury, North Scituate, Mass.

Mather K. and Jinks J. L. (1977) *Introduction to Biometrical Genetics*. Chapman and Hall, London. (For a more technical account, see Mather K. and Jinks J. L. (1982) *Biometrical Genetics* (3rd edn). Chapman and Hall, London.)

Sokal R. R. and Rohlf F. J. (1973) *Introduction to Biostatistics*. Freeman, San Francisco.

4 The diversity of invertebrates

Topics of this chapter

1 The particular merits of organisms such as bacteria, paramecia, nematodes and honeybees for behaviour genetic research.
2 The neurogenetic approach, using single gene mutations to dissect the nervous system, particularly of *Drosophila*.
3 Polygenic variation in *Drosophila*, its relation to neurogenetics and its role in analysing and comparing behaviours.
4 How insect behaviour relates to the ecological conditions of the environment and how the genetic architecture provides information about the action of natural selection on behaviour.

In Chapter 1 the importance of invertebrate behaviour was explained in terms of the unique advantages of particular species. Even though invertebrates appear to be far removed from humans and the determinants of human behaviour, they permit analyses at the molecular and cellular level which are impossible in higher organisms.

But the problem arises of what is 'behaviour'? In an early text on the behaviour genetics of 'simple systems' (Wilson 1973), Gamow describes the growth pattern of the fungus *Phycomyses* and the effects on it of stimuli such as light. Plant growth is a behavioural response to a stimulus, albeit a much more leisurely response than the behaviours we normally choose to study. But we can query whether there are any parallels to the behaviour of higher organisms, and contrast this work with the discussion in the same text by Feldman of circadian rhythms in the flagellate *Euglena* and the yeast *Neurospora*. He describes single gene mutations altering the rhythms of the production of asexual fungal spores, with specific reference to the aim of understanding biochemical systems affecting rhythms in a variety of animal species.

Rather than becoming involved in arguments over reductionism, we consider behaviours with a more typical time-course and examples where parallels can be made with higher organisms. Although *Drosophila* has contributed as much to behaviour genetics as to genetics as a whole, we first discuss some simpler species with more specific attributes.

Bacteria

Bacteria have obvious advantages for genetic research such as their
short generation time of 7-9 minutes and their one large chromosome
on which many specific loci are already known. For behaviour
genetics they have the added bonus of a clearly defined behavioural
response to chemicals. Bacteria such as *Escherichia coli* swim by
rotating their thin flagellar filaments rather like a propeller. As long as
the rotation is counterclockwise, the filaments stay together and
movement is smooth. As soon as rotation becomes clockwise, the
bundle of filaments disperses and the bacterium shows tumbling
movements and changes direction (Fig. 4.1a).

(a)

(b)

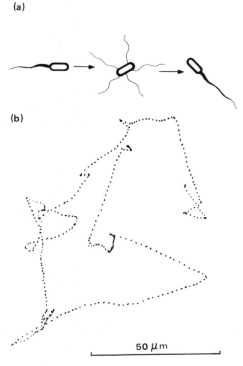

50 μm

FIG. 4.1   (a) The sequence of events in bacterial movement — smooth
swimming, dispersal of flagellar filaments leading to tumbling, before
reforming and moving off in another direction (From D. E. Koshland (1980)
Bacterial Chemotaxis as a Model Behavioral System Raven Press, New York (b)
The track of wild-type of *E. coli*. The points are 0.08 seconds apart, and the
tumbles and changes of direction roughly 1 second apart (From H. C. Berg
(1975) *Nature* **254**, 389).

This sequence forms the basis for studying the response of bacteria
to attractant and repellant chemicals. Chemoreceptors constantly
monitor the chemical concentration around the bacterium. In a spatial
gradient of increasing attractant or decreasing repellant, the
bacterium swims smoothly in an appropriate direction with few

'tumbles'. In the opposite situation of decreasing attractant or increasing repellant, the frequency of 'tumbling' increases until the bacterium is moving towards the attractant or away from the repellant. Figure 4.1b shows the typical path of wild-type *E. coli*. ('Wild-type' is the term used to refer to normal, non-mutant individuals for any phenotypic trait and is not a description of the animal's behaviour!) Other bacterial species (Berg 1975) use slightly different strategies to alter their direction of movement, depending on their flagellar organization.

So bacteria can recognize a stimulus, process the information and generate a behavioural response, much like higher organisms. To carry out genetic analysis it is necessary only to induce mutations usually by chemical means and to identify those mutant bacteria with abnormal behaviour patterns. Using this approach, Parkinson (1977) has shown how the behaviour can be broken down to five components, depending on where specific mutants act in the sequence:

$$
\begin{array}{c}
\text{Flagellar response} \\
\overbrace{\qquad\qquad\qquad\qquad}
\end{array}
$$

Stimulus detection	signalling	control	switch	rotor
$A \longrightarrow$	$B \longrightarrow$	$C \longrightarrow$	$D \longrightarrow$	$E \longrightarrow$

where A = mutants defective in the reception of attractants such as certain sugars and amino acids and of repellants such as ethanol. Each group of chemically similar compounds has a specific receptor and a mutation in one receptor need not affect other receptors.
B = mutants defective in signalling the response from one or more specific receptors to the flagella.
C = mutants of this type cause predominantly clockwise flagellar rotation and excessive tumbling. They do respond to chemical stimuli, but only in higher concentrations and for a shorter time than normal bacteria, suggesting that the defect is in a higher threshold for responding to signals. At lower concentrations the bacterium continues to tumble as it fails to perceive the stimuli and alters direction as it constantly searches.
D = mutants where the flagellar rotation is only counterclockwise. Such mutants lack the facility to switch direction since they can never tumble.
E = mutants affecting flagellar function or its capacity to rotate.

Far more detailed accounts of bacterial behaviour are provided by Parkinson (1977) and particularly by Koshland (1980). The above example suffices to show that mutations can efficiently dissect a behaviour into its components. Nor is bacterial behaviour such a rigidly programmed characteristic as it may initially appear since it shows
1 *Sensory adaptation* Even in extreme chemical concentrations which elicit smooth-swimming for several minutes, the bacterium will eventually adapt and return to its normal pattern of swims and tumbles. Parkinson (Fig. 4.2) explains this by a basal signal level around which the sequence of swims and tumbles fluctuates until a stimulus is received. After an initial response, the bacterium

gradually adapts to this new level, slowly if the stimulus suppresses tumbling, more quickly if it enhances tumbling.

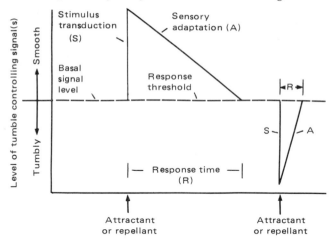

FIG. 4.2   The sequence of events in E. coli following an attractant (leading to smooth swimming and lengthy adaptation) or a repellent (leading to tumbling and rapid adaptation). Stimulus transduction (S) is the initial response which is followed by sensory adaptation (A) over a response time (R). (From J.S. Parkinson (1977). *Annual Review of Genetics* **11**, 397.)

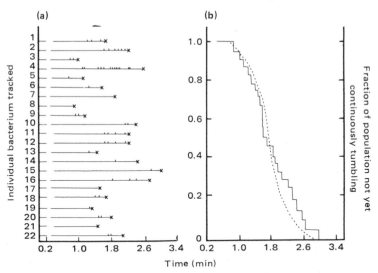

FIG. 4.3   (a) The time taken by individual bacteria from a tumbly mutant strain to begin tumbling continuously. Brief tumbles are shown by vertical marks. (b) The fraction of the population in (a) above not yet continuously tumbling. The broken line is a comparable cumulative normal distribution, showing how closely the data correspond to the expected values — a cumulative distribution gives the proportion of the entire population which have started tumbling by that point in time, whereas the conventional normal distribution gives the proportion which start tumbling at each point in time. (Both from D.E. Koshland (1980). *Bacterial Chemotaxis as a Model Sensory System* Raven Press, New York.)

2 *Memory* Most bacteria are so small that there can be no real concentration gradient from one end of the body to the other. To tell if it is moving towards an attractant or away from a repellant, it must be able to compare the concentration now with that at some point in the past, that is to remember. To confirm this memory hypothesis, Koshland has used an ingenious device where bacteria are suddenly exposed to a different chemical concentration. Only if the bacteria sense over time, rather than by differing concentrations along their bodylength, will they respond. Although the memory rarely lasts more than 60 seconds, it is finely tuned to the conflicting demands of long duration to improve accuracy and short duration for a rapid response. Even the asymmetry between tumble-enhancing and tumble-suppressing stimuli in Fig. 4.2 may be adaptive. Tumbling is suppressed for a longer interval when going to an attractant or away from a repellant than it is enhanced in the opposite situations.

3 *Environmental effects* Not all bacteria of a given species behave the same and they do exhibit individual non-genetic variation. Figure 4.3 shows the time taken for 22 individuals of a tumbly mutant strain to begin tumbling continuously after exposure to a sudden concentration change. If plotted appropriately (Fig. 4.3b), these figures show a normal distribution of response time within the strain which contains very little if any genetic variation.

Paramecia

While bacteria model many features of behaviour, they are so small that it is difficult to determine what happens within the bacterium when it responds. An alternative is a much larger unicellular organism *Paramecium* where intracellular recording is possible. *Paramecium* also have genetic advantages such as their capacity for both sexual and asexual modes of reproduction, so that they can be cloned as well as crossed. They even offer a third mode autogamy, which results in complete homozygosity within the one generation. With this method even recessive mutations become immediately obvious.

A *Paramecium* has a clearly defined behavioural response in avoiding objects or repellant chemicals (Fig. 4.4a). It normally swims forward by coordinated beating of its cilia, but in the avoidance reaction it briefly reverses the direction of beating, so that it moves backwards and can then progress in another direction. Intracellular recording demonstrates that cilia reversal is due to ion conductance changes across the external membrane of the *Paramecium*. In normal conditions there is a voltage difference (the resting potential) across the membrane, so that internally the *Paramecium* is electrically negative. Mechanical or aversive chemical stimulation causes a local influx of sodium ions and a more general influx of calcium ions, temporarily changing the polarization (the action potential) and reversing the direction in which the cilia beat. As the ions are pumped out, the cilia beat in the normal direction once again. Because this

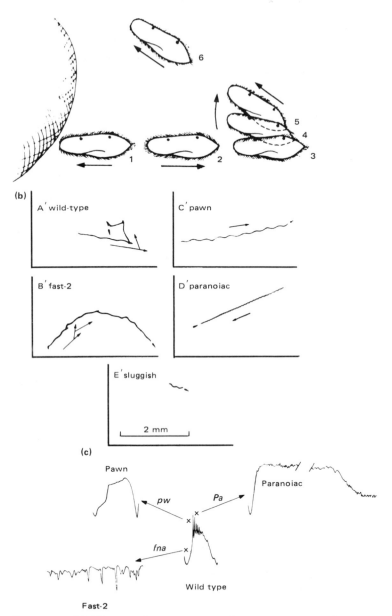

(a)

6

5

4

3

1

2

(b)

A' wild-type

C' pawn

B' fast-2

D' paranoiac

E' sluggish

2 mm

(c)

Pawn

Paranoiac

pw

Pa

fna

Wild type

Fast-2

FIG. 4.4 (a) The avoidance reaction of *Paramecium*, involving ciliary reversal, pivoting and moving off in a new direction. (From S. Dryl (1973). In A. Perez-Miravete (ed.) *Behaviour of Micro-organisms*. Plenum, London pp. 16–30.) (b) The tracks of individuals from different strains of *Paramecia* during 9.3 seconds of filming. (c) Electrophysiological recording from within individuals of different *Paramecia* strains, showing the effect of three different mutations. Xs mark where the mutant recordings peak relative to the wild-type. (b and c from C. Kung *et al.* (1975) *Science* **188**, 898.)

process is very similar to the mechanism by which neurons function in higher organisms, *Paramecia* may provide a key to understanding neuronal transmission at a molecular and genetic level.

Figure 4.4b demonstrates several mutants with altered avoidance reactions. Compared with the normal *Paramecium* which swims along in a helical path, often turning abruptly through 90° or more as a result of spontaneous avoiding reactions; *fast-2* makes frequent, rapid avoiding reactions through much less than 90°; *pawn* makes no avoidance reactions and can only move forward (like the pawn in chess); *paranoiac* makes greatly exaggerated avoidance reactions in which it moves backwards over a considerable distance; *sluggish* moves very slowly. Kung *et al* (1975) discuss in more detail these and other mutants with such intriguing names as *spinner*, *staccato* and *atalanta*. While some of their mutants may have the same behavioural phenotype, they can represent distinct genotypes. There are 151 *pawn* mutant strains representing at least 62 independent mutations over three unlinked loci.

Except for the *atalanta* mutant whose cilia reversal mechanism is damaged so that it can only swim forward or stop but not go backwards, the effect of the mutations are on the cell membrane not on the cilia. For example, *pawns* treated with the detergent Triton X-100 have no functional membrane and in certain circumstances can show a normal avoidance reaction, something impossible when their membrane is intact. One can be more specific and demonstrate that the mutants have altered action potentials (Fig. 4.4c). Compared with the wild-type which shows a rapid depolarization and a slow repolarization, *fast-2* is blocked near the start of depolarization so that it does not show full reversal of cilia direction, *pawn* shows a spikeless depolarization insufficient to reverse cilia direction, while *paranoiac* has a prolonged depolarization of up to 60 seconds consistent with its prolonged backward movement.

Quinn and Gould (1979) have sounded one note of caution about *Paramecium* as a model behavioural system. Detailed electrophysiological work has revealed such a variety of mechanisms that 'it is as if *Paramecium* tried to compact all the functions of a nervous system into one cell' (p.20) with consequent difficulties in disentangling them. But with the ease of chemically inducing mutations in *Paramecium* this organism has a future in integrating the fields of behaviour, genetics, physiology and biochemistry. Kung *et al* (1975) refer to screening for resistance or oversensitivity to drugs or other pharmacological agents, adding yet another dimension to the possibilities of this unicellular organism.

Nematodes

The nematode or round worm *Caenorhabditis elegans* was introduced to behaviour genetics by Brenner, as one of the few organisms suitable for genetic analysis of behaviour as he saw it:

Behaviour is the result of a complex set of computations performed by nervous systems and it seems necessary to decompose the problem into two; one is concerned with how the genes specify the structure of the nervous system, the other with questions of how nervous systems work to produce their outputs. Both require methods for analysing the structure of nervous systems. Thus, what has to be done is clear in general outline: i.e. isolate mutants affecting the behaviour of an animal and see what changes have been produced in the nervous system. (S. Brenner (1973) *British Medical Bulletin* **29**, 269.)

C. elegans has some obvious genetic advantages. It is a self-fertilizing hermaphrodite producing up to 300 progeny during its 3 day life-cycle. Thus in one week, a single adult can produce 10 000 or more second generation offspring. A small proportion (0.1%) are male, so that crossbreeding is possible for standard genetic analyses.

Of greater importance to Brenner are the small size of *C. elegans* (1 mm) and the total number of cells (600 of which 250 are neurons). These make it possible to observe the physical effects of mutations on the nervous system by slicing each nematode into some 20 000 serial sections, studying the sections by electron microscopy and using a computer to create a three-dimensional representation of the nervous system in its entirety. It is rather like determining the wiring diagram of a house in terms of where the switches are, where the cables go and to which appliances they are connected — do particular mutations cause lesions in specific neurons or prevent neurons branching properly or connecting to the muscles they normally innervate?

Mutations created by chemical mutagens are usually recessive and will segregate in the second generation. Apart from mutants altered in size or shape, the most easily detected mutants in *C. elegans* are behavioural ones affecting chemotaxis or locomotion and co-ordination. Because *C. elegans* is maintained on agar plates where they feed on bacteria, their tracks in the agar leave a permanent

(a)   (b)

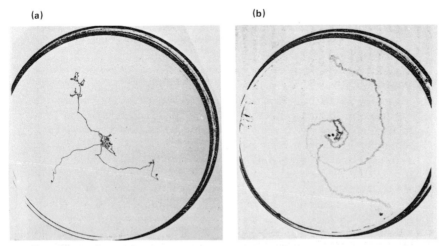

FIG. 4.5   The tracks of three nematodes responding to an attractant in the centre of an 85mm plate (a) wild-type (b) bent-headed mutants. The wild-type were observed over 15 minutes, while the mutants took 60 minutes to reach the centre (From S. Ward (1973) *Proceedings of the National Academy of Sciences USA* **70**, 817.)

record of their behaviour. For example the *roller* mutants move end over end, leaving distinctive craters in the soft agar.

Figure 4.5 shows the tracks of normal and of *bent-headed* mutant *C. elegans* moving towards an attractant. Whereas we saw earlier that bacteria use temporal cues to sense attractants, the analysis of different *C. elegans* mutants shows that they use a different system. A *slow* mutant taking eight times longer than normal to reach an attractant still tracks as accurately as normal, eliminating the temporal explanation. The comparison of concentrations over the body length is excluded by the adequate performance of mutants with blisters covering the sensory receptors on their tail. The *bent-headed* mutants reveal that orientation is by klinotaxis, a series of successive comparisons by receptors on the sides of the head. This searching requires side-to-side movements of the head which makes the bent-headed mutant follow a spiral path.

As in the case of bacteria (Fig. 4.2), *C. elegans* shows habituation. After reaching the attractant, nematodes fairly soon swim away to the periphery of the plate. The lower the concentration of attractant the sooner they leave. The nematode then returns to the centre and the cycle is repeated. There appears to have been little attempt to identify mutants with abnormal habituation despite the possibilities these would provide for detailed analyses of this process and its relationship to learning.

But what has the discovery of these mutants told us about behaviour? The coordination mutants arise from specific and quite distinct lesions in the nervous system. unc-30 has a set of neurons which do not synapse with other neurons and do not branch properly. In unc-5 there are no neurons to innervate the dorsal muscle cells, since the processes from the ventral nerve cord fail to develop in their normal direction. Similarly chemotactic mutants have observable defects in the terminals of the sensory neurons in the head. Mutants differ in the magnitude of the defects, some only having a few neurons terminating in the wrong place while others have almost all their sensory neurons misplaced.

In his detailed review of neurogenetics Ward (1977) concludes that almost half the nematode mutants identified by altered behaviour have some altered anatomy in their nervous system. Thus Brenner's aim has partly been justified. But there are limitations of the *C. elegans* studies. This nematode is too small for electrophysiology, so that the function rather than the structure of the nervous system cannot adequately be explained. One unsatisfactory solution (Ward 1977) is to combine the knowledge of *C. elegans* neuronal structure with electrophysiological recording from the much larger nematode *Ascaris*. There are also some fundamental differences in the organization and development of the nematode nervous system relative to other animals especially mammals, which may limit extrapolation from the nematode model. Nevertheless Brenner's approach is providing some unique information on the genetics of nervous system development (reviewed by Quinn and Gould 1979; Ward 1977), which is sufficient to justify this method, even if it has contributed less than expected to an understanding of behaviour (Gould 1974).

Honeybees

While the organisms discussed so far have specific advantages for genetic resarch, the honeybee's merits lie also in its behaviour. The genetic advantage of bees is that male bees (drones) are haploid, developing from unfertilized eggs. Hence all gametes from a particular male are identical, its female offspring have 75% of their genes in common rather than the usual 50% and there is higher genetic uniformity within a colony. In terms of behaviour the bee is a social insect, living in a colony which depends on such advanced attributes as communication, maternal care, the division of labour and the storage of food.

There is a surprising amount of behavioural variation between the three geographically distinct major honeybee races. Races are genetically and often geographically distinct groups from within the same species, the honeybee races being the Italian, the Caucasian (from the Caucasus mountains) and the Carniolan (from the south-eastern Alps and the northern Balkans). Rothenbuhler (1967) reviews this work on racial differences which is often overlooked, despite the fact that the social insects were the source of much early inspiration in sociobiology (Chapter 1). The sociobiologists have been concerned with explaining the evolution of insect society in general in terms of the consequences of the haplo-diploid mating system and have often ignored the variation between colonies and races. We shall consider the genetics of one social behaviour which is a classic example in behaviour genetics and also some of the differences in learning to illustrate how the behaviour of races of honeybees has evolved to fit their particular habitat.

Nest-cleaning behaviour

Rothenbuhler examined two inbred lines of honeybee, one selected for resistance and one for sensitivity to the bacterial infection,

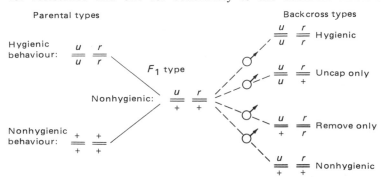

FIG. 4.6   Genetic hypothesis with two loci to explain the responses of bee colonies to larvae killed by American foulbrood. Non-hygienic behaviour is dominant to hygienic behaviour and the loci determining the two behaviour components segregate independently (From W.C. Rothenbuhler (1964) *American Zoologist* **4**, 111.)

American Foulbrood. The resistant line was quick to uncap the brood cells in the comb and then to remove foulbrood-killed larvae from the hive, but both the sensitive line and the F_1 between the lines usually failed to remove dead larvae, indicating that the 'hygienic' behaviour was recessive. The backcross of the F_1 to the recessive (i.e. hygienic) parent gave the result in Fig. 4.6 with two new behavioural pheno-types emerging, each showing only part of the hygienic behaviour. There must be two unlinked loci involved, one responsible for uncapping the brood cells and the other for removing the dead larvae from the uncapped cell.

As Ewing and Manning (1967) explain, this example remains one of the clearest cases where genes 'switch' on or off what is a very complex pattern of behaviour. It is one of the few situations where a behaviour pattern can be dissected into two clearcut components. Unfortunately the validity of this experiment has been questioned. It has proved difficult to replicate and anomalies exist in Rothen-buhler's data. The other backcross to the dominant (i.e. non-hygienic) parental line should give only non-hygienic offspring, but one colony exhibited the complete nest-cleaning behaviour.

There is also the question of why hygienic behaviour should be recessive. While hygienic lines restock the brood cells, non-hygienic colonies sometimes die out because of the accumulation of brood cells filled with dead young. One might expect that evolution would have made hygienic behaviour dominant, so that a far greater proportion of all colonies could have shown the nest-cleaning behaviour contri-buting to their survival.

Learning

The honeybee is dependent on good learning ability for spatial localization. On their very first flight bees must learn the location of their colony. If they cannot return to their colony, the only hope for survival is discovering one of the few other colonies which will accept intruders. Few other species are exposed to such intense pressure for learning and it is not surprising that learning in the honeybee shows many examples of the 'biological preparedness' discussed in Chapter 1. Bees can learn to distinguish food sites by odour cues after only one approach, whereas colour requires three rewarded approaches and form 20 or more approaches.

Lindauer (1975) demonstrates that learning is even more specific in that races of bees from particular localities are adapted to learning cues from their usual environment. In the examples of bacteria, *Paramecia* and nematodes, we saw that chemoreception was a major variable and the same applies to bees. Each race most easily learns those scents associated with its own native flora. The Indian *Apis cerana* most easily learns those scents of thyme, anise and oil of rose-wood, while Egyptian and European races find lavender and rosemary the easiest. The extent to which preferences can be varied also differ between races. The Indian *A. cerana* is most resistant to change while the mid-European *A. mellifera carnica* readily modifies its preferences. Lindauer suggests that this is because the more variable European climate demands adaptation to food plants which change with the season.

An even more extreme racial adaptation in learning is shown in Fig. 4.7. Bees were required to identify a food site by the presence of a

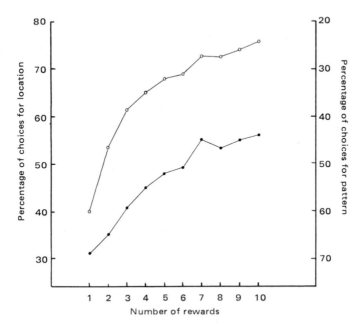

FIG. 4.7   The learning curves of two races of bee *Apis mellifera carnica* (o——o) and *A. m. ligustica* (●——●) on the basis of the choice of a pattern or its location on a food table. (From M. Lindauer (1975) In G. Baerends, C. Beer and A. Manning (eds) *Function and Evolution in Behaviour*, Clarendon, Oxford, p. 228.)

black star. The position of the star on the food table was varied, to determine if bees responded only to the pattern of the star itself or to the total location involving the star and other objects. The mid-European *A. m. carnica* relied more on the location and the Mediterranean *A. m. ligustica* on the specific pattern, a result Lindauer attributes to the climatic differences. Mediterranean bees can usually navigate by the sun-compass orientation and so need to learn only specific patterns, while in mid-Europe with more overcast skies, bees have to rely on a variety of landmarks associated with the food and hence attend to the overall configuration of the location.

None of these learning examples involve genetic analysis to any great extent. Comparable evidence on human racial differences in Chapter 8 is explained by cultural factors just as easily as by genetics and the same may apply to honeybees. The importance of honeybee learning is as a demonstration of how behavioural differences reflect adaptation and evolution to environmental differences, a topic to which we return with *Drosophila*.

Discussion

These four examples make it clear that behavioural research need not be confined to vertebrates. The range of behaviours possible in a bacterium and the influence of the cell wall on *Paramecium* behaviour show how much can be achieved without a nervous system. What other organism apart from *C. elegans* offers such possibilities for analysing nervous system structure or where else but in honeybees can we find such clear relationships between learning ability and the environment? There are many other examples from invertebrate behaviour genetics reviewed in Ehrman and Parsons (1981, Chapter 10) and in Gould (1974). Of particular note are the analyses of the calling song of male crickets as an example of how a polygenic system can be dissected into its genetically discrete components and the fine structure of the visual system of the water flea *Daphnia* as a means of analysing variation among non-mutant individuals. In *Daphnia* electron microscopy and computer reconstruction show that individual differences in the nervous system are not found in the gross organization of neurons which is similar in all individuals (unlike the mutation effects on nematodes), but in the shape of the dendritic trees (the fine processes) and the exact pattern of connections between the neurons.

Enthusiasm for the possibilities of such exotic species in behaviour genetic research must be tempered by caution. There are many problems in adequately defining and controlling behaviour and often even practical difficulties in maintaining the species, far less breeding them. The best known example of this was the controversy in the 1960s over whether the planarian (flatworm) could learn and whether memory could be transferred from one individual to another by cannibalism. One review of biochemical studies of the planarian work commences with a statement applicable to much behavioural research:

What typifies many investigators in the area is their treatment of the laboratory subject as a learning machine, a preparation that will display consistent and unequivocal learning under automated and somewhat restricted conditions. Their expectations allow for little consideration of the organism and when the beast rebels and dirties up the data, psychologists may prefer to conclude that learning is impossible in the particular animal and those of other disciplines may begin to express impatience with the science of behavior. This is sometimes followed by a negative review paper and a return to safer pastures, moves that could increase the unemployment rate among psychologists wanting to work in biochemistry. (W. C. Corning and D. Riccio (1969) in *Molecular Approaches to Learning and Memory* Byrne W. (ed.), p. 107. Academic Press, New York.)

Corning and Riccio go on to tabulate many of the mistakes made by scientists following up the initial reports of planarian learning. These range from such basic faults as keeping the planarians in distilled water, which is lethal, through to subtle effects such as not considering the presence of slime trails from other planarians, a factor which can halve the response to light. It is indicative of the state of psychology at that time that Corning and Riccio make only one brief mention of species differences and no mention at all of within species differences. We considered the possibility of using Australian planarians in behaviour-genetic research because as well as sexual reproduction, planarians will also regenerate into identical individuals after being cut into pieces (a sort of 'cloning by scalpel'). However, it took 5 years merely to establish for various species the optimum temperature and conditions for their laboratory maintenance and reproduction and to determine seasonal fluctuations in mortality which also affect behaviour.

It is easy to see why in such situations scientists retreat to the most convenient organism and the remainder of this chapter is devoted to *Drosophila*, the invertebrate on which so much of our genetic knowledge is based. As Benzer (1973) has pointed out, *Drosophila* is perhaps the best compromise out of all the animals one could choose. Its mass, number of genes, generation time and even the number of neurons are roughly the geometric mean between *E. coli* and humans.

Drosophila

Since one major recent behaviour genetics text (Ehrman and Parsons 1981) has a very comprehensive coverage of *Drosophila* and other detailed accounts of *Drosophila* behaviour genetics are available (Ashburner and Wright 1978; Parsons 1973), we shall only illustrate the different areas where *Drosophila* have been used. Although some 1500 *Drosophila* species have been described, behavioural research has been confined to a few species, most notably *D. melanogaster*.

Single genes

In the course of general genetic research on *Drosophila* many single gene mutations have emerged and one of the very first approaches to behaviour genetics involved *Drosophila* mutants with obvious effects on body or eye colour. The behaviour most widely studied in this context has been mating behaviour where *D. melanogaster* and many related species show a complex but consistent pattern (Fig. 4.8). The male (A) initially orientates to one side of the female, (B) vibrates one wing, (C) licks the female's genitalia, (D) mounts and grasps the female, spreading her wings and (E) copulates usually for 15-20 minutes. This sequence gives the female ample opportunity to reject any male with which she does not wish to mate (F).

Figure 4.9a summarizes how three components of courtship behaviour differ from wild-type *D. melanogaster* in four body colour mutants, *black, tan* and two ebony lines, *ebony°* and *ebony[11]*. *Black* is basically similar to wild-type except that it shows less vibration and licking, the later stages of courtship. *Ebony* and *tan* males show major

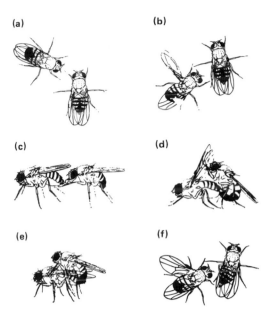

FIG. 4.8   Courtship in *Drosophila melanogaster*. The male (with solid black tip to its abdomen) in A orientates, B vibrates one wing, C licks the female's genitalia, D mounts, and E copulates. If a female rejects a male, F she turns her abdomen and extends her ovipositor. (From B. Burnet and K. J. Connolly (1974). In J.H.F. van Abeelen (ed.) *The Genetics of Behaviour*, North Holland, Amsterdam, p. 201.)

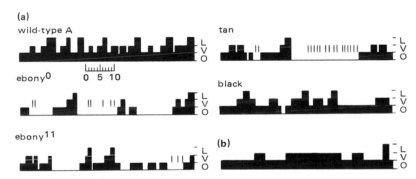

FIG. 4.9   (a) The sequence of courtship in five types of male *D. melanogaster* courting females of their own genotype. The sequence is from left to right with each unit being 1.5 seconds. L = licking plus attempted copulation, V = wing vibration (vertical dashes are wing flicks) and O = orientation. (From S. Crossley and E. Zuill (1970) *Nature* **225** 1064.) (b) A similar record from *D. simulans* (From A. Manning (1959) *Behaviour* **15,** 123.)

gaps in their courtship, often breaking away to run in a zigzag path, opening and closing their wings (wing flicking). Female mutants are relatively unaffected in their behaviour.

Although the mutants cause quantitative changes in the amount of time spent in the various components of courtship, there are few qualitative changes. That is, apart from the wing-flicking, they exhibit the same behaviour patterns as wild-type but in different proportions. Figure 4.9b indicates that the same happens when closely related species are compared. Compared with *D. melanogaster*, males of the sibling species *D. simulans* take longer to initiate courtship and are slow to change between the elements of courtship. Some of the species differences are attributable to the females since *D. simulans* females are relatively inactive. When required to court the more active *D. melanogaster* females, the *D. simulans* males speed-up to some extent. Both the mutant and the species differences show that genetic differences are expressed through differences in threshold for switching from one behaviour to another, not by introducing new behavioural components.

But how do the effects of the mutants occur? *Ebony* and *tan* must both be pleiotropic, in that much more than just body colour is affected. Their eyes have abnormal electroretinograms (ERGs), unlike *black* whose ERG is normal. Although *ebony* and *tan* do not rely sufficiently on vision during courtship and may lick or even attempt to copulate with a female's head, not all the effects on courtship can be attributed to poor vision. Kyriacou (Fig. 4.10) postulates three separate effects of the *ebony* allele on the nervous system in order to explain why ebony males are more successful than wild-type males in mating in the dark.

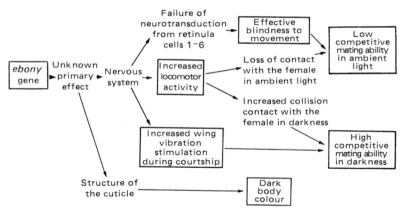

FIG. 4.10   Kyriacou's model of the pleiotropic effects of the *ebony* allele (From B. Burnet and K.J. Connolly (1981) *British Medical Bulletin* **37**, 107.)

Another approach to determining the action of the mutant has been to identify the biochemical abnormality responsible for the altered pigmentation in *ebony*, *black* and similar mutants. *Ebony* flies fail to incorporate into the developing cuticle β-alanine which both induces tanning and prevents blackening, while the levels of β-alanine within the body are elevated. The *black* mutant has decreased synthesis of β-alanine but can incorporate it if it is present. Jacobs (1978) assessed the role of β-alanine in courtship by creating phenocopies, making *black* flies phenotypically normal through the injection of β-alanine into newly-emerged flies. Such phenocopies do not blacken, they begin courting earlier than non-injected *black* flies and they are much more successful at mating. When above normal concentrations of β-alanine are injected, the males continuously court and chase females but with limited success. Jacobs attributes one cause of the poorer mating success of *ebony* and uninjected *black* flies to poor muscular coordination due to weakness in the exoskeleton which is normally stiffened by β-alanine. In contrast, the overinjected flies are so rigid that their wings are permanently dilated, again impairing courtship and mating.

FIG. 4.11   Song is an important part of *Drosophila* courtship.

The range of mutants available is such that mutants affecting particular organs can often replace the surgical techniques used to study particular systems. One example involves the courtship song of *Drosophila*. During vibration the male directs patterned auditory stimuli towards the female, his orientation ensuring that the sound is directed at her and not at other females in the area. This song enables the female to distinguish conspecific males and sexually stimulates her. Female flies process the information in the song after perceiving it through the arista, a feather-like structure evolved from the end of the antenna. The role of the arista has been studied using mutant alleles affecting its size and structure. Not surprisingly, sexual receptivity was reduced in the mutant females, but so also was mating ability in males with the double mutant *aristaless; thread* where only a vestige of the wild-type arista remains. It seems that males may need the arista to provide feedback and to control their wing vibration in courtship.

Neurological mutants

Rather than mutants such as *ebony* where obvious physical effects are first observed and then behavioural concomitants sought, there is an increasing interest in those mutants which directly affect the nervous system and behaviour. Hall *et al* (1982) provide a comprehensive review of this research into 'genetic neurobiology'.

The main approach is to screen mutants for abnormal behaviour and then seek a neurological explanation for the abnormality. Gould (1974) points out that making mutants is easy with the chemical mutagens available, but sorting the mutant from the many normal flies is another question. Benzer (1973) provides examples of some of the neurological mutants and the ingenious methods developed to detect them. Admittedly, it is stretching a point to call 'behavioural' mutants such as *drop dead* whose only behavioural anomaly is suddenly to become uncoordinated, fall on its back and die with major degeneration of the brain.

One method Benzer developed was the countercurrent apparatus which 'can fractionate a population of flies as if they were molecules of behavior' (Benzer 1973, p.29). Flies are separated into those which move along a tube towards a light source and those which remain stationary. This sequence is repeated several times dividing the poulation into groups ranging from the most to the least phototactic along a similar principle to Hirsch's maze discussed in Chapter 1. To separate flies that are positively phototactic from those that are merely active and those that are negatively phototactic from the sluggish ones, the separation is repeated, this time away from the light source. In the end the population is divided into five fractions (Fig. 4.12). These can then be analysed in more detail to find the neurological bases of the behavioural differences. Particular attention has been paid to the electroretinogram (ERG), the response to a flash of

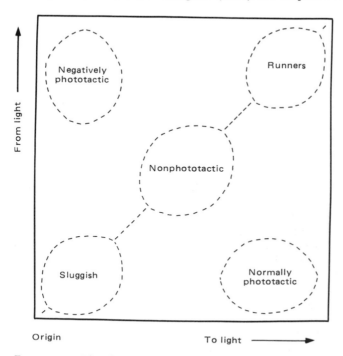

FIG. 4.12   The fractionation of a fly population by their phototactic behaviour (From S. Benzer (1971) *Journal of the American Medical Association* **218**, 1015.

light of the photoreceptor cells in the fly's eye. In some mutants the primary receptor cells are so affected that there is no ERG at all, while in others the receptor cells are normal but fail to trigger the next stage in the response and in others, the cells are normal in the young fly but gradually degenerate.

Another group of mutants involving three X-linked loci was detected by chance in the course of routine laboratory maintenance. In response to the ether used in sedating flies, these 'shakers' show rapid leg shaking and frenzied bursts of activity which can spread through a group of mutant flies in a chain reaction. Of the three mutants, the *shaker* also shows wing scissoring, the *hyperkinetic* a kinetogenic response (falling over and rolling about when a shadow passes over the tube) and the *Ether à go go* is temperature-sensitive, responding vigorously above 30°C but not at all below 20°C.

The neural mechanism of the shaking has been isolated to rhythmic bursts of activity from motor neurons within three specific parts of the thoracic ganglion. But is this the primary site of action or is it some secondary effect of a lesion elsewhere in the nervous system? It could even be some metabolite circulating throughout the fly which triggers the abnormal response. This question can be settled by a very important technique for analysing the *Drosophila* nervous system.

Gynandromorphs

Gynandromorphs are genetically mosaic flies, comprising male tissue carrying the mutant allele and female tissue which is heterozygous. The technique for producing gynandromorphs is described in detail elsewhere (Benzer 1973; Gould 1974) but centres on female *Drosophila* with a special ring X chromosome replacing one of the usual rod shaped X chromosomes. The first time an embryo cleaves, this ring X is lost, leaving half the cells with two Xs (phenotypically female) and half with one X carrying the mutant gene (unlike humans where such XO individuals are female, in *Drosophila* these are male). This distinction remains throughout development, the XX and XO cells populating different parts of the blastula (the single cell layer surrounding the yolk) according to the orientation at the first division (Fig. 4.13a). As different parts of the blastula are responsible for the development of different adult structures, a fly emerges that has some parts male, some parts female.

Figure 4.13b shows how this technique is applied to the problem of the shaking mutant. Even a single hyperkinetic leg on an otherwise non-mutant thorax will shake and an abnormal firing pattern is found in the motor neurons governing the movement of that one leg. Therefore the mutation is at this specific level, none of the rest of the fly or anything circulating within the fly being involved. This localization has been confirmed by electrophysiological studies of the muscles in *shaker* which showed an abnormal response to the flow of calcium ions.

Although the gynandromorph approach has been applied to many behaviours (Benzer 1973; Ward 1977), we shall consider only one of

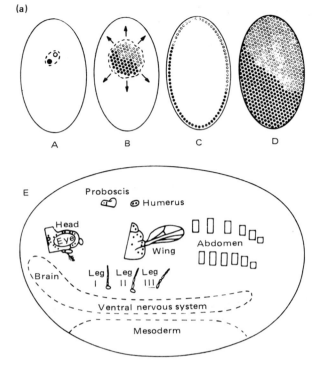

FIG. 4.13 (a) The formation of a gynandromorph — see text for details. A–D indicates the sequence of division following the loss of one X chromosome at the first division, leading to clones of male cells (open circles) and of female cells (black circles). E shows the location on the blastula of cells which will form the appropriate structures in the adult fly. (From Y. Hotta and S. Benzer (1972) *Nature* **240,** 527.)

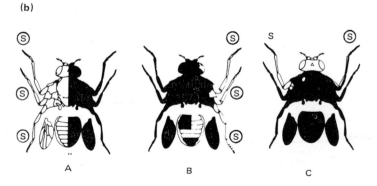

(b) Mosaic patterns in three *Hyperkinetic* gynandromorphs. Lighter parts are male tissue and Ⓢ marks the legs which shake and where an abnormal firing pattern in the motor region of the thoracic ganglion was recorded. For technical reasons, there was no record from the front left leg of fly C. (From W. D. Kaplan (1972). In J. A. Kiger (ed.) *The Biology of Behavior*. Oregon State University Press, p. 133.)

the most interesting questions. Given an individual which is part male and part female, which opposite-sex behaviours can or cannot coexist in the one individual and which parts of the body are required for behaviour to be predominantly male or female? If a fly has a male head but female external genitalia, how will it behave and how will other males and females respond to it? A study by Cook centred on those gynandromorphs with female external genitalia. He found that 65% of these would orientate, that is commence male courtship of females and some would even attempt to copulate. Even among those containing sufficient female tissue to be able to lay eggs, 45% showed male courtship.

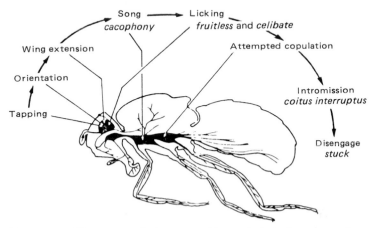

FIG. 4.14   The events during courtship, and their fate mapped sites of control in the nervous system. Several mutants are listed in italics at the points where they disrupt normal courtship. (From B. Burnet and K.J. Connolly (1981) *British Medical Bulletin* **37**, 197).

A technique called fate-mapping can be applied to the gynandromorphs to localize the control sites or 'foci' in the nervous system for particular behavioural components. Figure 4.14 shows the mapping for the various stages of courtship from the initial tapping of the female by the male through to disengagement. Mutations with suitably descriptive names influence different stages of the sequence. Analogous to the concept of dominant and recessive alleles, foci can be domineering, where only one side of the brain must be male for that courtship component to appear, e.g. orientation and wing extension, or submissive where male tissue must be present in both sides of the brain, e.g. licking. Courtship song is interesting in that though it depends on wing extension whose focus is in the brain, there must also be male tissue on at least one side of the meso-thoracic ganglion.

Learning

One of the most tantalizing possibilities is that the gynandromorph method could be applied to the processes of learning and memory to provide information impossible with any other organism or technique. There is one major problem, the limited ability of *Drosophila* to learn. Médioni (1977) describes the efforts which have been made to demonstrate learning in *Drosophila* and the difficulties and alternative interpretations which have dogged so many of the attempts.

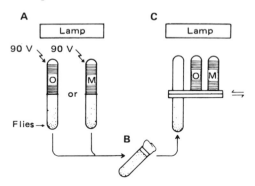

FIG. 4.15   The sequence in avoidance conditioning of *D. melanogaster*. In A, flies move towards light and are exposed to electric shock in the presence of one or another odourant (O and M). After a given time (B), they are tested (C) to see if they avoid O and M to different extents when no shock is administered. (From W. G. Quinn, P. P. Sziber and R. Booker (1979). *Nature* **277**, 212.)

One of the most widely used techniques is a development of Benzer's countercurrent fractionation where a population of 30–50 flies in a tube is attracted towards a light (Fig. 4.15). In the presence of one chemical odourant in the tube nearer the light, they receive an electrical shock but no shock in the presence of another odourant. Learning is demonstrated by more flies avoiding the shock-associated than the control odour. Benzer and his colleagues have excluded many biases which could explain the results, including phototaxis, social interactions and 'stampeding' of groups of flies, locomotor activity and inability to discriminate the odourants.

However several problems remain, most notably that learning is a property of the population not of the individual. In other words, about 30% of the flies learn on each trial but it is not the same 30% on successive trials. If one is to have a learning task suitable for genetic analysis, an individual must be capable of consistently showing behaviour which reflects its genotype. Apart from such practical problems as 25% of the flies being unaccounted for because the experimenter cannot see all parts of the apparatus, a more theoretical question is what should be learned in this situation. The optimal strategy may not be to move only in the presence of the control odour but simply to learn not to move at all, thereby avoiding any possibility of an electric shock. This response, a conditioned inhibition, was a frequent problem in many of the planarian learning experiments mentioned earlier and may well apply to *Drosophila*. The initial problem with mutants (Dudai 1977) is in demonstrating that the mutation affects learning directly, not by modifying phototaxis, activity or some other variable that indirectly impairs learning.

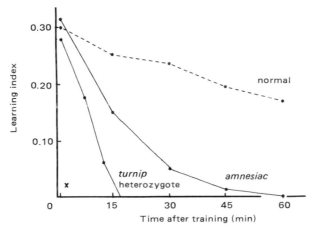

FIG. 4.16   Memory retention of normal, *amnesiac* and *turnip/wild*-type *D. melanogaster* heterozygotes on the task shown in Fig. 4.14. X marks the score of *dunce* and *turnip* mutants. (Modified from W.G. Quinn, P.P. Sziber and R. Booker (1979) *Nature* **277**, 212.)

Given these limitations, several X chromosome-mutants have been detected. Apart from *smellblind* which fails to learn because of a peripheral defect in the ability to discriminate the odours, mutants with effects on learning (Fig. 4.16) include:

dunce   Fails to learn the olfactory task due to a deficit in associative learning skills, not through any obvious sensory or behavioural handicap. However *dunce* can learn a visual discrimination task normally, implying that this task must involve a different genetic system.

amnesiac   Learns normally but memory decays rapidly, persisting for only 1 hour compared with 6 hours in normal flies. Even though memory cannot be observed it is still there. *amnesiac* flies trained a second time but to the opposite odour do not learn as well, showing that the memory can still interact with more recent experience. The memory is stored but cannot be retrieved.

turnip   While this mutant like *dunce* fails to learn, the heterozygote can learn but forgets very rapidly. It seems likely that the homozygote has such poor retention that it forgets immediately after training.

dunce has been most intensively studied and has a defect in the enzyme cyclic AMP phosphodiesterase. Cyclic AMP is involved in the control of neurotransmitters and the enzyme phosphodiesterase inactivates cyclic AMP so that information is carried for only a short time and not propagated endlessly in the nervous system. The animal has to have some phosphodiesterase or it would die, but one of the two forms of the enzyme is missing in *dunce*. To show that the absence of the enzyme was causative and not merely coincidental in the mutant, normal flies were fed caffeine, an inhibitor of the phosphodiesterases and this reduced their performance to the level of *dunce*.

Cyclic AMP is implicated in mammalian neurotransmission and so *dunce* may indeed turn out to be a useful model system for learning.

There are three other developments which make for more confidence in the avoidance learning task.

1 *amnesiac* also shows altered performance on a male courtship task involving experience. Normally male courtship towards virgin females is impaired after they have been confined with fertilized females which are unreceptive. The inhibition of courtship in *amnesiac males* is of much shorter duration.

2 When the learning task is modified for *Drosophila* larvae moving across an agar plate to or from airborne odourants, the same results are observed. *smellblind* larvae fail to respond to odourants, *dunce* and *turnip* and a similar mutant *cabbage*, all fail to learn and *turnip* heterozygotes learn but forget rapidly.

3 Exposure of adults to cold or to nitrogen just after learning on this task inhibits the consolidation of what has been learned into long term memory. The fly remembers normally for a short time but cannot store the information for long. The effects of cold- or nitrogen-narcosis parallel those observed in many vertebrate learning paradigms.

Another approach widely used in dissecting the components of vertebrate learning has been to administer the protein synthesis inhibitor cycloheximide just before training. Short-term memory is normal but cycloheximide inhibits long-term memory which depends on protein synthesis. Although cycloheximide did not have this effect on *Drosophila* avoidance learning there may be a good reason for this, as we can see from the example used to illustrate the next approach to *Drosophila* behaviour genetics.

Chromosome inversions and behaviour

One thing that attracted earlier geneticists to *Drosophila* was the fact that the chromosomes of the salivary glands in the larvae are so large that individual bands or segments could be readily identified visually. Natural *Drosophila* populations often contained polymorphisms for different inversion sequences. That is, there were large numbers of individuals in whom the arrangement of bands might have the order ABCDEFG and other individuals in whom the order of the segment CDE was inverted, so that their bands were arranged ABEDCFG. Due to the mechanisms of meiosis there is little crossing-over within such inversion sequences and they are transmitted from one generation to the next effectively as complexes of single genes, with the added advantage that they can actually be distinguished under the microscope.

Parsons (1973) reviews the extensive work on mating behaviour and inversions and we concentrate on the aspect of mating which demonstrates a very different approach to *Drosophila* learning from that discussed earlier. It involves two inversion types, Arrowhead (AR) and Standard (ST) of *D. pseudoobscura*. Normally AR females given a choice prefer to mate with AR males. However if forced to mate with ST males at 4 days of age, they will choose ST again at 11 days when they once more become receptive. One way of confirming that this situation has similarities with more conventional learning tasks is to feed the females cycloheximide to see if it inhibits memory as in vertebrates. Table 4.1 shows that while cycloheximide given shortly before mating has no effect, cycloheximide given within 30 minutes after mating with ST completely eliminates any memory of the mating partner.

TABLE 4.1   Effect of cycloheximide on choice of Arrowhead (AR) or Standard (ST) inversion-type mates by D. pseudoobscura AR females (derived from A. Pruzan, P. B. Applewhite and M. J. Bucci (1977) Pharmacology, Biochemistry and Behavior 6, 355).

| | Males chosen | |
Females	AR	ST
Virgin	45	11
Prior mating with ST males, no cycloheximide	11	48
Cycloheximide administered before earlier ST mating	5	28
Cycloheximide administered just after ST mating	30	2

While cycloheximide fails to disrupt the avoidance learning in this way, the reason may be the different duration of learning, since avoidance learning decays rapidly within 6 hours whereas one exposure in the mate recognition task leads to retention for 8 or more days. But the mate recognition task has also involved a different species, D. pseudoobscura, so the possibility of species differences in learning processes cannot be ignored.

Polygenic inheritance

Earlier chapters introduced examples of three different approaches to polygenic inheritance in Drosophila:

1 *Diallel cross*   The diallel cross where several inbred strains are mated in all possible combinations provides a good overview of the genetic variation within a species. Figure 3.12 gave the results of a diallel cross between nine D. melanogaster strains tested on the avoidance learning task with biometrical analysis indicating (a) a consistent dominance for high performance, 31 of the 35 F_1s scoring above the mean of the parents; (b) a fairly small narrow heritability (15%) but a much higher broad heritability (30%) reflecting the extensive dominance variation; (c) there are at least four loci involved. With the limitations of the analysis, this probably means that all four chromosome pairs of Drosophila have some influence on avoidance learning.

Thus the diallel cross shows that much more of the genome is involved in learning and indicates the limitations of the gynandromorph approach relying on X chromosome mutations. Not only are all three pairs of autosomes involved as well, but the heritability estimates are not very large, with at least 70% of the differences between flies being due to non-genetic causes. It might be more appropriate to search for environmental influences on learning than to concentrate on the limited range of genetic variation accessible with mutant studies.

2   Artificial selection   We saw in Chapter 1 that the geotaxis and phototaxis maze (Fig. 1.3) had many problems in terms of what it actually measured. Results obtained with this apparatus may bear little relationship to measures of the same behaviour made by different methods. But the artificial selection studies with this maze are so extensive that they provide the best example of selection in *Drosophila*.

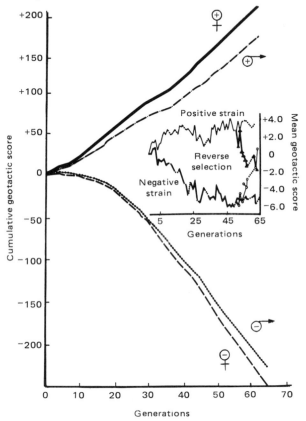

FIG. 4.17   The cumulative response of male and female *D. melanogaster* to selection for positive and negative geotaxis. Inset is the response to reverse selection from generation 52 to generation 64. (From L. Erlenmeyer-Kimling, J. Hirsch and J. M. Weiss (1962) *Journal of Comparative and Physiological Psychology* **55**, 722.)

The marked response to 65 generations of selection (Fig. 4.17) shows the considerable amount of genetic variation in geotaxis. Yet much genetic variation remained within the selection lines at this time. The inset in Fig. 4.17 shows the effects of reverse selection. Beginning at generation 52, the most negative pairs within the positive line were bred, as were the most positive pairs in the negative line. If no genetic variation remained within the lines this reverse selection would have been ineffective, but instead there was a rapid response over the next 12 generations.

Selection involves accumulating all the alleles for increased performance in one line, all those for decreased performance in the other. Clearly this had not been achieved at the time of the reverse selection and the lines had not reached homozygosity although they were getting there. As we saw in Chapter 3, two homozygous parental lines and their F_1s should have similar variances, while the F_2 variance is much increased through segregation. If the parental lines are not homozygous, the F_1 variance will be larger as it contains segregating alleles. The gradual approach to homozygosity in the geotaxis selection lines is seen in the changing relationship between the F_1 and F_2 generations. In generation 11, the F_2 variance was only 0.78 units greater than the F_1, but the difference increased to 3.33 by generation 15 and 4.40 by generation 53.

To look in more detail at the genetic changes accompanying selection, chromosome substitution lines can be created. These involve crossing the selection line (S) and the unselected base population (B) from which the selection line was derived with a stock of flies carrying inversions and dominant marker alleles on every chromosome (except the very smallest chromosome IV). The inversions inhibit crossing-over and the dominant markers allow each chromosome of a particular fly to be identified. By appropriate matings genotypes can be produced with all possible pairs of chromosomes. That is, each chromosome pair can be SS, SB or BB. Across all three major chromosome pairs, this gives 27 (3 x 3 x 3) female genotypes and 18 (2 x 3 x 3) male genotypes, since males have only one X chromosome. This approach was applied to the negative geotaxis line after 133 generations of selection giving the results in Table 4.2.

TABLE 4.2   Analysis of chromosome substitutions in the negative geotaxis line of D. melanogaster. (Derived from J. Hirsch and G. Ksander (1969) Journal of Comparative and Physiological Psychology **67**, 118.)

	Male			Female		
Chromosome	I(X)	II	III	I(X)	II	III
Effect of substitution of selected for unselected chromosome	− 1.16	− 0.70	− 0.15	− 0.63	− 0.89	− 0.32
% Contribution to additive genetic variance	21	10	4	15	16	3

All three chromosomes were involved, especially the X and the largest autosome. The substitution effect of the X was almost twice as large in males as in females, but the overall effect on the females was much greater. Taking into account that the selection line has two copies of each chromosome except for one X in males, the overall difference between unselected and selected lines can be calculated using the values in Table 4.2 as $1 \times -1.16 + 2 \times -0.70 + 2 \times -0.15 = -2.86$ for males and $2 \times -0.63 + 2 \times -0.89 + 2 \times -0.32 = -3.68$ for females. The chromosome substitution technique allows polygenic inheritance in a selection line to be divided into its genetic components and to see upon which chromosomes selection acts.

Selection lines are also ideal for identifying the ways in which a variety of behaviours are changed as a result of selection. This particular approach using selection has been most widely used in *Drosophila* because of one practical advantage, the short generation time (14 days at 25°C). This generation time both speeds up selection greatly, so that 26 generations can be achieved each year compared with 3–4 in rodents, and complicates alternative breeding programs such as the diallel. It becomes very difficult to test enough flies from the numerous different genotypes at roughly the same age.

Ehrman and Parsons (1981, p. 171) summarize the extensive selection studies on mating behaviour, addressing such questions as (a) if one sex is selected for slow mating, is the other sex affected? (b) which components of courtship are most influenced? and (c) are sexual and locomotor activity correlated positively, so that the most active flies are the fastest to mate or correlated negatively, so that flies which move a lot do not provide or receive sufficient courtship cues and hence mate slowly?

TABLE 4.3   Performance of negative geotaxis flies reinforced over three trials for choosing a smooth, rough or control (no cues) ascending maze pathway (40 flies in each group) (derived from O. W. Drudge and S. A. Platt (1979) *Behavior Research Methods and Instrumentation* **11**, 503.)

No. of flies in which the proportion of correct choices either	Smooth	Rough	Control
Increased over three trials	24	27	14
Did not change	6	7	10
Decreased over three trials	10	6	16

To continue the themes of taxes and learning, Table 4.3 is an ingenious demonstration of how selection lines can be extended into other areas of behavioural research. In an appropriately modified maze, the negative geotaxis line can learn to discriminate between smooth and rough substrates where the only reinforcement for a correct choice is being able to climb upwards in the maze, that is to display the taxis for which they have been selected.

3 *Strain differences*   Even without the elaborate mating schemes of the diallel and selection experiments, a great deal can be learned about behaviour simply through strain comparisons. Figure 4.18 demonstrates the use of isofemale strains. Each such strain derives from a single ('iso' in Greek) wild-caught inseminated female, whose emerging progeny are maintained and inbred in the laboratory for a

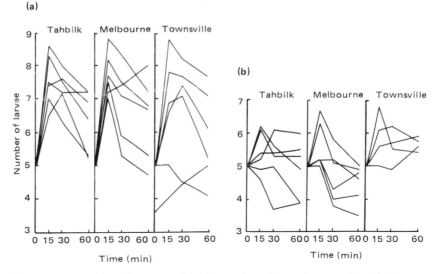

(a)

(b)

FIG. 4.18   Number choosing ethanol out of ten larvae from strains of (a)
D. melanogaster populations and (b) D. simulans populations both from three
different localities. Each line represents a different isofemale strain. (From
P.A. Parsons (1977) Oecologia **30**, 141.)

period of generations. The variation between strains gives some
indication of the variation in the natural population. For the
experiment in Fig. 4.18, two other variables were introduced. Females
of two species D. melanogaster and D. simulans were collected in
three localities — Melbourne and a nearby winery Chateau Tahbilk
both in the temperate south of Australia, and Townsville in the
subtropical north.

Adult D. melanogaster are more tolerant of alcohol in the environment than D. simulans, often being found around winery fermentation sites and this experiment was to determine if the same
preference applied in larvae allowed one hour to crawl towards an
area of an agar plate containing ethanol. The D. melanogaster but not
the D. simulans larvae almost always crawl to the ethanol at first,
paralleling the adult preference. Although the D. melanogaster
preference diminishes, they never avoid the ethanol as many of the
D. simulans strains are doing by the end of the hour.

This result in itself justifies the use of different strains since one
can be far more certain that a real species difference exists than if only
one or two strains had been used. But it is also possible to interpret the
variation between strains. The D. melanogaster strains from Townsville are far more variable than those from the south. Drosophila rely
on plant fermentation products as their food resources and the
difference between localities in behavioural diversity within
D. melanogaster probably reflects the decreasing plant diversity from
the subtropical to the temperate regions of Australia. Thus the isofemale strains demonstrate a relation between behavioural variation
in laboratory tests and ecological variation in the natural habitat.

The variation between strains can also be a useful tool in behavioural analysis. The taxis maze in Fig. 1.3 can be modified to study learning by forcing the flies to make a sequence of turns to move towards the food tubes and lights at the end of the maze. When given a choice, will they continue the sequence, having associated this response with progress to the end of the maze? Apart from learning, this response could be affected by such factors as activity, alternation of left and right turns, preferential bias towards left or right and flies following one another. Within any one strain it is impossible to distinguish between these factors.

TABLE 4.4   Mean performance of *Drosophila* crosses in two types of maze (terms are as defined in the text) (derived from D. A. Hay (1979) *Experientia* **35**, 310.)

Genotype	'Forced-choice' maze			Conventional maze	
	PL	CL	CR	Outer left	Outer right
Parental strains	51.5	22.9	23.8	22.2	11.1
Reciprocal F_1s	52.1	32.0	37.5	19.3	13.5
[h]	0.6	9.1	13.7	−2.9	2.4

Table 4.4 demonstrates one solution to the problem. The magnitude of [h], the measure of directional dominance (more correctly called 'potence' since it may be influenced by dispersed additive or ambidirectional dominance effects — Chapter 3) is calculated from the difference between parental and F_1 means for the measure PL, the percentage of flies showing an initial left-turning bias in the 'forced-choice' maze. [h] is also calculated for CL and CR, the percentage of flies which had initially turned left or right and which at the next choice-point (see Fig. 1.3) continue the sequence of left or right turns reflecting their learning of the maze. Significant [h] values are found only for CL and CR, the hypothesized measures of learning, not for PL, so learning is genetically distinct from the initial bias. [h] was also estimated for flies that had consistently repeated choices in the conventional maze where turns are not forced and that had ended up in the outermost left or right tubes of the maze through alternation or following effects. Since there are no forced sequences in the conventional maze, learning is unlikely to be involved in this behaviour. Again the values of [h] were small and insignificant. Thus the measures of learning in this maze are genetically independent from the other behaviours which might have explained the results, a conclusion confirmed by selective breeding (Hay 1980). Selection for learning did not influence the other behaviours and *vice versa*.

Utilizing the genetic architecture in this way has further benefits beyond dissecting the behaviour and demonstrating that the measure of learning is quite distinct from the other behaviours: (a) it indicates some similarities between this task and the avoidance learning task where there was also directional dominance; (b) it enables parallels to be drawn with other species. Table 5.9 and Fig. 6.11 show respectively that learning in rodents and human intelligence test performance are both characterized by directional dominance. While this does not prove that the *Drosophila* learning tasks are necessarily tapping the same phenomenon, it is at least reassuring.

But why should there be directional dominance for learning and not for other characteristics such as geotaxis? With the ease of population genetic studies in *Drosophila*, they can explain how behaviours evolve to have different genetic architectures.

Drosophila and the evolution of behaviour

Although species diversity in *Drosophila* courtship has been studied from the viewpoint of evolution (Fig. 4.9) an alternative evolutionary emphasis is on the advantages of a particular expression of behaviour within a species. Does the animal which, for example, learns the fastest or is more active, have a selective advantage over the rest of the population? This question can to some extent be answered by studying the genetic architecture of the trait.

The whole field owes much to Fisher's theory of the evolution of dominance and to the work of Mather and others at Birmingham which is summarized by Broadhurst and Jinks (1974) and which reflects the three main ways natural selection may act upon some trait in a population. Figure 4.19 illustrates that one extreme or the other of

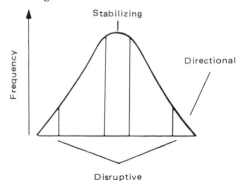

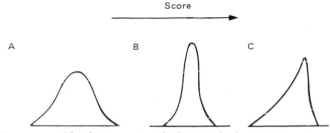

Fig. 4.19   The three ways in which natural selection may act upon variation and the effects on the distribution of scores in a population with (A) no dominance (B) ambidirectional dominance and (E) directional dominance for high performance. (From D. A. Hay (1980) *Neuroscience and Biobehavioral Reviews* **4**, 489.)

the population may be favoured (directional selection), an intermediate may be favoured (stabilizing selection) or two or more different levels may be favoured (disruptive selection). The concern here is with the first two as disruptive selection can lead to a stable polymorphism and even to speciation.

The difficulty is that there are two conflicting evolutionary needs, firstly to ensure that as many as possible of the population fall in that part of the distribution favoured by natural selection and secondly, to preserve variation in the population. If only the former mattered, then a simple solution would be to have the favourable alleles 'fixed' in the population, that is, every individual would be homozygous for these alleles. One implication is that there would be minimal additive genetic variance for those characters important for an organism's fitness. Thus by definition, any character with a large narrow heritability is part of what Thiessen called 'genetic junk' in the quote which concluded Chapter 1.

But this argument is not complete. The lack of genetic variability would make it difficult for the population to adapt genetically should circumstances change and some other level of expression of the behaviour be favoured, for example, low rather than high performance. Depending on whether natural selection is directional or stabilizing respectively, one compromise is to have dominance which is directional for all loci or ambidirectional (dominance for high performance at some loci, dominance for low performance at others), maintaining the variety of alleles but shifting the distribution of scores in the desired direction. The lower part of Fig. 4.19 illustrates schematically what happens to the distribution in a population where there is dominance for high performance, or ambidirectional dominance. (Recall Fig. 3.7 which demonstrated at the level of individual loci how such distributions arose.) Relative to the situation with no dominance, in the former case more of the population lie in the range favoured by natural selection for high performance and in the latter, more are in the range favoured by stabilizing selection. Certain types of epistasis may further enhance such effects.

The value of this approach is that instead of arguing on purely pragmatic grounds that a particular expression of behaviour has evolutionary advantages, one can now check this hypothesis by genetic analysis. If natural selection does favour high learning ability because in the wild animals with this ability are more likely to find food or mates or to avoid traps, then there should be directional dominance for high learning ability. Similarly, if natural selection favours an intermediate level for some other behaviour there should be ambidirectional dominance. The advantage of fast learning for *Drosophila* may not be obvious, but one possibility is suggested by the mutant *amnesiac* which is deficient in retaining altered courtship following exposure to unreceptive, fertilized females, as well as on the avoidance task. That is, slow learners may be more likely to waste time in unproductive courtship.

Except for *Drosophila* it is very difficult to find an organism that lends itself to experimental verification of this hypothesis and even in *Drosophila* most of the evidence comes from physical characteristics such as viability which, not unexpectedly, shows dominance for high performance. One of the few behavioural examples from *Drosophila* is locomotor activity following stimulation. Directional dominance for high activity was found and a study of the possible selective advantages revealed that the more active flies were less likely to become stuck to the walls of the culture bottles. Activity does have other selective advantages besides this rather mundane one as it is also involved in mating success, but nevertheless this provides a clear demonstration of the connection between the genetic architecture and the action of natural selection.

These results were confirmed by comparing flies reared in standard culture bottles with those reared in population cages where the population density is far higher and where there is greater competition for such limited resources as food, mates and egg-laying sites (reviewed in Hay 1980). If competition is more intense, selection pressures should be greater and higher levels of activity found if high activity really does have some selection advantage. Such turned out to be the case with cage flies being much more active only when activity was studied in the realistic situation of groups of flies and not in isolated individuals.

In the next chapter several examples from rodents show that the results are not confined to *Drosophila* and Table 2.4 has already provided one human example which can be interpreted in this evolutionary manner. If mental retardation is much more deleterious than many of the other human genetic disorders, having most disorders involving retardation inherited as autosomal recessive is one way to limit their occurrence while still maintaining the allele in the population.

There are two limitations to this technique, the first being that the genetic architecture depends on the way the behaviour is measured. Whereas *Drosophila* activity in response to stimulation shows directional dominance, there is no dominance for the level of spontaneous activity where there is no prior stimulation. So one has to be very precise about the behaviour on which natural selection is postulated to have acted. The second limitation is the population on which selection acts. In Fig. 4.18, the populations from which the strains came were clearly defined but in many biometrical analyses this is not the case. The nine strains used in the avoidance learning diallel (Fig. 3.12) came from all over the world, so how can one claim that they represent a particular gene pool influenced by natural selection? The best support comes from the consistency of the results. With the one exception of the cross 7 x 9, the results in Fig. 3.12 are remarkably uniform, suggesting that the different populations may be influenced by similar selection pressures.

Conclusion

In the 1970s, emphasis switched from the polygenic approach to *Drosophila* behaviour to the neurogenetic approach and particularly to the gynandromorph technique. The avoidance learning example shows that it is unwise to dismiss polygenic analysis, since it has demonstrated that there is much genetic variation not on the X-chromosome and that in any case, most variation is non-genetic in nature. It must therefore remain debatable whether the mechanisms of learning will ever fully be unravelled using the gynandromorph technique. Another less explored point of intersection of polygenic and single gene approaches is that the effects of a mutation may depend upon the particular genetic background of the animal in which the mutation was induced. Much of the gynandromorph work has been based on a single stock *Drosophila* line (the Canton-S stock) and there is no guarantee that the same mutation in another stock would have the same result. It is akin to the effects of the background genotype on Down's Syndrome (Fig. 2.13) or on PKU (Fig. 2.18).

The avoidance task also demonstrates how much less sophisticated is behavioural measurement than genetics. A situation where there is only a 30% chance that an individual fly will repeat the response on the next trial hardly justifies detailed genetic analysis. The importance of behavioural refinement is indicated by the fact that Tully (1983) has recently redesigned this apparatus, ensuring that all flies are shocked and improving the administration of the two odourants. In this new apparatus, 95% of wild-type flies learn and learning improves with number of training trials. While the task may now be better, it may not help our understanding of a central mechanism underlying learning. A comparison of 10 different strains on this olfactory learning task and on a visual learning task showed no relationship, leading Durcan and Fulker to conclude that rather than a central mechanism,

there are some quite specific factors controlling performance in both tasks, perhaps of a sensory nature, and that possibly the learning deficient mutants are mutants of these factors. (*Behavior Genetics* (1983) **13**, 189.)

The problem may come back to 'biological preparedness' and the fact that *Drosophila* never evolved to avoid electric shocks. A better way to tap learning potential may be through some of the effects of experience on courtship and mating behaviour (Tompkins *et al.* 1983) or through a recent demonstration that *D. subobscura* can learn to return to baits which mimic their natural feeding and breeding sites.

A more general observation is that the potential for genetic analysis in *Drosophila* may blind us to their behavioural limitations and that some other invertebrate may offer a better compromise between genetics and behaviour. Figure 4.20a presents the results of a selection experiment on learning in the blowfly *Phormia regina* where flies were classically conditioned to extend their proboscis for sucrose when chemosensory hairs on their legs were dipped in saline. In this insect which is large enough to be handled individually both for behavioural measures and for electrophysiology, the bright and dull strains have reliably separated after only seven generations. Note the somewhat greater and less variable response of the dull line, indicating like *Drosophila* some dominance for good learning ability.

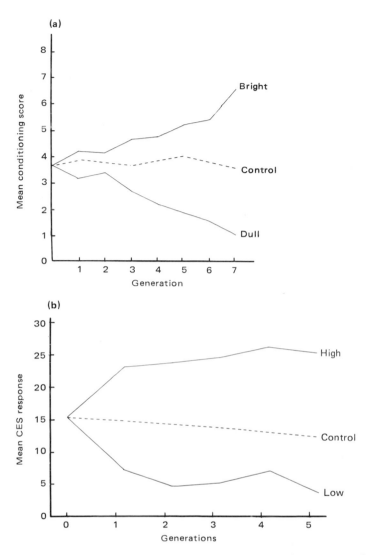

FIG. 4.20   (a) The response of *Phormia regina* to selection on a classical conditioning task. (From J. Hirsch and L. A. McCauley (1977) *Animal Behaviour* **25**, 784.) (b) The response in one generation to selection for central excitatory state. (Two other replicates produced identical results.) (Derived from T. R. McGuire (1981) *Behavior Genetics* **11**, 331.)

Crosses among similar selection lines has confirmed this result (Médioni 1977). The level of genetic analysis possible in blowflies is not as good as that in *Drosophila*, but at least a consistent learning task and selection lines in which to analyse the learning in more detail are available.

As the *Drosophila* learning example in Table 4.4 indicated, genetic techniques can be used to differentiate components of behaviour. In the case of blowflies, learning may be confounded with the central

excitatory state (CES), a proboscis extension in response to inappropriate stimuli or after prolonged immobilization of the flies. Figure
4.20b shows that CES is inherited quite differently from classical
conditioning. Whereas the selection response for classical
conditioning is gradual, implying several genes are involved, the
response for CES selection is complete in one generation, indicating
that only one locus is involved. Complex biometrical analyses of
crosses between the selection lines have confirmed this result. It is
rare that single locus inheritance is found in a normal (non-mutant)
population, but the situation in blowflies creates an exciting
possibility for the analysis of conditioning behaviour into its underlying components.

In summary, invertebrate behaviour genetics is currently a very
stimulating field with organisms and behaviours being offered as
models for what goes on in higher animals. Just how many of these
models will be successful in explaining behaviour remains to be seen.
Some of the initial enthusiasm for behavioural analysis has faded
particularly in work on nematodes and *Drosophila* gynandromorphs.
Instead the emphasis in these organisms has turned to development
and physiology of the nervous system (Ward 1977), which underlie
behaviour but which do not necessarily help to explain the
behaviour.

Discussion topics

1 "Quinn's law states that the convenience of genetics multiplied by
the quality of the behavioural repertoire is a constant for all animals.
Hence, bacteria have superb genetics and marginal behaviour, while
undergraduates have very interesting behaviour but cumbersome
genetics." (W. G. Quinn and J. L. Gould (1979) *Nature* **278**, 21.)

Are there any invertebrates which are an exception to Quinn's
law?

2 W. C. Corning, J. A. Dyal and A. O. D. Willows (eds) (1973)
Invertebrate Learning (three volumes), Plenum, New York, cover just
about every attempt to study invertebrate learning, but give little or
no attention to genetics. Choose an organism from their text and
describe how you would approach the genetic analysis of its learning
ability. (You may find useful R. C. King (ed.) (1974-76) *Handbook of
Genetics* vols 1-3, Plenum, New York, covering genetic systems in
invertebrates.)

3 "Whilst the ultimate aim of behaviour genetics is an
understanding of rich and complex behavioural repertoires of man
and the other higher animals, this is unlikely to be a realizable goal
until we have a clearer picture of how genes affect the 'nuts and bolts'
of such underlying cellular processes as sensory transduction and
neural transmission, and the mechanisms leading to the development
of connectivity in nervous systems. For this reason genetic
manipulation of invertebrate organisms, using them as model
systems, will continue to be valuable." (B. Burnet and K. J. Connolly
(1981) *British Medical Bulletin* **37**, 113).

Is this a realistic statement and what are the limitations of the approach?

References

Annotated bibliography

Ehrman L. and Parsons P. A. (1981) *Behavior Genetics and Evolution*. McGraw Hill, New York. (The general behaviour genetics text dealing most fully with *Drosophila* and with many other invertebrates.)

Ewing A. W. and Manning A. (1967) The evolution and genetics of insect behaviour. *Annual Review of Entomology* **12**, 471. (A review presenting a very different approach to behaviour with much more emphasis on evolution and species differences than the present chapter.)

Gould J. L. (1974) Genetics and molecular ethology. *Zeitschrift für Tierpsychologie* **36**, 267. (A provocative claim for the superiority of the molecular approach to understanding behaviour — is he justified, given some of the problems in behavioural measurement raised in this chapter?)

Hall J. C., Greenspan R. J. and Harris W. A. (1982) *Genetic Neurobiology*. MIT Press, Cambridge, Mass. (A detailed account of neurogenetics and of how behavioural mutants have become neurophysiological, neuroanatomical and neurochemical mutants. Note that they distinguish (p. 3) their focus from that of behaviour genetics.)

Koshland D. E. (1980) *Bacterial Chemotaxis as a Model Behavioral System*. Raven, New York. (A comprehensive account of bacterial behaviour genetics, but one with emphases on parallels to higher organisms with which not everyone agrees — see Berg's review *Nature* (1981) **292**, 870.)

Médioni J. (1977) Towards a genetic analysis of learning and memory in invertebrates. In *Genetics, Environment and Intelligence* Oliverio A. (ed.), p. 59 North Holland, Amsterdam (A description of the methods and problems involved in invertebrate learning. The concept of a 'genetics x learning intersection' is worth noting.)

Parsons P. A. (1973) *Behavioural and Ecological Genetics: A study in Drosophila*. Clarendon, Oxford. (Although predating the neurogenetic approach, this text is the most complete account of polygenic, ecological and evolutionary research on *Drosophila*.)

Wilson J. R. (ed) (1973) *Behavioral Genetics: Simple systems*. Colorado Associated University Press, Boulder. (Proceedings of a conference at the time behaviour genetics of lower organisms was just beginning and worth comparing with the subsequent developments.)

Additional references

Ashburner M. and Wright T. (eds) (1978) *Genetics and Biology of Drosophila*. vol. 2b. Academic Press, New York.

Benzer S. (1973) Genetic dissection of behavior. *Scientific American* **229** (6), 24.

Berg H. C. (1975) Bacterial behaviour. *Nature* **254**, 389.

Hay D. A. (1980) Genetics in the analysis of behaviour. *Neuroscience and Biobehavioral Reviews* **4**, 489.

Jacobs M. E. (1978) Influence of β-alanine on mating and territorialism in *Drosophila melanogaster*. *Behavior Genetics* **8**, 487.

Lindauer M. (1975) Evolutionary aspects of orientation and learning. In *Function and Evolution in Behavior* Baerends G., Beer C. and Manning A. (eds), p.228. Clarendon, Oxford.

Parkinson J. S. (1977) Behavioral genetics in bacteria. *Annual Review of Genetics* **11**, 397.

Quinn W. G. and Gould J. L. (1979) Nerves and genes. *Nature* **278**, 19.

Rothenbuhler W. C. (1967) Genetic and evolutionary considerations of social behavior of honeybees and some related insects. In *Behavior-genetic Analysis* Hirsch J. (ed.), p.61. McGraw-Hill, New York.

Tompkins L., Siegel R. W., Gailey D. A. and Hall J. C. (1983). Conditioned courtship in *Drosophila* and its mediation by association of chemical cues. *Behavior Genetics* **13**, 565.

Tully T. (1983) Properties of retention of a classically conditioned olfactory response in wild-type and mutant *Drosophila melanogaster*. *Behavior Genetics* **13**, 555.

Ward S. (1977) Invertebrate neurogenetics. *Annual Review of Genetics* **11**, 415.

5 Behaviour genetics in rodents and other vertebrates

Topics of this chapter

1 The problems in using single gene mutations to study behaviour, most notably the criticism that many mutations are trivial with only peripheral effects on behaviour.
2 Learning as a demonstration of the different levels of genetic analysis and the extent to which results on experimental animals may be related to humans.
3 The concept of the behavioural phenotype and the pooling of data from many different sources, especially neurochemistry, to better appreciate the ramifications of genetic differences in behaviour.
4 The relationship and possible extrapolation between laboratory and wild rodents and the role of behaviour in structuring wild populations.

Introduction

Genetic analysis in non-human vertebrates is the oldest branch of behaviour genetics in that man has always selected his domestic animals partly on their behavioural characteristics. The plethora of contemporary dog breeds is visual proof of behavioural as well as morphological genetic diversity. Dogs have been selectively bred for everything from bear-baiting to retrieving game, as well as for more exotic roles such as dachshunds for running along under carriages, and naked dogs used by the Chinese as hot water bottles in bed!

TABLE 5.1   Percentage of variance attributable to differences between dog breeds. (From J. P. Scott (1964) *American Zoologist* **4**, 161.)

Measures	Number of variables scored	Percentage of variance*
Physical size	10	27
Heart rate	5	38
Expressions of emotion	8	23
Social relationships with dogs and humans	3	21
Trainability	6	28
Problem-solving	12	27
All behavioural measures	34	27

*These do not correspond to either the narrow or the broad heritability, since dog breeds are not inbred and at least 12% of the total variance is genetic variation within breeds.

More structured behaviour genetic analyses on dogs are limited. The major effort has been the 20 year study at the Jackson Laboratory, Maine, concerned mainly with the emotional and social behaviour of five breeds of medium-sized dog (reviewed in Scott and Fuller 1965). Table 5.1 indicates the sort of behaviours measured and makes an important point about the potential for selective breeding for behaviour. There is as much genetic variation for the behavioural as for the physical features, so that if breeding can create dogs as different physically as the Chihuahua and the Great Dane, it should be possible to breed also for similar behavioural diversity.

Research on dogs has the potential to develop into an applied area of behaviour genetics in breeding better guide-dogs for the blind. Currently in Australia dogs are rejected for guide-dog training principally for behavioural defects (fearfulness, ease of distraction by other dogs and excitability) rather than for such physical reasons as hip dysplasia. Labradors have been bred specifically for guide-dog work to be less fearful and less dog-distracted, but Labradors can be particularly affected by excessive attention and affection in the period they spend in private homes before formal training begins. So there are not only genetic differences within and between breeds but also genotype–environment interactions to consider. Attempts are being made to determine if particular hybrids such as labrador x German shepherd combine the best characteristics of the different breeds or introduce some other undesirable features such as aggression.

But dog breeds are unlikely to play a major role in experimental behaviour genetics. Beyond the practical problems of housing and maintaining such large and slow-breeding experimental animals, they have genetic disadvantages. The breeds are not inbred and attempts at inbreeding usually increase the incidence of infertility and malformations. Following Quinn's law introduced in the discussion topics in Chapter 4, laboratory rodents provide the best compromise. Their behavioural repertoire is extensive and well defined, so that it is easy to demonstrate behaviours such as learning without the controversy that accompanies similar attempts in invertebrates (Chapter 4). Many different types of genetic analysis are possible and there is a long history of inbred strains (Fig. 3.3) and of mutants often bred initially by pet fanciers for their bizarre behaviour or coat colouration. The waltzing mutant mouse which makes exaggerated circling motions can be traced back to 80 BC in China.

Chapter 3 introduced the relevant methodology and provided many rodent examples principally from the areas of learning and open-field behaviour. The present chapter demonstrates how the information about genetic variation for a particular behaviour can then be used to understand more about the behaviour and its biochemical and physiological underpinnings. For example, by 1975 the Maudsley Reactive (MR) and Non-reactive (MNR) rat lines, bred for open-field defaecation (Table 3.1) had been used in 131 comparisons on aspects of open-field behaviour, and 39, 23 and 83 comparisons on physiological, endocrine and pharmacological measures respectively (Broadhurst 1975). Against this array of behavioural data the genetic details of the lines' response to selection are of minor consequence.

We consider only single gene and polygenic inheritance in rodents since surprisingly little attention has been paid to behaviour in chromosomal variants beyond limited studies of a few of the more common translocations. Apart from the ease of alternative approaches to rodent behaviour genetics it is unclear why this should be so. There are no difficulties in identifying rodent chromosomes and chromosome differences have been the bases of many studies of their population dynamics, but not at the behaviour level.

Single-genes

As with *Drosophila* there are two approaches to single gene effects. One is to examine the effects on behaviour of an allele with observable, non-behavioural effects, e.g. on coat colour, while the second is to follow the segregation of behavioural phenotypes where behaviour is itself the genetic marker. In mice the latter group are represented by the 300 neurological mutants so far identified, where the first clue to the mutation often comes from altered behaviour. The way in which behaviour is affected is indicated by the names of such mutants as *gyro*, *quaker*, *reeler*, *staggerer*, *tipsy*, *waltzer*, and *zig-zag*.

Determining the influence on behaviour of a single gene affecting, say coat colour, involves much more than just comparing two inbred strains differing in coat colour. Differences between them in behaviour may be due not to the single locus but to the many other loci at which inbred strains differ. Figure 3.8 was an example of this point. There the albino mice of the BALB/c strain showed a much higher rate of intracranial self-stimulation (ICSS) than the pigmented DBA/2 strain. But reference back to the backcross in Fig. 3.8 shows that ICSS and albinism are unconnected. In the backcross where there is segregation into pigmented and albino mice, mice with very high and very low rates of self-stimulation occur equally among the albino and pigmented. Thus the connections between albinism and high ICSS were purely fortuitous, not functional and were disrupted through segregation and crossing-over when the backcross generation was produced.

In an important critique of single gene research, Wilcock (1969) reviews the methods by which single gene effects can be reliably isolated and introduces three other problems for this type of research:

1   A confusion between causation and consequence. It is too easy to find a mutant with a behavioural anomaly, detect some biochemical or physiological defect and then claim that this defect is the means through which the mutant affects behaviour. One example would be the *obese* mouse which grossly overeats and shows abnormal enzyme activity. However, the abnormal biochemistry is not the cause of the abnormal behaviour but its consequence. Normal mice, made to

overeat by chemical or surgical means, also show the same abnormal enzyme activity.

2 Most single gene defects are trivial and can be explained by peripheral physical or physiological differences, rather than by complex behavioural mechanisms. A simple example would be two strains of mice differing in temperature preference not through any behavioural or metabolic mechanism but merely because they differ at a locus influencing skin thickness and hair density. The behavioural difference is an inevitable consequence of the physical difference and of no great importance.

A more sophisticated example comes from Fig. 3.4 where three inbred mouse strains and a heterogeneous stock were tested on an avoidance learning task, where the mouse could avoid shock by moving in a shuttle box as soon as a light signal came on. The CBA strain hardly learned at all on this task and it was hypothesized that they differed from the good learners in terms of consolidation of the information into long-term memory (rather like the *amnesiac Drosophila* in Fig. 4.16). However CBA is homozygous for the retinal degeneration (rd) allele and may do badly not because of any memory defect but only because these mice cannot see the warning light before the shock. If they are tested instead in a shuttle box where the warning stimulus is a buzzer rather than a light, they do much better. While Wilcock would regard this result as a trivial mutation, Thiessen in a reply points out: '... at the very least they provide a more parsimonious explanation of learning defects and warn investigators against premature neurologizing' (*Psychological Bulletin* (1971) **75**, 104.)

3 In his rejoinder to Thiessen, Wilcock makes a most cogent observation about the limitations of single gene research:

Some reservations about regarding single genes as treatments in this sense, however, should be expressed. Such treatments must be considered in developmental perspective. For example, an rd mouse is not a normal mouse with enucleated eyes, nor is an albino a normal mouse with unpigmented skin. These deficiencies have a developmental history in which background genic interaction and genotype-environment interaction may have compensated for the defects in particular ways. Thus, the spatial ability of congenitally blind people may be quite different from that of people blinded in later life... (*Psychological Bulletin* (1971) **75**, 107.)

If Wilcock is correct, then in theory it is impossible to analyse the effect of any single gene mutation relative to its normal allele.

The albino locus

We can use the albino locus to illustrate three approaches to single gene research on mice. All homozygous recessive albinos (including humans with this condition) lack melanin pigmentation, not only in the coat but also in the iris of the eye, leading to the characteristic 'pink-eyed' condition.

Classical Mendelian breeding schemes   In this approach, segregating generations are created to determine if the linkage between albinism and behaviour remains after segregation and recombination. The ICSS situation in Fig. 3.8 was one example, but most work has been done with albinos in the open-field where albino strains were initially observed to be less active and to defecate more. To isolate the effects to the albino locus, DeFries crossed two inbred mouse strains, C57Bl/6J (black pigmented, high activity and low defecation in the open-field) and BALB/cJ (albino, low activity and high defecation). The F_1 were crossed to get the F_2 and these crossed in turn for the F_3 giving the results in Table 5.2a, where the pattern of reduced activity and increased defecation in the albinos is continued.

TABLE 5.2   Open-field activity and defecation in albino and pigmented mice. (a From J. C. DeFries (1969) *Nature* **221**, 65. b From J. C. DeFries, J. P. Hegmann and M. W. Weir (1966) *Science* **154**, 1577.)

	Activity		Defecation	
	Albino	Pigmented	Albino	Pigmented
(a) Genotype				
F_2	10.7	12.7	1.97	1.76
F_3	10.0	12.3	2.24	1.81
S_8	12.7	15.2	2.41	1.97
(b) Illumination				
White	8.8	12.9	2.10	1.95
Red	13.3	14.1	1.73	1.76

DeFries then began to breed selectively from this F_3 generation for high and low activity, *irrespective* of the coat colour. In the F_2 there should be 25% albinos according to genetic theory, but by the eighth generation of selection (S_8), there were virtually no albino mice in the high activity selection lines, whereas one low activity selection line had only albino and a replicate selection line had 60% albinos. The behavioural differences between the albino and pigmented mice are still found in S_8 (Table 5.2a), including the defecation difference even though selection was only for activity.

This result could mean two things. One is that the effect is not on the albino locus, but on one closely linked to it on the same chromosome. This possibility had been suggested earlier where one experiment on inbred strains and their crosses found albino mice to be much slower on a task involving escape from water, while another experiment involving a randomly bred albino stock showed no albino-pigmented difference. An explanation for this discrepancy could be another locus near the albino one, controlling the behavioural differences and differing between the inbred and the randomly bred population. In DeFries's selection experiment, this explanation is unlikely since by S_8 there would have been so many opportunities for recombination that any connection between the albino and this other allele would probably have been lost.

The remaining possibility is that the albino locus has pleiotropic effects, other than on coat colour. This pleiotropism could be purely 'trivial' in Wilcock's terms, if it were through something as obvious as the increased sensitivity to bright light of mice with unpigmented irises — they cannot reduce the visual input in the way possible with normal animals. Table 5.2b supports this view in that the difference between albino and pigmented is almost abolished when the mice are tested in low levels of red light rather than in bright white light.

TABLE 5.3   Scores of pigmented and albino coisogenic C57Bl/6J mice. (From J. L. Fuller (1967) *Animal Behaviour* **15**, 467.)

	Water escape time (in seconds over four trials)	Open-field activity over 15 minutes	Activity wheel for 24 hours (in 1000s of revolutions)	Water maze (number of correct trials out of 40)	Swimming time in water maze (in seconds)
Pigmented	51	317	13.1	36	2.6
Albino	112	266	14.5	24	2.8

Coisogenic strains   A more satisfactory method which lessens the possibility of any linkage effects is to find a pigmented strain in which there has been a mutation at the albino locus. One can compare such *coisogenic* pigmented and albino strains which differ only at this one locus, since presumably the rest of the genotype is unaltered. Table 5.3 compares pigmented C57Bl/6J mice and an albino mutant of the same strain. Albino mice escaped more slowly from water, were less active in an open-field and made more errors on the water-maze (the last involved a black–white discrimination, so visual acuity may play a part). There were no differences in swimming time (during the last trials in the water maze) or in activity measured over a full 24 hours. Fuller points out that the differences observed may be due to peripheral mechanisms or to neurochemical effects through an imbalance of catecholamines (hormones secreted by the adrenal gland and involved in stress reactions), but concludes

It appears that only an intensive investigation into the neural effects of the gene for albinism will determine whether its behavioural effects are a curious fact or a clue to the way in which genetic information programmes the behaviour of a complex organism. (*Animal Behaviour* (1967) **15**, 470.)

Congenic strains   Whereas coisogenic strains rely on a mutant being found in an existing strain, *congenic* strains involve the introduction of a chromosome segment containing the specific allele into another strain. One then compares animals with this newly introduced allele with their congenic strain which retain the original allele at this locus. The introduction of a specific chromosome segment is possible by testing for genetic compatibility using skin grafts which are only accepted by genetically compatible animals. Skin grafts are used because the histocompatibility loci involved are so diverse that they provide markers throughout the entire genetic complement which can identify the strain from which every chromosome segment comes.

Consider transferring the albino allele from a donor strain to the agouti C3H/HeJ (agouti mice are pigmented with brown coats due to an allele at a coat-colour locus other than the albino one leading to each black hair having a yellow tip). Some 10–14 generations are made of backcrosses to C3H of the albino × C3H F₁, with constant checking by skin grafts to see that only those with particular donor histocompatibility alleles are mated. Intercrossing among the final backcross progeny produces individuals identical to C3H except that

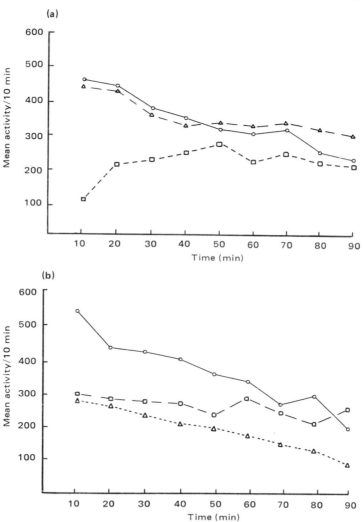

FIG. 5.1   (a) Activity of coisogenic pigmented (o—o), albino (□--□) and heterozygous (△--△) C57Bl/6J mice. (b) Activity of congenic pigmented (o—o) and albino C3H/HeJ (□--□) mice and of the albino donor strain (△--△) which provided the chromosome segment. (From R. J. Katz and R. L. Doyle (1981) *Behavior Genetics* **11**, 167.)

they are homozygous for specific histocompatibility alleles from the donor strain. It only remains to find one albino strain among such congenic strains and you have a C3H mouse in every respect, bar the presence of a small chromosome segment containing the albino allele.

Figure 5.1a compares activity of coisogenic homozygotes and heterozygotes recorded over 90 minutes. Albino mice are much less active initially but gradually the difference disappears. This result would explain why Fuller (Table 5.3) found an activity difference over 15 minutes but not over 24 hours or at the end of a bout of swimming. The performance of congenic strains (Fig. 5.1b) is similar but not exactly the same, probably because the pigmented strain in Fig. 5.1a was the black C57Bl/6J and in Fig. 5.1b the agouti C3H/HeJ. That is, although the genetic background was consistent *within* experiments it differed *between* experiments.

Such sophisticated approaches are relatively new and except in the area of drug effects (Eleftheriou 1975) have not been widely used. Although analysing whether the albino locus affects activity by these means may seem excessive, it serves as a model for what can be done with other loci. None of these methods help answer the original question of how the albino locus actually affects behaviour but there are two lines of evidence to suggest that not all the effects are as trivial as Table 5.2b would suggest.

Firstly albinos may be less sensitive to a wide range of stimuli other than visual ones. In a large-scale study of many coat colour mutants on very diverse behavioural tasks, Thiessen, Owen and Whitsett (in Lindzey and Thiessen 1970) found albinos had a smaller geotaxis response, and hesitated to descend from heights, to avoid light, to escape from water and to avoid sounds. This slowness to respond provides a common explanation for the different effects of albinism on the behavioural tasks in Table 5.3 and Fig. 5.1 in that only those involving a short-term response are affected. It also explains Erlenmeyer-Kimling's suggestion discussed in Chapter 3 that the albino BALB/c is less affected by specific trauma than the pigmented C57Bl, with long-term maintenance variables such as cage illumination having the opposite effect.

Secondly, albinism does have neurological consequences. In the normal animal nerves from each eye project to both the right and left halves of the visual cortex, that is

$$
\begin{array}{ccc}
\text{L} & \text{R} & \text{eye} \\
\downarrow \;\; \times \;\; \downarrow & & \\
\text{L} & \text{R} & \text{cortex}
\end{array}
$$

But in the albino, connections are predominantly contralateral,

$$
\begin{array}{ccc}
\text{eye} & \text{L} & \text{R} \\
& \times & \\
\text{cortex} & \text{L} & \text{R}
\end{array}
$$

Work on humans has demonstrated that this is an effect primarily of the lack of pigmentation not necessarily of the specific allele. Human albinos were compared to those suffering from the Waardenberg syndrome where pigmentation is absent only in the irises and hair but not in the skin. This syndrome is quite different from albinism, being autosomal dominant rather than recessive, but the effect on the brain is identical. That miswiring is due to the pigmentation deficit itself rather than to the allele specifically responsible for this deficit was confirmed in mink where many alleles create an almost continuous distribution of pigment. The degree of visual abnormality was correlated precisely with the degree of eye pigmentation across the range of alleles.

This miswiring means that whereas normal individuals have similar inputs to both sides of the brain whether one or both eyes are open, the albino animal's input is asymmetrical when one eye is shut. For this reason, pigmented rats learn a visual discrimination task better than albinos when one eye is covered, but other behavioural effects of the miswiring are unclear. Albino strains such as A/He, A/J and BALB/c still learn a maze better than some pigmented strains (Fig. 3.6) and so can use visual cues.

Wahlsten's method

While the effects of albinism on behaviour initially appear to be peripheral and of little interest (Table 5.2b), the neurological evidence suggests otherwise. It is possible that other alleles have been dismissed as trivial merely because neurological abnormalities have not yet been detected. Wahlsten (1978) proposed three possible effects of a mutation on behaviour (Table 5.4). (He used albinism and learning as an illustration but any other mutation and behaviour could be substituted.) Effects may be (a) purely peripheral, where albino animals fail to learn because they cannot see properly, (b) central as well as peripheral, so that albinism has some direct effect on learning as well as the peripheral effect, or (c) purely central, where the direct effect on learning is paramount and the effects on visual acuity unimportant. As Wahlsten says, this third possibiilty could provide important information about the nervous system and behaviour.

These three possibilities can be distinguished without having to wait for neurological evidence to emerge, by comparing the performance of normal and mutant animals on one measure of visual acuity and two learning tasks, one dependent on visual discrimination and one on discrimination in some other sensory modality. Each of the possibilities yields a distinct profile of performance (shown in Table 5.4) across the three tasks.

Do any examples exist of case (c), the purely central effect? Wahlsten proposes the brindled mouse, an X-chromosome disorder of copper metabolism. The absence of copper affects many enzymes including lysyl oxidase (leading to whiskers being curly rather than straight), tyrosinase (leading to reduced pigmentation) and dopamine-beta-hydroxylase (leading to tremors). Thus abnormalities of pigmentation do not *cause* behavioural abnormalities but they are *correlated* since both are affected by a common biochemical mechanism.

TABLE 5.4 Wahlsten's method of studying possible relations between albinism, eye pigmentation, visual acuity and learning ability. (from D. Wahlsten (1978) in *Psychopharmacology of Aversively Motivated Behaviour* Anisman H. and Bignami G. (eds), p. 63. Plenum, New York.)

		Consequences for albino (A) and normal (N) mice	
		Discrimination learning	
Model	Visual acuity	Tactile cues	Visual cues
I *Peripheral effects only* albinism other mechanisms ↓ Lack of pigment ￨ ↓ visual acuity → learning ability	N > A	N = A	N > A
II *Peripheral and central effects* albinism other mechanisms ↓ \ ￨ Lack of pigment \ ↓ ↘ visual acuity → learning ability	N > A	N > A	N > A
III *Central effects only* albinism other mechanisms ↓ \ ￨ Lack of pigment \ ↘ visual acuity → learning ability	N = A	N > A	N > A

Arrows indicate causal relations between characters and the absence of arrows no causal connections although correlations may still be found.

The brindled mouse is of particular interest because it parallels the human disorder Menkes Syndrome, also called Kinky Hair Disease because of the steel-wool appearance of the hair. There is genetic similarity in that the locus is at the same relative position on the X chromosome of both species. In both species the defects are very

TABLE 5.5 Parallel defects in Menkes Syndrome boys and brindled mice

Menkes Syndrome	Brindled mice
Hair abnormal in both structure and pigmentation	Coat wavy, whiskers curly, fur light
Weakness and spontaneous rupture of walls of the arteries	Same
Skeletal defects	Same
Convulsions and twitching	Uncoordinated movements and tremors
Hypothermia	Abnormal body temperature during development
Brain atrophy, demyelination and mental retardation	Demyelination likely but other effects uncertain, except for very low levels of many neurotransmitter substances
Death before 4 years	Death by about 14 days

similar (Table 5.5). Apart from tremors, it is difficult to compare the species behaviourally since male brindled mice die at 14 days, unless given a large injection of copper at about 7 days after birth. This injected copper enables the enzymes to function sufficiently well for the pigmentation to appear and the whiskers to straighten.

Neurological mutants

Although the albino and brindled mice have neurological or neurochemical defects, neurological mutant mice are normally detected by gross abnormalities of locomotion. Sensory abnormalities are rarely detected as is clear from the routine use of CBA and C3H mice despite the retinal degeneration present in some substrains. The major categories of neurological mutants compiled in Table 5.6 demonstrate the large array of genes influencing the morphology and action of the nervous system. However, these mutants suggest

TABLE 5.6   Categories of neurological mutant mice

Category	Example
Embryonic inner ear malformation	*Dancer* runs in circles with jerky head movements. When lifted by their tail, they contort violently.
Postnatal inner ear degeneration	*Waltzer* hyperactive, runs in patterns such as figure-of-eights, especially after handling. Syndrome is most marked in young adults but is suppressed during voluntary activities such as feeding
Dysraphic (congenital fissures of the skull and/or spinal column)	*Loop-tail* twisted tail and a twitching which at first is a diffuse fluttering of the whole body but later becomes localized with development of the abnormal motor system in the forebrain
Neural crest disorders	*Dreher* circles ventrally when held by the tail or placed on its back, due to lesions in the central nervous system
Cerebellar malformations	*Reeler* lack of muscular coordination, disorders of balance and tremors when the mouse is active. When swimming the hind legs twist
Nervous system pathology	*Quaking* severe deficiency of myelin in entire CNS. When moving, tremor especially in the hind portion so that the animal appears to bounce
Epileptic seizures	*Epileptiform* seizures triggered by minor stimuli and when the mouse awakens
Probable metabolic disorders	*Spastic* rapid tremor of limbs and tail. If stimulated continually, muscles can become so rigid the mouse is thrown onto its back and cannot right itself

relationships between morphology and behaviour which do not always survive closer scrutiny. Although the inner ear is responsible for the sense of equilibrium, this is not the primary effect of the mutation in cases such as *waltzer*. The inner ear develops normally, but then degenerates because the central nervous system does not send the right processes to it.

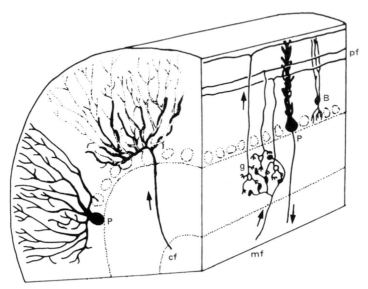

FIG. 5.2   A simplified diagram of some cell types and their connections in the cerebellum. Arrows indicate the direction of nerve impulses. P, Purkinje cell. B, Bergmann glial fibre. pf, parallel fibre. mf, mossy fibre. cf, climbing fibre. g, granule cells. (From S. H. Chung (1975) *Nature* **257**, 86.)

Development of the nervous system is the major thrust of neurological mutant research in mice (Pak and Pinto 1976). Most work has been done on the cerebellum (Fig. 5.2) where the six main neuronal types are particularly well understood. There are layers of three neuronal types, the molecular layer on the outside, then the Purkinje cell layer (P) and finally the granular layer (g) whose axons, the parallel fibres (pf) extend back towards the molecular layer. Inputs to the cerebellum and its Purkinje cells come from the climbing fibres (cf). In the *weaver* mutant the granule cells die within 2 weeks of birth because of abnormal growth of the Bergmann glial cells (B). Glial cells are the other major component of nervous tissue and are needed developmentally to guide the migration of the granule cells from the molecular layer down to the granular layer. With the granule cells not in their normal place, the mossy fibres (mf) and ultimately also the Purkinje cell fibres grow abnormally.

Is such abnormal growth of Purkinje cells a product of the environment (the consequence of the granular cell deformities) or is it itself determined by the mutant allele? This question can be answered with the *staggerer* mutant and a technique which is the closest thing in mice to the gynandromorphs discussed in Chapter 4.

In *staggerer* the granule cells die early and the Purkinje cells are smaller, fewer and are located in abnormal positions. Cells from early *staggerer* mutant embryos were introduced into normal embryos to form chimeras, mosaics of the two cell types. Whereas gynandromorphs result from chromosome loss, mouse chimeras are produced mechanically. Normal and mutant embryos at an early stage of division are washed to remove the outer layer (the zona pellucida) and then cells of each type are mixed together, cultured and finally transplanted into females made 'pseudopregnant' by hormone injections. When the chimeric mice are fully developed, a biochemical marker can distinguish between Purkinje cells from the normal and mutant strains. Those cells with the *staggerer* allele were deformed, while those with the normal allele were identical to cells from normal mice even though they had developed in a cerebellum where half the cells were abnormal. Thus the growth of the Purkinje cells depends on its genotype not on its environment.

One limitation to such neurological analyses is the genetic background. Bergmann glial fibres of the *weaver* mutant on a C57Bl/6J x CBA rather than the C57Bl/6J background used in the study mentioned in relation to Fig. 5.2 are almost as numerous as wild-type, although slightly different in morphology. There are significant differences between inbred strains in brain structure and very little is known about the effects that these may have on the expression of neurological mutants.

One striking example is the difference of BALB/c from most other strains in the size of the corpus callosum, the major bundle of fibres connecting the right and left hemispheres (Fig. 5.3). Some differences exist among other strains but many BALB/c mice have little or no corpus callosum while others are quite normal. The effect is not due to albinism since the other albino strain A/J in Fig. 5.3 is similar to the pigmented strains. Another single gene locus has been postulated but its effect must be very variable to produce the large range within BALB/c. An alternative is that the small corpus callosum is another result of the unstable development common in BALB/c for many characteristics and does not reflect a particular locus. Other genotypes are not affected. Introducing the formal nomenclature of mouse genetics Fig. 5.3 presents data from other strains and their crosses with BALB/c where except for one individual they all have a corpus callosum longer than 3 mm.

Recombinant inbred strains

While demonstrating again that single gene effects are influenced by the genetic background, the discussion of albinism shows that the opposite also applies. Rodent behaviour genetics is unique in that studies of 'normal' variation routinely use albino animals with neurological abnormalities, or specific strains such as BALB/c or C3H where some substrains respectively lack a corpus callosum or have retinal degeneration. While it is convenient to have strains 'colour-coded', very little attention has been paid to the impact of the

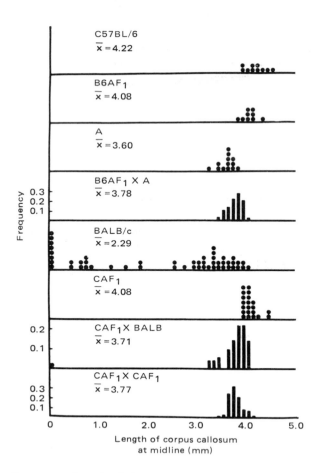

FIG. 5.3   Length of the corpus callosum in different inbred strains and crosses. B6 and C are standard abbreviations in crosses involving C57Bl/6 and BALB/c respectively, so that B6AF$_1$ means the F$_1$ between C57Bl/6 and A etc. Dots represent individual animals and solid bars proportions of the sample in groups where many animals were tested. (From D. Wahlsten (1974) *Brain Research* **68**, 1.)

coat-colour or other single gene effects in polygenic variation. Indeed such effects are rarely even detected. Wilcock (1969, Fig. 1) gives one example where in a diallel analysis of avoidance conditioning, those crosses involving BALB/c were quite distinct from those involving any of the other strains. He concluded that a recessive allele in BALB/c was responsible but its nature is still not known. It was not the *albino* allele since another albino strain A/J responded like the pigmented mice.

Some distinctions between the single gene and polygenic approaches to rodent behaviour have disappeared with the development of recombinant inbred (RI) strains which can determine if differences between strains in a particular behaviour are due to a single locus. RI strains are inbred strains derived from the F$_2$ between two inbred strains so that they represent new combinations of the alleles from the parent strains (Table 5.7). They have been used in genetic studies of drug effects on behaviour (Eleftheriou 1975) and Table 5.7 demonstrates their application to the changes in activity following a morphine injection. Seventy-five minutes after the injection, the activity of the RI strains falls into two categories — three strains (D, E and K) are similar to the BALB/c parental strain, while four (G, H, I and J) resemble C57Bl/6, the pattern expected if only a single locus determines the strain difference for that particular behaviour. Which parent each RI strain resembles depends on segregation in the F_2 at this locus. Some variation exists within each grouping of RI strains reflecting the other loci at which these strains differ, but this is inconsequential compared with the difference between groups.

TABLE 5.7   The derivation of seven recombinant inbred (RI) strains from BALB/c and C57Bl/6. The effects of morphine on activity are shown below each genotype — top line is exploratory activity 60 minutes after morphine injection, the bottom line activity 75 minutes after injection (data from L. Shuster (1975) in *Psychopharmacogenetics* B. E. Eleftheriou (ed.), p. 73. Plenum, New York).

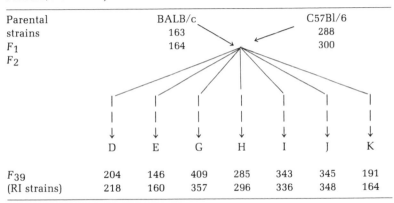

Parental	BALB/c					C57Bl/6	
strains	163					288	
F_1	164					300	
F_2							
	D	E	G	H	I	J	K
F_{39}	204	146	409	285	343	345	191
(RI strains)	218	160	357	296	336	348	164

If other major loci are involved, RI strains exist which are not homozygous for the same alleles as the parents. For example, if the parental strains were aabb and AABB some RI strains will have the combination aaBB and AAbb and do not resemble either parent. At 60 minutes after injection, such a situation arises with RI strain G differing from all the others, in a way which Shuster interpreted as the effects of an additional locus.

RI strain analysis has also been applied to avoidance conditioning where the difference between BALB/c and C57Bl/6 was attributed to a single locus. Using the histocompatibility markers mentioned earlier, this locus was mapped to chromosome 9, which is interesting as the albino locus is on chromosome 7. Thus this result supports Wilcock's diallel analysis that there is some single gene effect apart from albinism differentiating BALB/c from other strains on avoidance conditioning. Perhaps there really is an allele affecting the corpus callosum of BALB/c (Fig. 5.3).

While RI strains are an important development, interpretation becomes difficult when the results do not follow a single locus hypothesis, a point discussed by Broadhurst (1978) and clear from some of the other examples in Shuster's paper from which Table 5.7 was derived.

A final question is whether any of these mutations provide models for human disorders. Apart from the *brindled* mouse and Menkes Syndrome, the *spastic* mouse has obvious human parallels and the course of the *epileptiform* seizure is very similar to *grand mal* epilepsy. The *jimpy* mutant fails to develop myelin (the fatty sheath surrounding the axons of nerve fibres). This resembles the Pelizaeus-Merzbacker disease in humans, in both cases probably because a precursor of myelin accumulates and is not converted into myelin. There are also two mutants *pallid* and *fidget*, whose viability lessens with maternal age, which has possibilities for some of the human disorders such as Down's Syndrome whose incidence is also age-related.

One of the most widely used animal models has been the audiogenic seizure response. Some strains, notably DBA/2J respond to intense noise by a sequence of wild running, a clonic seizure (spasms), a tonic seizure (muscle extension and rigidity) and finally stupor or even death as a result of anoxia in the tonic phase. C57Bl/6J does not respond this way unless 'primed' by exposure to a loud bell rung sometime between 15 and 24 days of age. A wide variety of genetic models have been proposed but with little agreement (Fuller and Thompson 1978).

DBA/2J carries the coat colour mutation *dilute* leading to the fur being blue-grey. *dilute* mice are deficient in the enzyme phenylalanine hydroxylase, the same enzyme missing in human phenylketonuria. An 'attractive, reasonable and evergreen' hypothesis, (Collins 1972, p.365) is that since human phenylketonurics frequently have diluted hair colour and suffer from epilepsy, the *dilute* mouse is a potential model for epilepsy. However a coisogenic DBA/2J line with a mutation back to normal coat-colour behaved exactly the same as normal *dilute* DBA/2J on every aspect of audiogenic seizures, ruling out this hypothesis.

Polygenic inheritance

To demonstrate the uses of biometrical genetic analysis we consider learning, one of the most intensively studied aspects of animal behaviour genetics and one with many parallels to the human abilities discussed in subsequent chapters. Chapter 3 provided examples of genetic variation for both maze learning and avoidance conditioning. To apply this knowledge, we examine firstly three cases where critics of aspects of human behaviour genetics used animal research to support their views.

1 "g is a property which emerges from multifactorial statistical analysis, and what Jensen, Eysenck and others do is merely to reify a statistic. Because this apparently quantitative approach to the

question of intelligence essentially ignores what neurobiologists have been doing in the last decades in the field of learning, all the multitude of factors known to be involved in behavioural performance are discounted. To suggest that all of these may be combined to produce a single 'general intelligence factor' is as lacking in biological rationale as to talk of 'high IQ genes'. To go on from there to try to locate the 'habitation' of intelligence in the brain is, in terms of a scientific research programme, comparable to astrology or scientology. Unfortunately, it has much greater social resonance than these follies." (S. P. Rose (1975) in *Racial Variation in Man* Ebling F. J. (ed.), Symposia of the Institute of Biology **22**, 197.)

Recall that the concept of g was introduced in Chapter 1. To find a common trend running through the animal learning research so that a strain performing well on one learning task also does well on other learning tasks would support the idea of g. But one obvious problem hinders attempts to find a general factor in animal learning and means

FIG. 5.4　　One problem with finding a general factor in animal learning is that unlike human intellectual skills, not all abilities are measured in the same test situation.

that, whatever the result, it may have little relevance to humans. This point is illustrated by the cartoon in Fig. 5.4 and more seriously in Fig. 5.5. In humans, the same sorts of verbal or written responses in the same situation are used to assess quite different types of ability. If humans had to solve a maze by physically running round it rather than by merely tracing the pathway, performance would depend on athletic prowess as well as on intellectual ability.

So it is with rodents. In Fig. 5.5 Henderson views learning behaviour as a network of processes starting with what he calls Level III subcharacters arising from the specific situation e.g. the effects of foot-pad thickness on the perception of electric shock or the effects of being grabbed out of the maze at the end of a correct run. As the *Drosophila* maze example in Chapter 1 showed, most behaviour genetic research operates at the next level, Level IV, observing the response which is the outcome of Level III factors both relevant and irrelevant to learning. Above Level IV are the supercharacters defined

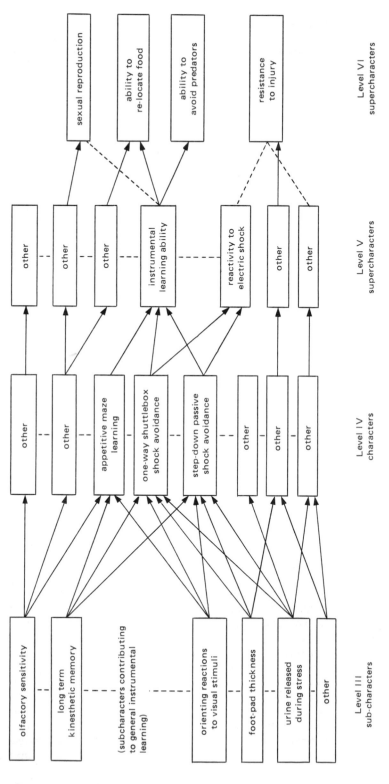

FIG. 5.5   Henderson's model of four levels of factors influencing learning (omitted are Levels I and II which refer respectively to primary gene action and to neural and chemical factors). Dotted lines are possibly spurious correlations, e.g. reactivity to electric shock may be related to resistance to injury, because both relate independently to foot-pad thickness, other than through any causal connection. (From N. D. Henderson (1979) in *Theoretical Advances in Behavior Genetics* Royce J. R. and Mos L. P. (eds), p. 234. Sijthoff and Noordhoff, Alphen aan den Rijn.)

in terms of the goals of any analysis. Level V is where a general learning factor would be found, while Level VI introduces evolutionary effects.

This model shows the limitations of simply studying the correlation between learning tasks in a variety of strains or selection lines. Different tasks may depend upon different Level III factors or may relate to quite different Level VI evolutionary factors, so that low correlations occur. On the other hand, albino or retinal degeneration strains may be greatly impaired on all visual acuity tasks, with this Level III factor leading to spurious but quite consistent patterns of strain differences at the Level V level.

Fuller and Thompson (1978, ch. 9) and Wahlsten (1972, 1978) review the data on the generality of learning. One approach is to see if lines selected for learning ability on one task do well on others. The Tryon maze-bright (TMB) and maze-dull (TMD) rats turn out to be exactly that — on Tryon's original 17 unit and on a similar but shorter 14 unit maze, TMB do better than TMD, but the differences disappear on both a 16 and a six unit maze and TMD actually are better at water-escape learning. In avoidance learning, TMB is better in a shuttle box but worse on a jump-out task (where the animal jumps out of the apparatus rather than shuttling into the compartment in which shock was administered previously). In the course of selection, TMB and TMD have been selected for a range of characteristics associated in some way with specific features of the maze. TMB rats are more motivated by hunger but less motivated to escape from water and are less distractable, while TMD are scared by the noisy mechanics of the original maze. The rats even respond to different cues — TMB solve problems with spatial cues, while TMD depend more on specific visual cues.

Results with other selection lines are not much more promising. The Roman high and low shuttle box avoidance lines of rats (RHA and RLA) differ also in their sensitivity to electric shock. Taking this Level III factor into account by adjusting shock intensities to elicit equal escape responses from both lines reduces but by no means eliminates the line difference. What does affect it is the exact response the rat has to make. Using a circular runway to permit responses other than the usual reversal of direction in the shuttle box, line differences are reduced if the animals can avoid by always going in the same direction. With a different and simpler response where the rat has to jump out of the box to escape shock, differences between the lines are eliminated.

An approach directed more at Level III factors is to take lines selected for some other characteristic and to see if this affects learning. The possible connections between emotionality and learning can be explored by comparing the Maudsley reactive and non-reactive (MR and MNR) lines on learning tasks. While differences occur with MNR doing better at shuttle box avoidance and on one specific maze task, MR does better in other situations. One difference relating closely to the selected characteristic is that MNR are better at learning an active avoidance response, and MR at learning a classically conditioned emotional response where the animal learns to associate a stimulus with subsequent, non-avoidable shock.

The comparison between selection lines usually involves only two or three genotypes. A better approach is to compare inbred strains where a greater number of genotypes can be contrasted. A more important advantage of inbred strains is that they are likely to differ by more than just one Level III factor which may have been the basis for the selection response, e.g. the different effects of the noisy apparatus on the TMB and TMD rats or of the specific avoidance apparatus on RHA and RLA rats. Therefore comparisons of strains across different learning tasks will not hinge on differences in this one variable.

Such a variety of strains have been ranked across such a variety of tasks that it is less confusing to quote the conclusions of Wahlsten (1978):

(a) when relatively minor variations in the task are used, strain rank orders are not greatly changed; (b) as greater differences are introduced into the training apparatus and procedure, large reversals of strain rank orders are commonly observed (this is especially true when different sources of motivation, such as electric shock, hunger or water immersion, are compared); and (c) when many tasks and strains are studied, certain strains do exhibit consistently good or poor performance across tasks within a single experiment, whereas other strains vary widely from task to task. (p. 81.)

Certain differences between the types of task involve factors other than learning. The order of strains consistently reverses between active avoidance (where the animal makes the response to avoid the shock) and passive avoidance (where it is shocked only if it moves). Strains active in stressful situations do well at active avoidance but poorly at passive avoidance, because of the different demands placed on their activity not their learning.

So far we have ignored the actual amount of genetic variation for different measures of learning, but Table 5.8 shows that in mice genetic variation is greater on more complex tasks. That is, it is low where animals have only to learn a classically conditioned emotional response (where the shock is inevitable whatever they do), greater when they have to learn to press a lever a certain number of times and much greater in a complex maze or in shuttle avoidance. As Wahlsten (1972, p. 152) puts it, 'genotypes which are all sufficient for learning simple tasks may not be equally effective when the demands for processing information are increased'. We return to this particular

TABLE 5.8   Heritabilities for different measures of learning in mice (derived from D. Wahlsten (1972) *Behavioral Biology* **7**, 143).

Task	h^2
Conditioned emotional response to shock	0.21
Running straight alley	0.30
Fixed ratio lever-pressing	0.34
Complex maze	0.40
Shuttle avoidance (two separate experiments)	0.48
	0.50

point in Chapter 8, since it parallels Jensen's concept of racial differences in ability appearing only on complex tasks (Fig. 8.4). However, the results in Table 5.8 should be interpreted with caution since the heritabilities were calculated in different ways from different crossing programs. Some are narrow and others are broad heritabilities including dominance which we see below is an important factor in animal learning.

Rose's criticisms can be answered in that animal data may not always show a general learning ability, but this has less to do with learning than with all the other variables apart from learning which differ between the various learning situations. We should also be cautious in extrapolating from rodents to humans because of the high degree of assortative mating in humans for *all* aspects of cognition (Chapter 6). Even without the genetic arguments this may be enough to indicate a general factor in intelligence with no parallels in experimental animals.

2 "It seems to us therefore that the attempt to estimate from family resemblance a variance component due to dominance for a trait like IQ is poorly motivated. We subscribe to the view of Falconer: 'The *additive variance*, which is the variance of inbreeding values, is the important component ...' If this is true for the experimental organisms considered by Falconer, it holds *a fortiori* for man." (N. E. Morton (1978) in *Genetic Epidemiology* Morton N. E. and Chung C. S. (eds), p. 169. Academic Press, New York.)

TABLE 5.9   The number of reports of dominance in three rodent behaviours. Numbers in parenthesis refer to reports of dominance for high performance. (derived from P. L. Broadhurst (1979) in *Theoretical Advances in Behavior Genetics*, Royce J. R. and Mos L. P. (eds), p. 43. Sijthoff and Noordhoof, Alphen aan den Rijn).

	Dominance present		
	Yes	Weak	No
Avoidance learning	19 (15)	3	1
Maze learning	5 (5)	3	1
Activity in experimental situation	9 (7)	1	11

Despite Morton's denial of the role of dominance, it has already been noted in Chapter 4 for learning in both *Drosophila* and blowflies. Should rodents be any different? Table 5.9 is compiled from an extensive review of genetic architecture in many behaviours and clearly indicates how common is dominance for high performance on learning tasks. That the dominance is not merely due to inbred animals being less active is confirmed by the studies of activity in situations such as runways or in response to electric shock where dominance often occurs for low activity.

Henderson (in Royce and Mos 1979) used dominance as a means of unravelling the relationship between measures of learning. If one takes very different measures, the composite score across all measures will depend far less upon task specific Level III effects and give a better indication of the mode of inheritance. Henderson used two water-escape maze discrimination tests (one where the discrimination was black versus white and the other where it was vertical versus horizontal) and a T-maze (where the animal had to reverse choices on each trial to avoid electric shock). While the genetic analysis of a composite score on the two water-escape mazes resembled the analysis of each maze separately, combinations of the water-escape and the T maze showed a very small additive variance and a larger directional dominance effect. Henderson interprets this result as meaning that task specific factors are largely additive genetic or environmental and are of little evolutionary significance, while the learning ability is far more closely related to fitness and hence shows much more directional dominance — the former cancel out when different tasks are combined while the latter becomes more obvious.

Applying the same argument to intelligence test performance, which is the composite of scores on many individual items each with slightly different determinants, would suggest that directional dominance should be a major consideration, contrary to what Morton claims to be shown by animal studies.

3 "A similar explaining-away is used in respect of the observations that in certain cultures, Black babies tend to be more advanced than White babies in terms of sensori-motor coordination; here Eysenck and Jensen claim that this illustrates a so-called 'biological law' that animals of lower final range of learning potential mature faster than those of a higher final range. The dubious validity of this proposition is demonstrated clearly by comparison of two closely interrelated species, rat and guinea pig. The rat is born immature — blind, naked and with much of its brain development still to come. The guinea pig by comparison is much further developed at birth — its eyes are open, it has a coat of hair, can run well and its brain is much closer to the adult. Yet one may doubt whether even Eysenck and Jensen would wish to maintain that the rat has, in parallel to its relative prematurity, a substantially greater learning potential than the guinea pig. Perhaps psychologists should be more careful in arguing about 'biological laws'" (S. P. Rose *loc. cit.*, p. 196).

Since White and Black people belong to the same species such arguments about differences between species may be irrelevant. But the work of Oliverio (reviewed in Oliverio 1977 and in Royce and Mos 1979) links the course of development within a species with later learning. Based on a decade of studies of strain differences in mice, Oliverio characterizes C57Bl/6 relative to SEC/1ReJ (Table 5.10) as being more advanced in infancy, but responding less to environmental effects and being 'rigid' rather than 'plastic' in adult behaviour. The reduced flexibility of adult C57Bl/6 behaviour is typified by poorer learning ability and by their circadian rhythm being little affected by alterations of the light–dark cycle. Oliverio interprets C57Bl/6's higher activity as representing more stereotyped exploratory behaviour which interferes with the purposeful activity required in situations such as maze-learning.

TABLE 5.10   Early development and later behaviour of C57Bl/6 mice relative to SEC/1ReJ (derived from A. Oliverio, C. Castellano and S. Puglisi-Allegra (1979) in *Theoretical Advances in Behavior Genetics*, Royce J. R. and Mos L. P. (eds), p. 139. Sijthoff and Noordhoff, Alphen aan den Rijn).

Infancy		Adulthood
Earlier reflex development	→ Less effect of enriched or impoverished environments	→ Lower brain weight and cortical thickness
Faster development of electrical activity in the cortex		Higher activity
Earlier imprinting of circadian rhythm to set light-dark cycles		More pronounced circadian rhythms of activity and sleep
		Poorer active and passive avoidance and maze learning

Corresponding to precocial species with rapid early development, C57Bl/6 should show more imprinting, learning which occurs at an early, often specific, developmental stage and which is very enduring. The example Oliverio uses is that 15–18 day old C57Bl/6 reared under a 12 hour on, 12 hour off light–dark cycle develop a clear circadian rhythm whereas SEC/1ReJ do not. Denenberg (in Oliverio 1977) provided an example more akin to conventional imprinting by rearing C57Bl/10J and random-bred Swiss-albino mice with rat mothers. This rearing experience had no effect on aggression in the Swiss-albino, but it reduced the incidence of fighting from 44.8% in C57Bl/10J reared by mice to 8.5% when reared by rats. That is, C57Bl is particularly sensitive to early experience, even though in later life it is much less affected.

Another feature of the 'rigid' genotype such as C57Bl is a high standard of maternal care which helps achieve rapid maturation of the young animal. Just how effective maternal behaviour can be at this

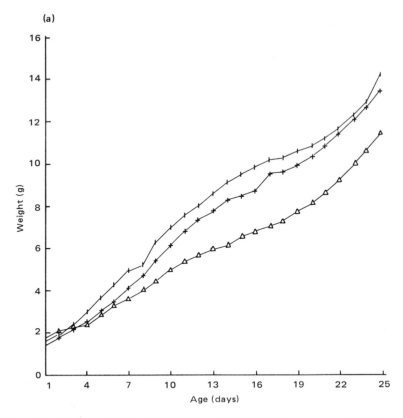

(a)

FIG. 5.6   Daily weights for (a) BALB/c and (b) C3H/HeJ young reared by their own mother (control, I), by another female of the same strain (simple-fostered, +) or by a female of the opposite strain (cross-fostered, △). (From D. A. Hay (1980) *Neuroscience and Biobehavioral Reviews* **4**, 489.)

stage is seen in Fig. 5.6 where the growth of BALB/c and C3H young is compared. Like C57Bl/6, this particular BALB/c strain is a 'good mother' making a compact nest, allowing the young long, unbroken feeds and retrieving the young when they leave the nest. C3H is the opposite and it shows in Fig. 5.6 where C3H fostered to BALB/c grow much faster and *vice versa*. Reflex development is similarly affected. Using eye-opening and the final disappearance of a rooting reflex as indicators that a mouse has reached the adult stage of reflex development, cross-fostering BALB/c to C3H delays maturity by 1.33 days. A BALB/c mother even increases brain weight (Wahlsten 1983).

As we saw earlier, BALB/c generally is a poor learner for reasons other than its albinism, so the faster physical and reflex development of BALB/c pups may mean Oliverio's model generalizes beyond C57Bl. Further evidence that maternal behaviour is negatively related to learning is the study by Fulker (in Royce and Mos 1979), of the RHA and RLA high and low avoidance lines. Not only does RHA spend less time with their litter after the first few days, but so also do F_1 mothers. That is, both additive and dominance effects for low

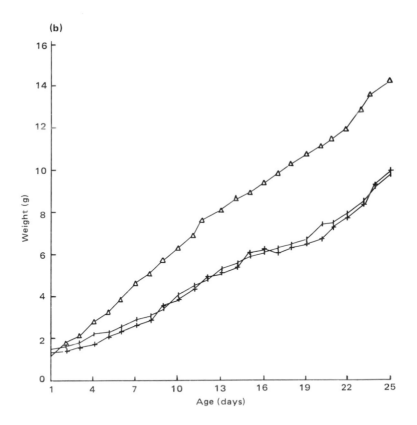

(b)

maternal care have parallels in the high learning ability discussed earlier.

Mechanisms of learning

Genetic analyses of learning provide information about how learning occurs and the internal processes involved. Three aspects of these are the time-course of learning, its biochemistry and the effects of drugs on learning.

The *Drosophila* learning mutants (Fig. 4.16) showed that learning and retention consist of a short-term or labile memory followed by consolidation into long-term storage. TMB rats are faster than TMD rats at consolidation, in that their retention is disrupted less by electroconvulsive shocks to the brain given shortly after learning or by learning trials given close together. Such massed trials interfere with each other unless consolidation occurs before the next trial.

Bovet *et al* (1969) made detailed analyses of strain differences in short- and long-term memory. Their C3H/He strain exhibited excellent short-term memory for avoidance conditioning but had forgotten all by the time the next day's trials commenced and was little affected by electroconvulsive shock and massed trials. DBA/2J showed the opposite pattern and the one most frequently observed in other strains. Given that C3H/He often has retinal degeneration, the explanation for its behaviour may lie in its vision as well as in its good short-term and poor long-term memory. Similar problems confound other attempts to make a genetic distinction between short- and long-term memory (Fuller and Thompson 1978). It remains an interesting question given that the two stages of memory involve different biochemical mechanisms, with protein synthesis being involved only in long-term memory.

As Will (1977) points out in his review on biochemistry and learning, it seems unlikely that the same biochemical mechanisms will apply to all learning tasks and biochemical effects on activity and emotionality must be distinguished. The TMB and TMD strains demonstrate how there is more to biochemical analysis than just finding a difference between the lines. TMB have higher levels of acetylcholinesterase (AChE), one enzyme which inhibits the activity of the neurotransmitter acetylcholine (ACh). But when lines selected for AChE levels were compared on the mazes, the low AChE line performed better than the high line, the opposite of what the TMB results would predict. The alternative hypothesis is that the ratio of ACh to AChE is involved. Other neurological differences between these lines are as complex — TMB have lighter brains than TMD but the cortex constitutes a greater proportion of the total brain weight in TMB.

AChE and a related enzyme choline acetyltransferase (ChA) have also been implicated in differences between mouse strains (Table 5.11) in that they are low in the temporal lobe of poor learning strains such as C57Bl/6J and high in good learning strains such as DBA/2J and SEC/Re1J. ACh and related enzymes involved in the cholinergic system play a role in habituation to novelty and in suppressing exploratory behaviour in the learning situation. ChA is more important than AChE since the F_1 of C57Bl x DBA in Table 5.11 has high AChE but still learns poorly and is active. Similarly a connection can be drawn between performance and norepinephrine levels in the pontine medulla, the region of the brain containing the reticular formation whose role is in short-term emotional arousal.

Will (1977) stresses that these results are very much correlations and not causations. One way to verify them is to use drugs which inhibit the particular systems. The anticholinergic drug scopolamine increases the activity of DBA/2 and SEC/1ReJ but decreases activity in C57Bl (recall the arousal model in Fig. 3.16 to explain such opposite effects). Nicotine facilitates consolidation so that poor avoidance learning strains such as C57Bl improve relative to the better learning strains. The anticholinesterase physostigmine, impairs avoidance learning less in RLA than in RHA, not by affecting learning but because RHA becomes very inactive, while the antiacetylcholine compound, N-ethyl-3 piperidyl benzilate, greatly enhances avoidance conditioning in RLA but not RHA. To some extent these may merely be scaling effects since RLA's avoidance learning is so poor that it cannot deteriorate much more while RHA is the opposite. Broadhurst (1978) provides a very full account of genetic analysis of

TABLE 5.11 Behavioural characteristics of three mouse strains and two hybrids with some neurochemical correlates (from P. Mandel, A. Ebel, G. Mack and E. Kempf (1974) in *The Genetics of Behaviour* van Abeelen, J. H. F. (ed.), p. 397. North Holland, Amsterdam).

	C57Bl/6J	DBA/2J	F1(C56×DBA)	SECRe/1J	F1(C57×SEC)
Maze learning ability	low	high	low	high	high
Avoidance behaviour performance	low	high	low	high	high
Wheel-running activity	high	low	high	low	low
Acetylcholinesterase in temporal lobe	low	high	high	high	high
Choline acetyltransferase in temporal lobe	low	high	low	high	high
Norepinephrine in pontine medulla	high	low	high	low	low

drug effects on behaviour and the almost ubiquitous nature of drug–genotype interactions.

Limitations of the polygenic methods

Several other behaviours apart from learning could have been used to illustrate the scope of polygenic analysis. There have been selection lines, strain comparisons and biochemical studies for alcohol preference and the sedative effects of alcohol, as well as for aggression although this last-mentioned is fully covered elsewhere (Simmel et al 1983). Not all results are consistent, as indicated by this extract from a recent review of aggression.

However, not even the rank ordering of strains on aggressiveness is consistent. The reader has no doubt noted that in some studies, C57Bl males are ranked above BALBs in aggressiveness. In others, no difference is found and in still others, BALBs are reported as more aggressive than C57Bl males. Further, DBA males are variously described as high, intermediate, or low in aggressiveness. While differences in method of testing may underlie some of the discrepancies, this explanation cannot account for disparities which have been noted within the same basic design (e.g. panel of testers, where C57Bl males are described as either more or less aggressive than BALB males).

Disparities also are found when the proposed genetic mechanisms underlying aggression are compared. For example (1) aggressiveness follows the paternal line or it does not; (2) it is regulated by two loci with dominant alleles having additive effects or it is a single gene, autosomal recessive character; or (3) there is one incremental and one suppressor factor for aggressiveness, both of which are autosomal. (From N. G. Simon (1979) Neuroscience and Biobehavioral Reviews 3, 100.)

The first paragraph of this quotation indicates that the problem lies partly not in the genetics but in the definition of the phenotype. As pointed out with the Drosophila avoidance conditioning paradigm in Chapter 4, it is unrealistic to expect consistent and useful genetic information when the behaviour is very variable within the one individual (or strain in the case of the aggression example). Neither aggression nor genetic analysis are unique in this respect and there are many other examples of inconsistent results in the field of animal behaviour, e.g. drug studies (Silverman 1978).

Fortunately, learning has sufficient consistency to demonstrate the use of strains and selection lines in comparing behaviours, their underlying biological mechanisms and their genetic architecture. What is apparent is how well some of these genotypes are known behaviourally, so that learning is not studied in isolation from maternal behaviour, development, activity, emotionality, drug and environmental effects. Oliverio's model (Table 5.10) demonstrates how strain comparisons enable commonalities to be found among a diversity of behavioural and neurological information. Parsons (1974) developed the concept of a 'behavioural phenotype' to describe the set of behaviours corresponding to a particular genotype and extended the situation to include morphological and physiological characteristics of the mice.

Knowing one's animals genetically has implications throughout psychology. The dispute between the early learning theories of Hull and Tolman may have resulted from Hull and Spence working with rats from Hall's original non-emotional strain which learn efficiently, whereas Tolman used rats from the emotional strain which show random trial-and-error behaviour (Eysenck in Royce and Mos 1979). To take a more recent learning example, Fulker (in Royce and Mos 1979) compared RHA and RLA rats on a measure of Seligman's theory of 'learned helplessness'. Seligman considers that inescapable aversive stimulation leads to an individual regarding all aversive situations as inescapable and displaying the apathy characteristic of depression. Not surprisingly Fulker found that RHA always escape if possible while RLA becomes more unable to escape as trials progress, that is, shows learned helplessness. But the strain difference suggests Seligman's theory needs to be revised to consider different strategies of response.

There are three limitations on the polygenic approach to analysing behaviour.

1 The emphasis is largely on finding correlations and very rarely are segregating generations used to check that these correlations are anything but fortuitous. Yet the examples of ethanol preference in blood-pressure selection lines in Chapter 3 and of the Tryon lines and AChE in this chapter show how easy it is to jump to conclusions before adequate breeding programs are carried out.

The same situation applies with strain differences. As a change from learning, consider the work of Batty (1978) on male sexual behaviour in relation to levels of the hormone, plasma testosterone. Among five mouse strains and three hybrids she found those genotypes with the highest testosterone levels had the longest latency until mounting and the lowest proportion of males mounting. To check if this relationship was spurious, she looked at the F_2 generation from C57Bl/6 female x DBA (called BDF_2) and from BALB/c female x C57Bl/6 (CBF_2). The BDF_2 showed no correlation between the hormone and behaviour but this is an unusual genotype, deriving from BDF_1 the only genotype to retain sexual behaviour after castration, that is, independently of any testicular secretions. In the CBF_2, the albino allele segregated and the 25 % albinos showed no hormone-behaviour relationship while the pigmented mice did. So what appeared to be a neat relationship in the inbred strains and hybrid became much more complex in the F_2, being absent in one cross and related to a single gene effect in another.

2 The same strain does not always behave the same way in every situation. The quote from the aggression review shows just how bad things can be. Wahlsten (1978) reports some experiments where C57Bl/6J was superior on learning, whereas Oliverio's model is based on his own C57Bl/6J being a poor learner. Some of the variation is due to environmental factors. Wahlsten describes how two experiments in the same university, one done in old, dilapidated laboratories and one in new facilities produced opposite ranking of strain differences on avoidance conditioning with fairly minor apparatus modifications.

There may also be substrain differences which limit the pooling of data from different laboratories. The C3H/HeJ strain available in Melbourne is a much poorer mother than BALB/c as Fig. 5.6 showed. When fresh stocks of the UK C3H/HeJ from which it was derived 25 years ago were imported to Melbourne and tested under the same conditions, they were found to be good mothers and to differ from the Melbourne substrain in certain skeletal features and, more significantly, on three biochemical genetic markers (Lavery and Hay 1984). In the 25 years since the C3H/HeJ were first brought to Melbourne, there has either been mutation or accidental breeding with other strains to make them genetically distinct. Wahlsten (in Oliverio 1977) gives examples of substrain variation at the neurological level in BALB/c.

FIG. 5.7   Genetic contamination of inbred mouse strains is usually first noticed when one animal stands out from the uniformity of the others.

3 Mice from varied origins are combined in the same study, irrespective of the different evolutionary pressures under which they may have developed. van Oortmerssen (1970) expanded this argument in a way which helps explain particular behavioural phenotypes but also casts doubt on the ability to generalize from particular strains. In comparing three strains of mice he found:

CPB (an albino closely related to BALB/c) is territorial (will win a fight in their own territory and will fight until one dies); shows elaborate courtship; makes spherical nests without wall-supports by fraying the nesting material; explores the open area in the centre of an open-field.

C57Bl is not territorial (any fights are indecisive); shows little courtship; cannot make spherical nests but usually builds in a corner; digs holes where possible and explores corners more than open spaces.

CBA is intermediate in territory formation and in courtship; builds poor nests and destroys them; digs at random.

van Oortmerssen's explanation is that CPB developed from ancestral, surface-living mice and C57Bl from hole-dwelling ones. The behaviour of CBA is fragmented, because CBA arose (Fig. 3.3) from a cross of an ancestor of BALB/c with DBA which is similar in behaviour to C57Bl. Habitat preference tests confirm that CPB and C57Bl still differ with CBA being intermediate.

van Oortmerssen (1970) provides the best account of the potential importance of such information:

> Such knowledge is of importance in experimental studies using house mice; e.g. if we artificially select for fast and slow maze learning using descendants from a 'hole living' population (or a cross between two 'hole living' strains) we may expect that selection will give a good response in the direction of slow learning and only little response in the direction of fast learning. For as the animals are adapted to living in holes natural selection will have deleted most or all of the additive genetic variance, which works in the direction of fast maze (hole) learning. In a similar selection experiment in which 'surface living' mice or a mixture of both types are used, one may get results which are quite dissimilar from the results of the first selection experiment mentioned. (p. 83)

Not all animal models are helpful

Animal models of human behaviour provide situations where rigorous control of both environment and genetics are possible. But at least two examples now exist where animal models in behaviour genetics have proved of limited benefit despite these advantages.

The fetal alcohol syndrome   The fetal alcohol syndrome is gaining increasing recognition as a severe risk in children born to chronically alcoholic women. It consists of characteristic craniofacial, limb and cardiovascular defects with growth deficiency, prenatal onset of developmental delay and childhood learning difficulties. But the woman who drinks sufficiently during pregnancy may pass on genes or environmental influences which are unrelated to the alcohol but which still impair behavioural development. Apart from depression

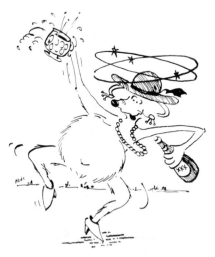

Fig. 5.8   The drunken, pregnant animal depicted here may look amusing but raises the serious question of whether rodents provide an adequate model of the major effects on humans of prenatal exposure to alcohol.

and frequent gynaecological and obstetrical problems, such women have many other difficulties including often an alcoholic partner. In one study, 12–17 year old children with an alcoholic father were down 2.4 grades in reading and 21% of them were mentally retarded. Adding on the many other family problems accompanying an alcoholic parent makes it difficult to accept the poor performance of children born to alcoholic mothers as necessarily part of the fetal alcohol syndrome. The action of alcohol on the fetus is complicated by the report (Kaufman 1983) that ethanol given to recently mated female mice can induce non-disjunction in the female but not the male-derived chromosome set of the fertilized egg. Chapter 2 indicated the severe effects of human autosomal non-disjunction so it is not yet clear if Kaufman's observation explains only the high rate of spontaneous abortions associated with fetal alcohol or also the behavioural anomalies seen in the survivors.

While non-disjunction would suggest that the cause of all fetal alcohol effects lies in straightforward genetic damage, the other question is why some individuals are more affected than others. Here animal models are vital with their potential to examine polygenic inheritance and to control environmental variables. Good models are available of the physical features of the fetal alcohol syndrome and strain differences exist with CBA being much more sensitive than C3H. But Table 5.12 shows the range of effects which experience with

TABLE 5.12   Some variables in the fetal alcohol syndrome (derived from D. A. Hay and J. E. Cummins (1981) *Australian and New Zealand Journal of Medicine* **11**, 148).

Ethanol administration	Prenatal variables	Preweaning variables	Adult variables
1 Dose	1 Maternal fertility	1 Maternal care	1 Genetics of the behaviour without ethanol
2 Duration			
3 Time during development	2 Maternal ethanol metabolism and preference	2 Maternal ethanol metabolism and (if ethanol still available) preference	
4 Method of administration			2 Physical effects of ethanol
	3 Progeny ethanol metabolism		3 Effects of ethanol on other behaviours
		3 Progeny ethanol metabolism	
	4 Maternal stress		
	5 Maternal nutrition	4 Maternal nutrition	
	6 Prior history of maternal ethanol exposure	5 Litter size	
		6 Progeny genetics	
	7 Mutagenic effects on parents and progeny	7 Laboratory procedures	
		8 Infection and parasites	

mouse strains suggests may be important when it comes to behaviour. For example, the stress of being kept in isolation and being regularly administered sucrose is enough to make BALB/c females lose their litters, over and above any effects alcohol may have; prolonged alcohol consumption increases susceptibility to infections and parasites. Hyperactivity is one consequence of the fetal alcohol syndrome in both rodents and humans and can make it very difficult to tell if learning has been affected by the alcohol or just by the increased activity. Some of the effects are very complex — normal BALB/c in certain mazes rarely enter paths not travelled before. One effect of prenatal alcohol is to reduce this inhibition so that BALB/c explore more and learn better, while other strains which lack this inhibition and normally do much better are adversely affected.

The fetal alcohol syndrome is a situation which takes the behavioural phenotype to such a level of complexity that any behaviour genetic experiment has so many variables it is in danger of becoming completely unworkable. The contribution of Randall and Riley (1981) at a symposium on fetal alcohol is recommended as an indication of many specific questions which can only be addressed accurately with animal models. Others at the symposium pointed out the enormous variability in the effects of alcohol on children of women with comparable levels of alcohol consumption and it is only through the animal model that the role of genetics in this variability can be adequately assessed.

FIG. 5.9   A left-pawed C57Bl/6J female reaching for food. Her parents and grandparents were all right-pawed. (From R. L. Collins (1969) *Journal of Heredity* **60**, 117.)

2 *Laterality* Although there is some evidence of genetic determinants in human hand preference (Fuller and Thompson 1978), the situation in rodents is very different and of no help in understanding what happens in humans. Using the apparatus shown in Fig. 5.9 where mice must use one paw to reach for food, Collins (1972) has found that while almost all mice have a distinct paw preference, the following lines of evidence suggest that it is not genetic: (a) Some 54% of mice from most inbred strains are right-pawed. If all mice within a strain are genetically identical, they should all have the same paw preference if genetics matters. (b) Selection for left or right paw preference has not been successful. (c) Mating mice of particular paw preferences does not affect the distribution of preferences among their progeny.

Collins altered the apparatus in Fig. 5.9 and moved the food tube so that it was flush with the wall of the test chamber, forcing mice to use a particular paw. He found that the paw preference of some mice was easily changed while others would not alter and he has since been able to select for strength of preference irrespective of which paw is the preferred one. His argument is that the decision as to which paw is preferred is a random event, but then a genetic process is involved fixing this paw preference more firmly in some mice than in others. While Collins argues for parallels with the determinants of human hand preference, there are many differences. In humans there is a considerable excess of right-handers and definite localization of skills in particular hemispheres of the brain (something only now being investigated in animals).

Evolution and behaviour

Inbred mice have been isolated in laboratories for so long that they are often considered unrepresentative of 'real' mice and independent of the selection pressures which act in the wild. Thus Bruell (cited in van Oortmerssen 1970, p. 82) said '... cancel all your orders for inbred mice, drive out to the country and get your own'.

There is no doubt that wild mice show much more genetic variation than laboratory mice and it has been proposed that all the major inbred strains derive from a single female alive some time between 1200BC and 1920AD. But as Henderson (in Royce and Mos 1979) points out, studies of the process of domestication in newly-trapped rodents indicate fairly small changes, usually only in such obvious characteristics as resistance to handling and ability to escape. For example Australian wild mice placed in an open field may spend an entire 25 minute test period jumping high in the air, trying to clear the perimeter. Inbred strains move around more and explore the environment rather than trying to escape from it.

There are now three rodent examples which suggest that the genetic determinants of behaviour in inbred strains need not be that different from wild populations.

1 Wilcock and Fulker (summarized by Fulker in Royce and Mos 1979) found in an 8 x 8 diallel cross of laboratory rat strains that dominance in a conditioned avoidance response is initially for poor performance but changes progressively over trials to dominance for high performance. In evolutionary terms they interpret this change as meaning the optimal response to aversive stimuli is initially to freeze and then to flee only when it becomes apparent that the aversive stimuli are continuing.

To test this hypothesis, Hewitt, Fulker and Broadhurst caught 22 wild male rats from a refuse disposal centre by digging out the nests. This is a much more satisfactory technique than trapping adults since only a selective group of rats may enter traps. They carried out a genetic analysis using the Triple Test Cross, a particularly convenient breeding scheme which does not involve the difficult task of trying to mate laboratory males with wild-caught females. In the Triple Test Cross, each wild male is mated to females of two inbred strains and their F_1, thus:

		laboratory females		
		strain 1	strain 2	F_1
	1	x	x	x
	2	x	x	x
wild males	.	.	.	.
	.	.	.	.

In this design, females must come from strains differing as much as possible for that behaviour, so in this case RHA and RLA (after a few generations of inbreeding to reduce heterozygosity further) were the obvious choice for a study of conditioned avoidance. The genetic architecture of conditioned avoidance in this situation corresponded

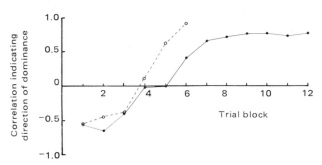

FIG. 5.10   Parallel directions of dominance in a triple test cross using wild and laboratory rats (●—●) and in a diallel cross with laboratory rats alone (o--o). (From J. K. Hewitt, D. W. Fulker and P. L. Broadhurst (1981) *Behavior Genetics* **11**, 533.)

remarkably closely (Fig. 5.10) to that from diallel analysis of the inbred strains alone. Not only does this mean similar influences on both the wild and laboratory rats but also that the inbred strains reflect the same selection pressures even though they were obtained from different localities. The evolutionary interpretation of the reversal of dominance is that the optimal response to pain is to 'freeze' at first and then, if the noxious stimulus continues, to learn quickly to escape it.

2 van Oortmerssen's CPB mice demonstrate that selection still goes on within the laboratory. CPB mice are very variable in the extent to which they fray the nesting material in order for it to be woven into compact nests. Considering they might be heterozygous, van Oortmerssen selected successfully for high and low fraying mice but found the selection lines died out. Table 5.13 shows that there is a clear selective advantage in fraying about an average amount. Both high and low fraying mice produce fewer young and these suffer a higher mortality.

TABLE 5.13   The relation between fraying scores and breeding results in CPB mice (derived from G. A. van Oortmerssen (1970) *Behaviour* **38**, 1).

Breeding pairs*	H × H	H × M / M × H	M × M	M × L / L × M	L × L	L × H / H × L
Average fraying	66.8	44.4	37.3	25.4	6.3	36.9
Total number of young per pair at birth	12.6	16.0	18.0	5.3	12.3	12.4
Mortality (%) at 60 days of age	26.3	18.8	4.6	12.5	40.5	37.9
Mean number of healthy young per pair at 60 days of age	5.7	9.5	17.0	4.7	1.7	4.8

*H, L and M represent high, low and medium fraying scores, respectively.

If there were little directional dominance, this disadvantage of the high and low fraying mice would be an ideal way to maintain heterozygosity in an otherwise inbred strain. In his genetic analysis van Oortmerssen found no consistent dominance, providing one of the few examples where the absence of dominance or possibly the presence of ambidirectional dominance is associated with an advantage to intermediate performance (see Fig. 4.19). Broadhurst's review of genetic architecture (in Royce and Mos 1979) shows that every other component of nesting behaviour involves dominance for high performance as one might expect if good nests aid survival of the young. Fraying is a vital component of making good spherical nests but its genetic architecture is different for the reason given by van Oortmerssen:

... because mice which do not fray in the proper way cannot build good spherical nests at the surface. If they do not fray, the nesting material does not stick; if they fray too much, they will bite their nesting material to pieces, so that the material becomes too small. Only intermediate fraying may lead to a proper nest. (p. 68)

3 The analysis of how dominance in laboratory mice changes in different situations can be interpreted in an evolutionary context (Table 5.14) and supported by studying wild mice in the same situation. When 4 day old wild pups were tested they were even less active than hybrids between the inbred laboratory strains, suggesting that the selection pressure for low infant activity, while still present, has been reduced during the domestication process. In contrast, by the age of 15 days, wild mice and to some extent hybrids frequently exhibit a hopping or jumping response in a novel environment, which also can be interpreted as adaptive at this age.

TABLE 5.14 Dominance and activity in young mice (1–3 derived from N. D. Henderson (1979) in *Theoretical Advances in Behavior Genetics*, Royce J. R. and Mos L. P. (eds), p. 263. Sijthoff and Noordhoff, Alphen aan den Rijn. 4 from N. D. Henderson (1981) *Developmental Psychobiology* **14**, 459).

Situation	Dominance	Evolutionary explanation
1 4 day old mice removed from nest but still in cage with mother	For low activity	With poor senses and coordination, best to wait for mother to retrieve them
2 10–11 day old mice still in cage but 15 cm away from litter mates and with parents absent	For fast return	Best response at this stage when sensory cues can guide them back to the nest and parents are not there to retrieve them
3 10–11 day old mice in new environment	No dominance	A 'nonsense' situation which never occurs in real-life
4 15 day old mice in new environment	Dominance for an explosive hopping response	When the young are first exploring away from the nest but still lack the visual-motor skills of adults, this is one way of escaping from predators

Behaviour genetics in wild populations

Population biologists have developed a very different approach to the analysis of behaviour in wild rodent populations. Only rarely is it possible to track or observe the behaviour of rodents in the wild and therefore much of the knowledge of behaviour comes from inference. The most important tool in this process of inference has been the technique of 'electrophoresis' by which individuals can be identified as homozygous or heterozygous for genetically determined enzyme systems (Fig. 5.11). When individuals from populations living in large barns or other defined areas are captured and their pattern of electrophoretic markers plotted, two things generally have emerged: (a) the different genotypes are not distributed at random but clustered in discrete groups throughout the barn, and (b) there are fewer heterozygotes than expected.

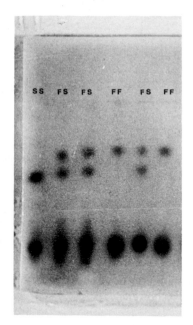

FIG. 5.11   Detection of genetic differences in protein by gel-electrophoresis. In the presence of electric current, protein samples in the gel migrate at different rates (in this case, from the bottom of the photograph towards the top). Calling the alleles F (fast) and S (slow), the FF homozygote shows up as one band further along the gel than SS, while the FS heterozygote has both bands. (Examples of esterase-3 in wild mice, courtesy of Dr G. R. Singleton, CSIRO Division of Wildlife and Rangelands Research, Canberra.)

The explanation is that the mice form small breeding groups called demes with little gene flow between the demes. With the inbreeding which inevitably accompanies such small populations, the proportion of heterozygotes decreases and each deme develops a distinctive set of genotypes. The number of mice in a deme and the physical size of the deme vary a lot — demes as small as 1 m² with only four mice contributing to the breeding have been reported but much depends on the habitat. Island populations have larger, more loosely defined demes than those in barns and similar localities. But whatever the size of the deme, if such inbreeding can occur in nature it is no wonder that inbreeding mice in the laboratory has been so successful. As a result of prior evolutionary pressures, far fewer deleterious alleles are likely to emerge through their inbreeding than in humans (Table 2.8).

What maintains the demes? They are generally unrelated to geographical features and depend on territorial behaviour, e.g. one study reported that 95% of the mice intruding into another territory were attacked and 91% of the males and 78% of the females were killed. So there is little chance for a male but some chance for a female to contribute to the breeding in another deme. It is more difficult than it might seem to analyse such territorial behaviour and its consequences. In small groups of laboratory mice the situation is clearcut — one male is aggressive and this dominant male sires the majority of the offspring. The situation is less certain in larger, more natural areas where the subordinate mice can escape more easily and where there are territorial boundaries which the dominant male patrols to fight off intruders. One of the other males may well do the mating while the dominant male is fighting or recovering from his wounds.

Busser *et al* (in van Abeelen 1974) studied mice in a small clearly defined habitat and found that they formed a three-tier hierarchy. Studies of mice establishing territories after being introduced together into an experimental enclosure support such a hierarchy (Fig. 5.12). The *dominant* male emerges as the one which generally

FIG. 5.12   A three-tiered dominance hierarchy in male mice. There are two dominant males (L1x2, PR), one subdominant male (HS14) and three subordinate males (HS5, HS12 and HS17). (A representation of data from G. R. Singleton and D. A. Hay (1983) *Behavioral Ecology and Sociobiology* **12**, 49.)

initiates aggressive interactions and electrophoresis confirms that this male sires most of the young. Next come the *subdominant* males which are aggressive initially but which thereafter consistently retreat from encounters with the dominant male. As the dominant male ages, subdominant males can quickly claim his territory and mating rights. At the bottom are the *subordinate* males which initiate very few aggressive interactions and are very submissive to both the dominant and subdominant males. As juveniles they may leave the territory. Once such a hierarchy is established it can remain stable for many months. Females generally remain with the dominant male and form much less of a hierarchy, so that movement of females across territories is permissible in such situations.

Very little is known of the genetics of behaviour in this situation but it indicates the value of considering behaviour outside the laboratory. The development of techniques such as the Triple Test Cross means that the genetic architecture can be established in such settings and should lead to a greater balance in rodent behaviour genetic research between laboratory investigations and the study of animals in a natural or seminatural population.

A new perspective on population genetics and behaviour has come with the discovery that animals can identify conspecifics which differ genetically only at the cellular level. Female mice use odour to distinguish male mice with t alleles and are less likely to mate with them. (The t locus is close to one part of the complex of major histocompatibility loci in the mouse and is important in embryonic development. Various t alleles are lethal when homozygous but are maintained by selective advantage of sperm carrying such alleles.) In general mice are likely to associate with mice of different histocompatibility types and perhaps also to choose them as mates. In a review of this work Jones and Partridge (1983) consider its potential in ensuring outbreeding and genetic diversity and suggest its relevance to humans where some immunological differences between fetus and mother may contribute to a successful pregnancy. However the existence of many thriving inbred mouse strains where no immunological differences exist and the limited genetic diversity in wild mouse demes, suggest that this phenomenon may not be quite so fundamental to population genetics.

Conclusion

The emphasis here on laboratory rodents mirrors the interests of most researchers. But there are other vertebrate groups which receive little attention. Chapter 1 mentioned the work of Immelmann and Kovach on the genetics of imprinting in birds. Imprinting is particularly important for many precocial bird species in the wild and deserves greater attention in behaviour genetics. It lends itself to studies in natural populations where Cooke has demonstrated the role of such early learning in the social behaviour and choice of mates among the blue and white colour morphs of the Lesser Snow Goose. The population consequences of such assortative mating are analysed by O'Donald (1983) using the Arctic Skua as a model.

Birds are not the only vertebrates to be ignored. There has been little attention to domestic animals despite their practical significance and the availability of many breeds often with very detailed mating records. The rodent examples in this chapter show just what can be achieved nowadays in behaviour-genetic analysis, but should not be taken to mean that other species can be ignored just because they are less familiar to the laboratory scientist.

On the other hand, changes in the emphasis of behaviour genetics are putting even more reliance on rodents. Lieblich (1982) demonstrates how much effort is now being applied to genetic analysis of the brain at all levels. The lesion, self-stimulation and drug approaches to the study of the brain which experimental psychologists have been using for years are increasingly involving genetic analysis because of the appreciation of the extent of genetic variation in the normal brain, as distinct from the many single gene mutations discussed in this chapter. While this trend is to be applauded, it must rely on mice and rats on which so much of the methodology of brain research is based.

Discussion topics

1 A. Oliverio and P. Messeri (*Behavioral Biology* **8**, 771) examined 31 single gene mouse mutants on a variety of behavioural tasks and found that on every task some mutants differed from their non-mutant siblings. They conclude:

"In general the present findings point out that, though these behavioral traits are influenced by genes at many loci, a single gene may substantially contribute to the additive and dominance genetic variance of the behavior." (p. 781).
Do you agree with this interpretation?

2 J. H. F. van Abeelen carried out detailed analyses of behavioural development in two types of neurological mutant mouse — *looptail* (*Animal Behaviour* (1968) **16**, 1) and *Nijmegen waltzer* (*Animal Behaviour* (1970) **18**, 711). Wilcock (1969, p. 23) explains such effects as *looptail* spending less time cleaning its face as obvious consequences of the damage to motor systems in the brain. Is there more to the behavioural development of neurological mutant mice than just this?

3 Apart from the Tryon lines, another set of maze-bright and maze-dull lines were established at McGill University — the MMB and MMD lines. The most widely cited result with these lines is the effect of rearing environment on number of errors in the maze.

	Environment		
	Normal	Enriched	Restricted
MMB	117.0	111.2	189.7
MMD	164.0	119.7	169.5

Data from R. M. Cooper and J. P. Zubek (1958) *Canadian Journal of Psychology* **12**, 159.

In a paper on racial differences in intelligence, Bodmer and Cavalli-Sforza say "This experiment is particularly relevant to differences in IQ because of the structure of human society." (*Scientific American* (1970) **223** (4), 24). Do you agree?

4 "The use of a single outbred stock in drug testing is sometimes justified on the grounds that humans are random bred, so in order to model humans realistically, an outbred stock of laboratory animals should be used. However humans are not random bred." (M. R. W. Festing (1975) *Food and Cosmetics Toxicology* **13**, 370).

Given the complex structure of wild mouse populations and the many factors acting upon laboratory populations, how best can extrapolation to humans be achieved?

References

Annotated bibliography

van Abeelen J. H. F. (ed.) (1974) *The Genetics of Behaviour*, North-Holland, Amsterdam. (Nine chapters cover every aspect of rodent behaviour from neurological mutants to populations. Chapters 12 and 13 deal with aggression, a behaviour not dealt with in the present text.)

Behavior Genetics (1981) **11** (5), 427. Special issue on '*Rattus norvegicus* as a tool in mammalian behavior genetics'. (A very useful summary of different aspects of research with rats. Particularly noteworthy are the descriptions of the different selection lines in contrast to the limited data on inbred strains and Boice's account of differences between wild and domesticated rats.)

Broadhurst P. L. (1978) *Drugs and the Inheritance of Behavior: A survey of comparative psychopharmacogenetics.* Plenum, New York. (A demonstration of just how useful strain and selection line differences can be in analysing drug or other treatment effects.)

Lindzey G. and Thiesen D. D. (eds) (1970) *Contributions to Behavior Genetic Analysis: The mouse as a prototype* Appleton-Century-Crofts, New York. (The only text dealing exclusively with rodent behaviour genetics — Bruell's chapter on wild mice and behavioural population genetics is particularly recommended.)

Oliverio A. (ed) (1977) *Genetics, Environment and Intelligence.* North-Holland, Amsterdam. (Part II dealing with rodents has full accounts of different aspects of the research by Oliverio and his colleagues and demonstrates the value — and the risks — of concentrating on a few strains.)

Royce J. R. and Mos L. P. (eds) (1979) *Theoretical Advances in Behavior Genetics.* Sijthoff and Noordhoff, Alphen aan den Rijn, Netherlands. (As well as other chapters on rodents, those by Broadhurst, Fulker and Henderson present complementary but distinct views on the adaptive significance of behaviour.)

Wahlsten D. (1978) Behavioral genetics and animal learning. In *Psychopharmacology of Aversively Motivated Behavior.* Anisman H. and Bignami G. (eds), p.63. Plenum, New York. (This review along with Wahlsten's earlier review — *Behavioral Biology* (1972) **7**, 143 — are the most complete accounts of the extent and limitations of genetic research on vertebrate learning.)

Additional references

Batty J. (1978) Plasma levels of testosterone and male sexual behaviour in strains of the house mouse (*Mus musculus*). *Animal Behaviour* **26**, 339.

Bovet D., Bovet-Nitti F. and Oliverio A. (1969) Genetic aspects of learning and memory in mice. *Science* **163**, 139.

Broadhurst P. L. (1975) The Maudsley reactive and non-reactive strains of rats: a survey. *Behavior Genetics* **5**, 299.

Collins R. L. (1972) Audiogenic seizures. In *Experimental Models of Epilepsy — A manual for the laboratory worker* Purpura D. P., Penry J. K., Tower D. B., Woodbury D. M. and Walter R. D. (eds), p.347. Raven, New York.

Eleftheriou B. E. (ed.) (1975) *Psychopharmacogenetics.* Plenum, New York.

Fuller J. L. and Thompson W. R. (1978) *Foundations of Behavior Genetics.* Mosby, St. Louis.

Jones J. S. and Partridge L. (1983) Tissue rejection: the price of sexual acceptance. *Nature* **304**, 484.

Kaufman M. H. (1983) Ethanol-induced chromosomal abnormalities at conception. *Nature* **302**, 258.

Lavery K. and Hay D. A. (1984) Electrophoretic, skeletal and behavioral divergence of two C3H substrains of mice. *Journal of Heredity* **75**, 171.

Lieblich I. (ed.) (1982) *Genetics of the Brain.* North Holland, Amsterdam.

van Oortmerssen G. A. (1970) Biological significance, genetics and evolutionary origin of variability in behaviour within and between inbred strains of mice (*Mus musculus*): a behaviour genetic study. *Behaviour* **38**, 1.

O'Donald P. (1983) *The Arctic Skua: A study of the ecology and evolution of a seabird.* Cambridge University Press, Cambridge.

Pak W. L. and Pinto L. H. (1976) Genetic approach to the study of the nervous system. *Annual Review of Biophysics and Bioengineering* **5**, 397.

Parsons P. A. (1974) The behavioral phenotype in mice. *American Naturalist* **108**, 377.

Randall C. L. and Riley E. P. (1981) Prenatal alcohol exposure: current issues and the status of animal research. *Neurobehavioral Toxicology and Teratology* **3**, 111.

Scott J. P. and Fuller J. L. (1965) *Genetics and the Social Behavior of the Dog.* University of Chicago Press, Chicago.

Silverman P. (1978) *Animal Behaviour in the Laboratory.* Chapman and Hall, London.

Simmel E. C., Hahn M. E. and Walters J. K. (eds) (1983) *Aggressive Behavior: Genetic and neural approaches.* Erlbaum, Potomac, Md.

Wahlsten D. (1983) Maternal effects on mouse brain weight. *Developmental Brain Research* **9**, 215.

Wilcock J. (1969) Gene action and behavior: an evaluation of major gene pleiotropism. *Psychological Bulletin* **72**, 1.

Will B. E. (1977) Neurochemical correlates of individual differences in animal learning capacity. *Behavioral Biology* **19**, 143.

6 Methods of human behaviour-genetic analysis

Topics of this chapter

1 The family, adoption and twin methods of genetic analysis of human behaviour and the limitations of each method.
2 The model-fitting approach which considers data from many relationships and which thus avoids most of the problems specific to each method.
3 Complications for genetic analysis such as assortative mating and genotype–environment correlation and interaction.
4 The importance of adequate and representative data and of testable hypotheses compatible with a variety of data rather than *ad hoc* explanations of particular results.

Introduction

In theory, many of the methods of studying familial relationships described in Chapter 3 should be appropriate for humans. However there are many more practical problems centering around the fact that humans cannot be controlled in the same way as laboratory animals — researchers cannot determine the breeding patterns of their sample, the choice and number of tests may be limited and the whole project must often rely on volunteers, bringing in yet another bias. These limitations open the way to attacks on the methodology and in particular Kamin (1974) has criticized many earlier projects on the genetics of intelligence on the bases of specific defects in each experiment.

The situation is aptly summarized by Henderson.

Investigators undertaking a large-scale study of a human population, in spite of the most careful preparations, can fully expect a series of *post hoc* criticisms about their assumptions, sample selection and size, and general forms of analyses and their interpretations. Indeed, for every hardy soul undertaking such an experiment, I expect that there are ten critics, often with legitimate and constructive ideas who are ready to point out flaws in the research program. Further, for each of these critics, who provide a useful and necessary function in the field, there are probably ten additional persons who will use such criticism as a basis for misinterpreting or altogether dismissing the entire research effort. I think it is fair to say, for example, that virtually no study presently available on the genetic basis of intelligence cannot be criticized on several grounds. Nevertheless, one cannot help but be impressed by the apparent consistency that seems to occur in the overall picture relating degree of kinship and similarity on various IQ measures, ... (In K.W. Schaie, V.E.

Anderson, G. E. McClearn and J. Money (eds) (1975) *Developmental Human Behavior Genetics: Nature-nurture redefined*, p.5. D. C. Heath, Lexington, Mass.)

In essence, this quotation summarizes our approach to human genetic analysis. Every study is open to some criticism, but there is an underlying consistency between the different studies which not only justifies them but also invalidates the criticisms. If two studies are each criticized on different grounds but still arrive at the same conclusion, then the criticisms cannot be of much consequence. Urbach (1974) takes this argument even further and attacks the critics for their *ad hoc* approach. Anyone can examine a genetic study in detail, find some flaw and use it to condemn the whole project. We see in Chapter 8 that such an 'environmental factor X' as Urbach calls it is a frequent *ad hoc* explanation of ethnic group differences. What Urbach feels is needed for an adequate criticism of genetic analysis is to predict anomalies and their consequences for the results and then carry out the necessary research to test this prediction.

We use some of the better-known sets of data on intelligence test performance to illustrate the three main methods of genetic analysis and then consider how these can be combined in one more informative analysis. Chapter 7 provides a more detailed interpretation of genetic analyses of the structure and development of intelligence. Although we criticized intelligence tests in Chapter 1, there are few other measures on which such a range of genetic methods have been tried and the problems so thoroughly analysed. With the renaissance of interest in intelligence (Sternberg 1982), there are now many more aspects of research on intelligence which lend themselves to genetic analysis.

Family studies

The most obvious genetic analysis is the study of the relationship between parents and children using the methods of correlation and regression discussed in Chapter 3. The largest recent study of this type is the Hawaii Family Study of Cognition where 731 Caucasian families were assessed on a battery of 15 cognitive tasks. Table 6.1 summarizes the familial relationships for the major component underlying performance on the 15 tests. (These tests do not comprise a measure of

TABLE 6.1   Heritability estimates for the first component of test performance in the Hawaii Family Study (data from J. C. De Fries *et al* (1979) *Behavior Genetics* **9**, 23)

Mid-child on mid-parent	Father– son	Mother– daughter	Mother– son	Father– daughter	Brother– sister	Brother– brother	Sister– sister
0.60*	0.60	0.80	0.70	0.70	0.52	0.72	0.50

*Heritability calculated as the regression of mid-child on mid-parent and as twice the correlation in the case of the other relationships.

IQ as such, but since they all measure some facets of intellectual ability, statistical techniques can be used to analyse the interrelations between scores on the different tests and to extract the component which reflects that variance common to all tests. This composite measure relates closely to standard IQ scores.) These results indicate that:

1 *Assortative mating* between people of like phenotype is of little importance. If it were, the correlations between children and a single parent would yield a much higher estimate of heritability than the regression of children on midparent. Positive assortative mating inflates the heritability estimated from the resemblance to one parent, since the other parent and hence the progeny are more similar than expected. In the midparent resemblance, the average of the two parents takes into account how similar they are. Intelligence test performance usually exhibits a higher degree of assortative mating than most other behavioural measures but in the Hawaii study the correlation between spouses was only 0.23. This indicates a small degree of positive assortment, much less than the values of 0.4–0.6 often reported from earlier studies. DeFries and his colleagues believe that such higher estimates were inflated by the fact that spouses are usually of similar ages. If this is taken into account by statistical adjustment, the degree of assortative mating is much reduced.

2 Both daughters and sons have higher correlations with their mother than with their father, suggesting a maternal environmental effect.

3 The genetic variation is predominantly additive. If there were dominance variation, the correlations among siblings would be inflated, relative to the parent–child relationships (see Table 3.4).

A considerable degree of familial resemblance is therefore indicated by the Hawaii Family Study. As DeFries stresses, 'familial' is a function of both genetic factors and those environmental factors common to members of the same family, so that the finding of a large familial resemblance is *necessary* but *not sufficient* evidence for genetic variation. (In theory this is not so, as there could be genetical and environmental effects cancelling each other out, rather like the rodent maternal buffering in Table 3.1. While such genotype–environment correlations are discussed later in the chapter, in practice they have never been observed to act this way.)

In the Hawaii Family Study the tests were administered *en masse* to many people from different families on the same occasion, eliminating the bias of many earlier studies where members of the same family were often tested together by the one tester. The danger then arises that any familial resemblance may partly be a product of how the tester interacts with that particular family.

However problems do remain for any family study due to the obvious fact that parents and children are of different ages. There are a few longitudinal studies such as the Fels Study (Chapter 7) where children were tested and years later their own children tested at the same age, but in general family studies are complicated by the parents and their children representing two distinct *cohorts*. The parent group grew up in an era differing from their children in education and in other aspects of the environment.

Most studies measure parents and children at the same point in time not at the same age and McAskie and Clarke (1976) review major behavioural problems which can then arise. One issue is longitudinal stability — if the same individual is assessed at different points in time, the resemblance between the test scores decreases the longer the interval between the tests. Some of the change will be unreliability of the test, some may be due to differences in what the tests measure at different ages or to the use of different tests. For example, in the well known Wechsler series of tests, the Preschool and Primary Scale of Intelligence (WPPSI), the Intelligence Scale for Children (WISC) and the Adult Intelligence Scale (WAIS) cover successive parts of the age spectrum. If these tests fail to measure the same abilities, then the relationship between test scores at different ages will be affected.

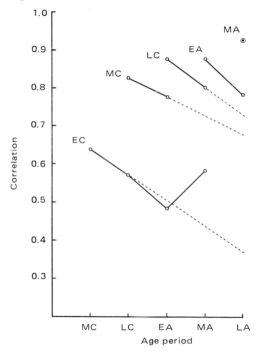

FIG. 6.1   Longitudinal correlations for IQ scores of the same people tested at different ages. Projected correlations for which data are unavailable are shown by dashes. EC, early childhood (3–5 years). MC, middle childhood (6–9 years). LC, late childhood (10–14 years). EA, early adulthood (15–21 years). MA, middle adulthood (22–42 years). LA, late adulthood (43 years +). (From M. McAskie and A. M. Clarke (1976) British Journal of Psychology 67, 243.)

Figure 6.1 shows how the correlation between IQ scores of the same individuals changes with age. For example, if a child is assessed during early childhood the correlation with his/her score at middle-childhood will be 0.64 and with early adulthood 0.48. If this can happen within the same individual, then the resemblance between the scores of parents and children would also be likely to be affected by the age difference, as would be the relationship between siblings of different ages. The age effect is complicated by the fact that tests differ in reliability and reliability changes with age, often being lower in younger children. Therefore the practice has developed of *correcting for attenuation*, adjusting the results to take into account reduced familial resemblance due to test unreliability rather than to any genetic or environmental differences.

These variables would least affect the Hawaii Family Study which used the same highly reliable tests on all ages (the minimum age being 13), which used a detailed system of standardizing results at different ages and which reanalysed the data taking into account the ages of the children (but not the parents). McAskie and Clarke estimate that in less well-designed family studies, the parent–offspring relationship would be reduced by some 30% and up to 65% if very young children and their parents are used. Therefore it is clear that parent–offspring relations are hazardous in research areas such as the genetics of behavioural development (Chapter 7) where repeated measures of behaviour throughout childhood are required and may be suspect in other situations.

Whatever the sophistication in experimental design, familial resemblance cannot provide unequivocal evidence for genetic effects and it is necessary to use situations such as the adoption or twin study, where the normal relationships within the nuclear family are altered.

Adoption studies

For a variety of reasons including more adequate contraception and reduced social stigma attached to having an illegitimate child, the number of children being offered for adoption has declined to such an extent that few major adoption studies are likely in the future. There remains a considerable emphasis on adoption studies in the arguments about the genetics of intelligence (Kamin 1974; DeFries and Plomin 1978) and adoption studies will continue to be important in studies of mental illness, alcoholism and criminality, since these are reasons why parents may still offer children for adoption.

We shall illustrate the adoption data with the classic adoption study by Skodak and Skeels and the more recent Texas Adoption Study. Other adoption studies are described by DeFries and Plomin (1978). The Skodak and Skeels study uses the traditional adoption design (Fig. 6.2a) where adoptive families are 'matched' with biological families rearing their own children. Since matching introduces many possible biases, the more recent approach used in the Texas study (Fig. 6.2b) is to involve also families with both their own biological children and adopted children. As indicated in Fig. 6.2, the aim of the adoption technique is to get away from the usual family situation where parents provide both genes and environment for their children to one where the biological parents provide only genes and

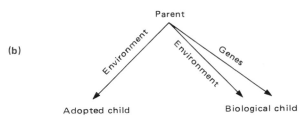

FIG. 6.2   (a) The traditional adoption design involving the matching of two groups of parents, one transmitting only environment, the other environment plus genes. (b) the more recent adoption design where each family has both an adopted and a biological child. (Derived from H. J. Eysenck versus L. Kamin (1981) *Intelligence: the battle for the mind*. Pan, London.)

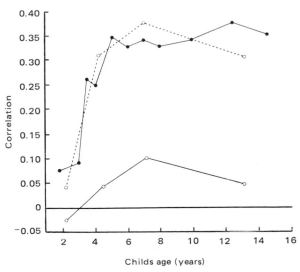

Childs age (years)

FIG. 6.3   The correlation between mother's education and child's IQ. This figure is a composite of Skodak and Skeel's adoption data and Honzik's data on children reared by their own mother. o- - -o, child reared by own mother (Honzik). ●—●, adopted child's IQ × biological mother's education (Skodak). o—o, adopted child's IQ × adoptive mother's education (Skodak). (From K. Honzik (1957) *Child Development* **28**, 215.)

prenatal environment and the adoptive parents only the postnatal environment. The problem of Fig. 6.1 still exists where one is comparing adoptive and biological parents with children who are much younger and often tested in different ways.

In the Skodak and Skeels study, 100 illegitimate children were adopted before 6 months of age and assessed at 2, 4, 7 and 13 years. Honzik combined these data with her own study of parents rearing their own children to demonstrate (Fig. 6.3) that once these children were 4–5 years old and could be adequately assessed by intelligence tests, the correlation of the adopted children's IQ with the biological mother's educational attainment was over 0.3 and similar to the correlation of children reared with their biological mother. In contrast the correlation with the adoptive mother's education was much less, fluctuating between 0.05 and 0.10. The mother's educational attainment level was used since IQ scores were not available for all women and educational attainment correlates fairly well with IQ (Chapter 8).

This result indicates a small environmental component (the correlation with the adoptive parent) and a large genetic component (the correlation with the biological mother). Given that the heritability is $2\times$ the parent–offspring correlation (Table 3.4), this gives a value very similar to those obtained in the Hawaiian Family Study (Table 6.1) where children were reared with their biological parents. But a variety of problems can arise in adoption studies (Munsinger 1975) and these include the following.

1 *Characteristics of the mothers who offer children for adoption* In the Skodak and Skeels study, the average IQ of the mothers was 86. These assessments are questionable since they were made shortly after the birth and the trauma of the woman agreeing to give the child up for adoption. In contrast Munsinger cites studies where the more able mothers are more likely to give up their children for adoption.

Mothers who give children for adoption may differ from normal in other related respects. In the Texas Study many more than expected showed high scores on personality measures related to schizophrenia and paranoia. While such scores are associated with thought disturbances and have been linked with low IQ, in practice they had no effect on the genetic analysis of IQ in this study.

2 *Children are more likely to be adopted by couples of above average ability* If IQ, educational attainment, income and socioeconomic status (SES) are related (Chapter 8), more couples of high ability can afford to adopt children.

3 *More able children are more likely to be adopted* On the assessments at different ages, the average IQs of Skodak and Skeels' children range between 107 and 117. Munsinger favours the explanation that more able children are selected, but against this is the fact that the Skodak and Skeels children were adopted before the age of 6 months. Given the very poor predictability of any assessment at this age, it seems improbable that the adoption agency or prospective parents were making a subjective decision which is more accurate than formal assessment! An alternative explanation is that adopted children are wanted children. Since people make a positive decision to adopt children and frequently have no other children, adopted children may receive more care and stimulation than other children. In addition they are less likely to grow up in a large family where the later-born children receive less attention and often are lower in IQ (see Figure 6.7).

4 *Biases in the data collection*   Not all adoptive parents will permit the children to be assessed for research purposes. As in the Skodak and Skeels study, IQ measures may not be available on all biological mothers and are even less likely to be available on the biological fathers. If the missing data are spread across the entire range of individual variation, no problem arises. But if certain groups are less likely to participate, such as less able biological mothers or adoptive families who have problems with the children, then the results can be distorted.

5 *Selective placement*   A frequent complication is that adoption agencies match the biological and adoptive parents on a mix of economic, educational and social factors in deciding where to place the child. The outcome of this is to increase the correlations between the children and both the biological and adoptive parents. The adoptive parents are providing an environment like the one the biological parent would have provided, while the two sets of parents are no longer randomly chosen from the population gene pool but may well have some genetic similarity. The effect is to inflate the resemblance to the adoptive parents and hence the apparent effect of environment rather than genetics.

Even though it is easy to find flaws in adoption studies, genetic estimates need not be distorted as long as only the mean performance and not the relationship between individuals is affected. For example, Kamin (Eysenck versus Kamin 1981) cites a recent French study by Schiff where children of low SES parents adopted into high SES homes had IQs 16 points higher than their siblings who remained with the low SES parents. Apart from the fact that such an environmental effect is quite consistent with the polygenic model (Fig. 3.11), it is also irrelevant to the genetic analysis that is concerned with the relationship between individuals. As long as the children with the higher IQs in the adopted group have siblings with the higher scores amongst those remaining at home, the genetic argument is supported despite any mean difference between the two groups.

It is interesting that the elevated scores of adopted children can also be used *against* the environmental argument. In Skodak and Skeels' study the IQs of the adopted children declined progressively with successive tests and did so equally in children adopted into upper or lower class homes. If performance hinged on the beneficial adoptive environment and not on the below average ability of the biological mothers, such a result is difficult to explain.

No one adoption study will have all these complications and the better ones such as the Texas Adoption Project have fewer problems and so many statistics that effects such as selective placement can be estimated as part of the analysis. The necessary statistical techniques are too complex to consider in detail here, but in essence involve the method of *path analysis* which defines all the possible paths by which the various genetic and environmental factors including selective placement and assortative mating can create a relationship between different individuals. Then models can be fitted and parameters calculated, analogous to the animal examples in Fig. 3.8 and Table 3.1. Table 6.2 gives eight of the 28 correlations used by Loehlin in

TABLE 6.2   The observed* and expected correlations for eight familial relationships in the Texas Adoption Project (from J. C. Loehlin (1979) in *Theoretical Advances in Behavior Genetics*. Royce J. R. and Mos L. P. (eds), p. 303. Sijthoff and Noordhoff, Alphen aan den Rijn)

Relationship	Expected	Observed
Adoptive father, biological son	0.29	0.27
Adoptive father, biological daughter	0.29	0.34
Adoptive mother, biological son	0.29	0.23
Adoptive father, biological daughter	0.29	0.19
Adoptive father, adopted son	0.12	0.18
Adoptive father, adopted daughter	0.12	0.04
Adoptive mother, adopted son	0.12	0.24
Adoptive mother, adopted daughter	0.12	0.04

*These correlations have undergone a procedure of z-transformation which makes the distribution more normal. With this range of values, any differences from the real correlations are so small they can be ignored.

analysing the Texas Adoption Study and their predicted values on the genetic and environmental model. Only one of the eight correlations, that between the adoptive mother and her adopted son is grossly discrepant, the observed value being twice that expected. Although critics may seize upon this point as a failure of the genetic analysis, Loehlin emphasizes that it is the overall fit to the data of the model as a whole that counts, not one discrepancy that may just be due to chance. In any case it is equally difficult to explain on environmental grounds why only the adoptive mother's correlation with the adopted son is discrepant and not that with the adopted daughter. The only effective criticism would be one which proposes an alternative model predicting all 28 correlations more accurately than Loehlin's model.

The conclusion from Loehlin's analysis and from more detailed analyses paying specific attention to the possible problems with adoption studies (Horn *et al* 1982) is that the narrow heritability is 0.38 and the broad heritability is 0.5. This leaves 0.5 of which 0.4 is due to the environment, comprising 0.18 attributable to factors common to all family members and 0.22 to factors specific to the individual and not shared with the others. Chapter 3 introduced the notation of E_b and E_w for these terms respectively. The remaining 0.1 is due to *genotype–environment correlation*, which results from parents providing not only the child's genotype, but also the environment in which that genotype develops. For example, a high ability parent may provide a favourable genotype and also access to quality schooling and to other resources such as an extensive library which enables that child's ability to flourish.

The heritability is lower than that observed in the Hawaii Family Study (Table 6.1) perhaps because the family method gives inflated values. There is also a higher correlation with the adoptive parents than in the Skodak and Skeels study (Fig. 6.2) where some children were not adopted until 6 months of age. In the Texas study, adoption was within 1 week of birth and many biological mothers never even saw their babies. Thus the influential early environment was provided by the adoptive not the biological parents with appropriate consequences for the pattern of correlations.

Other major adoption studies are in progress, supplanting the older and less adequate studies. The Minnesota Adoption Studies include surveys of adolescents and of transracial adoptions which are discussed in Chapters 7 and 8 respectively. A full account of this project and of different methods of analysis available with adoption data is provided by Scarr and Carter-Saltzman (in Sternberg 1982). Another major program, the Colorado Adoption Project has so far concerned only infants and is mentioned in Chapter 7.

Twin studies

The study of identical and fraternal twins has been an important part of human genetics ever since the pioneering work of Galton in the 1870s. With a different emphasis, twins were also widely used by psychologists in the 1930s in co-twin control studies. If one identical twin child is taught a particular skill while the other is not, does the first one acquire the behaviour sooner than the second? That is, are changes in behaviour attributable to learning or to maturation? Although Mittler (1971) summarized the results of such research as neither impressive nor conclusive, the co-twin method is being used again in other contexts, e.g. assessing alcohol-induced brain damage by comparing identical twins who differ in alcohol consumption.

Twinning has been described as 'the most common congenital abnormality', something which twins may find insulting but which at least suggests that twins should be studied not merely as a means to genetic analysis but as a significant, unique portion of the population (approximately 2% given 1 birth in 100 results in twins). Both here and in Chapter 7 we shall consider some problems specific to twins since they greatly complicate the deceptively simple twin method of genetic analysis.

The rationale behind the twin method is shown in Fig. 6.4. Identical twins result from the splitting of a single fertilized egg some time after conception and hence are monozygotic (MZ). Genetically these twins must be identical to each other, but may differ due to environment. Hence we shall call them MZ rather than identical twins to indicate that they are not always phenotypically identical. Fraternal twins result from the independent fertilization of two eggs by two sperm and are called dizygotic (DZ) twins. In terms of their genetics, DZ twins are only as alike as any brothers and sisters, although they have shared the same intrauterine environment.

Monozygotic twins

Are products of a single sperm and a single egg

In an early stage
the embryo divides

The halves go on to become
separate individuals

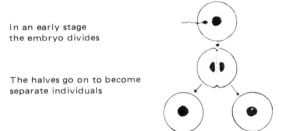

Having come from the same sperm and the same egg,
these two twins carry the same chromosomes and alleles,
thus they are always of the same sex

Dizygotic twins

Are products of two different eggs, fertilized by two
different sperms

They go on to develop into two different individuals

They carry different assortments of chromosomes and
alleles. They may be of the same sex or opposite sexes

FIG. 6.4   The different processes by which monozygotic and dizygotic twins
are produced. (Derived from A. Scheinfeld (1973) *Twins and Supertwins*.
Penguin, Harmondsworth.)

The factors determining twinning differ between the DZ and MZ
situations. Multiple ovulation and hence DZ twinning is more
common in older women and in women taking fertility drugs such as
Clomiphene which has a twinning rate of one in 10. DZ twinning may
partly be under genetic control, since it is more common in families
where the mother and to a lesser extent the father has a family history
of DZ twinning. DZ twinning rates also differ between ethnic groups,
being lowest in Asian populations (2–8 per 1000 maternities),
intermediate in Caucasian groups (7–11 per 1000 maternities) and
highest in Africa (10–40 per 1000 maternities), the Yoruba of Western
Nigeria having a uniquely high frequency (MacGillivray *et al* 1975). In
contrast MZ twinning rates are relatively constant across maternal
ages and ethnic groups.
    The advent of reliable contraception has brought considerable
changes in the relative proportions of MZ and DZ twins. Older women
are having fewer children so that the DZ rate is declining. At the same
time, there are indications of a higher incidence of MZ twinning

among women who conceive within 3 months of ceasing use of the contraceptive pill. In 1951, 71% of twins born in Australia were DZ but by 1977 this figure had dropped to 53%. The changing nature of the twin population may limit the parallels we can make between current studies of young twins and studies of twins born in the 1950s or earlier. For example in Australia in 1951, 18% of twins were born to women 35 years or over, while in 1977 only 7% of twins were born to women in this age range with consequent differences in child-rearing.

The resemblance between twins

In the twin method, MZ and DZ twins are compared for the degree of similarity of scores on the particular test. If performance is partly under genetic control MZ twins should be more similar than DZ twins. Although similarity is measured by the correlation coefficient, the usual formula (Table 3.3) is often considered inappropriate in the twin case. We have no X and Y variables as such, but only the scores of each twin of a pair, either of whom could equally well be classed as X or Y. Instead we use a different correlation, the *intraclass correlation* developed from the *analysis of variance*.

Analysis of variance is one of the most widely used statistical procedures in experimental research and centres on the comparison of variation within and between groups. If the groups represent different families, the variance between the groups should be larger than the interindividual variance within the groups in the situation where families really do differ in performance. In analysis of variance, the variances are called *mean squares*, abbreviated to MS_b and MS_w for the between and within group terms. These estimate the extent of family resemblance through the intraclass correlation which is

$$\frac{MS_b - MS_w}{MS_b + (n - 1) MS_w}$$

where n = number of individuals in each group. For twins $n = 2$, and the intraclass correlation is simply a comparison of whether there is more variation between rather than within twin pairs. If the only variation between groups (MS_b) is the individual variance (MS_w), the top line and the correlation will be zero. In the opposite case if all the variation is between groups, MS_w will be zero and the correlation will be +1.0.

When MS_b and MS_w are condensed into one figure, the correlation coefficient, information is lost since many combinations of MS_b and MS_w can yield the same correlation. The Birmingham approach favours MS_b and MS_w over correlations for genetic analyses because they provide a better check on some of the assumptions underlying the analysis, whereas another major research group based in Honolulu (Rao and Morton 1978) prefers correlations because they feel that environmental factors can be handled more realistically. Although Kamin (Eysenck versus Kamin 1981) emphasizes the disagreements between the two approaches, the results obtained by the two methods are generally quite similar (Loehlin 1978).

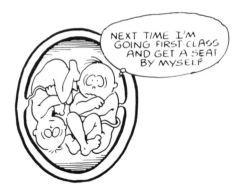

FIG. 6.5   The problems of twins begin early in life. (From N. Matterson (1982) *It's a Baby!* Marion, Greensborough, Australia.)

The mean squares within and between twin pairs are a convenient means of understanding how we specify genetic effects on twin similarities and differences. First we must distinguish the mean squares between and within groups from the *components of variance* between and within groups, which we term V_b and V_w respectively. Remembering that the variance of a mean is 1/nth the variance between individuals, then groups are going to differ even if the only source of difference is that between individuals (V_w) and not between the groups (V_b). In the case of twins where $n = 2$, the variance between pair means $= \frac{1}{2}V_w + V_b$ and the variance between pairs (MS_b) $= V_w + 2V_b$, where $V_w = MS_w$ and $V_w + V_b = V_t$, the total variance between individuals.

Although we have already seen how V_t can be defined genetically as $\frac{1}{2}D + \frac{1}{4}H + E_w + E_b$, Jinks and Fulker (1970) introduced yet another notation which is particularly convenient for describing different familial relationships. They extended the usual equation P (or V_t) $= G + E$ to $P = G_1 + G_2 + E_1 + E_2$ where G_1 and G_2 are the genetic effects within and between families. (As mentioned in Chapter 3, $G_1 = \frac{1}{4}D + \frac{3}{16}H$ and $G_2 = \frac{1}{4}D + \frac{1}{16}H$). E_1 and E_2 correspond to the E_w and E_b we introduced earlier. Some authors use a slightly different definition of E_2, but for the present the distinction is irrelevant although it matters in any actual analysis.

The outcome of all this is that we can specify V_w and V_b for every relationship. In MZ twins brought up together (MZ_T), the only possible factor contributing to V_w, differences between co-twins from the same family, is E_1. Therefore, for MZ_T

$$V_w = E_1$$
$$V_b = G_1 + G_2 + E_2$$

If MZ twins had been separated and brought up apart in different homes (MZ_A), E_2 is an additional cause of environmental differences between the co-twins, making

$$V_w = E_1 + E_2$$
$$V_b = G_1 + G_2$$

In the case of DZ twins brought up together (DZ_T), there are now genetic differences between the co-twins and

$$V_w = E_1 + G_1$$
$$V_b = E_2 + G_2$$

Other relationships such as full sibs and unrelated children brought up together or apart can be specified in the same way.

The method is equally applicable to correlations, where $V_t = 1.0$. Substituting $V_w + 2V_b$ and V_w for MS_b and MS_w respectively in the intraclass correlation formula shows that the correlation equals V_b. That is $r = V_b$ and $1 - r = V_w$. So, for example, an MZ_T correlation (rMZ_T) of 0.87 means

$$G_1 + G_2 + E_2 = 0.87$$
$$E_1 = 0.13$$

and a DZ correlation (rDZ_T) of 0.53 means

$$G_2 + E_2 = 0.53$$
$$E_1 + G_1 = 0.47$$

This approach has general implications to which we return later when we see how family, twin and adoption data can be combined in the same analysis. But it also explains one problem which plagued earlier methods of deriving heritabilities from twin data. Although such formulae were based on logical assumptions (Mittler 1971, Appendix B), Jinks and Fulker (1970) substituted G_1, G_2, E_1 and E_2 in these formulae to demonstrate that discrepant results arose because the formulae were measuring different things. As shown in Table 6.3, Holzinger's H involved only genetic and environmental effects within and not between families, while Nichols' HR included everything but E_1.

TABLE 6.3   Two heritability estimates from twins reared together and their application to four hypothetical data sets (derived from A. R. Jensen (1967) *Proceedings of the National Academy of Science* **58**, 149)

		Holzinger's H	Nichols' HR
Formula		$\dfrac{r_{MZ} - r_{DZ}}{1 - r_{DZ}}$	$\dfrac{2(r_{MZ} - r_{DZ})}{r_{MZ}}$
Biometrical model		$\dfrac{G_1}{G_1 + E_1}$	$\dfrac{2G_1}{G_1 + G_1 + E_2}$
Data sets			
	r_{MZ} r_{DZ}		
1	1.00 0.50	1.00	1.00
2	1.00 0.50	0.25	1.00
3	0.40 0.20	0.50	0.22
4	1.00 0.99	1.00	0.02

Other proposed indices of heritability can be analysed in the same way and found wanting. Vandenberg modified the *F-ratio test* which is normally the statistical technique used in analysis of variance to test if MS_b is larger than the MS_w and made it a test of genetics, by testing the ratio MS_w for DZ pairs/MS_w for MZ pairs. While this may seem valid — the DZ MS_w has a genetic component in addition to the environmental component found in the MZ MS_w — substitution of the biometrical model shows that this ratio is $(G_1 + E_1)/E_1$. Hence the F ratio test like Holzinger's H tests only effects within the family.

MZ twins reared apart

The G_1, G_2, E_1 and E_2 model shows us that it is difficult to combine MZ and DZ data in a way which provides an adequate measure of the heritability. To give some feel for such data, Table 6.4 summarizes several results from an early study of the different types of twin. For standing height the rMZ_T and rMZ_A are very similar, indicating that postnatal environment is not very important in determining height. In contrast environment matters for weight and for performance on the achievement test since rMZ_A is less than even the rDZ_T. In other words having the same genetic potential has less effect on the similarity between twins than being reared in the same environment. The two IQ measures show intermediate results with rMZ_A being less than rMZ_T but a little larger than rDZ_T.

TABLE 6.4   Correlations for physical and behavioural measures in MZ_T, MZ_A and DZ_T twins in the 1937 study by Newman, Freeman and Holzinger (from P. Mittler (1971) *The Study of Twins*. Harmondsworth, Penguin)

	MZ_T	MZ_A	DZ_T
Standing height	0.98	0.97	0.93
Weight	0.97	0.89	0.90
Binet IQ	0.91	0.67	0.64
Otis IQ	0.92	0.73	0.62
Stanford achievement	0.96	0.51	0.88

Before one concludes that rMZ_A by itself is a reasonable way of estimating the heritability for IQ, consider some unique problems which may arise in the separated MZ twin situation. Firstly not every family with MZ twins sends them out to be reared by different families. Several of the families involved in the early studies were poor, could not carry the financial burden imposed by a multiple birth and therefore gave one twin to relatives in the neighbourhood to raise. In such a case it is probably not justified to accept without closer scrutiny the premise that the twins differ in all aspects of the between family environment. On the other hand the similarity of IQ scores of MZ_A twins cannot be explained just by environmental similarity. One study showed such twins separated at an early age to be more similar in IQ score than those separated later which contradicts the purely environmental explanation.

The researcher still has to find twins and obtain their consent to being tested which raises the issue of biased sampling and what has been called 'the rule of two-thirds'. In volunteer samples of adult twins (not necessarily separated twins) roughly two-thirds are female and two-thirds MZ. In his extensive studies of French twins, Zazzo has shown that MZ females tend more often to remain in close contact throughout life and to regard their similarity as an asset. The consequence for genetic analysis is that males and DZ twins in the sample are less representative of the population and may have a smaller variance between pairs (MS_b) resulting in a smaller intraclass correlation and an overestimate of the heritability when MZ and DZ are compared.

The situation is compounded in MZ_A since the twins may have been adopted and know little about their co-twin. It is easy to envisage how only those most interested in twinning may try to find their co-twin, how only those most similar in appearance or mannerisms may be put in contact with each other by people who note their similarities or how only those who find remarkable parallels in their lives may seek out both publicity and participation in research. Those MZ_A cases who are not very similar or who care little about being a twin will not be included in research.

Apart from these biases there is the problem of sample size. Biometrical genetics demands large samples and MZ_A twins are rarely present in sufficient numbers to warrant their inclusion in any genetic analysis. The situation of MZ_A is adequately summarized by a quote from Shields (1978), reprinted in a report of a Minnesota study where some remarkable similarities in lifestyle are being found between separated twins:

I doubt if MZAs will ever be numerous and representative enough to provide the main evidence about environment, or about genetics, but . . . they can give unique real-life illustrations of some of the many possible pathways from genes to human behaviour — and so will always be of human and scientific interest. (Science (1980) **207**, 1328.)

We cannot leave the topic of MZ_A without mentioning Sir Cyril Burt who in the 1950s and 1960s reported MZ_A data based on a uniquely large sample. These results were frequently used in biometrical analyses until the early 1970s when it became apparent that many anomalies existed in the data. The question of how much data Burt actually collected and how much he fabricated remains obscure (Hearnshaw 1979) and in any case is irrelevant to continued progress in behaviour genetics. Some critics have seized upon Burt, arguing along the lines (a) MZ_A are fundamental to genetic analysis of intelligence, (b) Burt had the largest sample of MZ_A, (c) Burt's data are hopelessly flawed or are non-existent, and (d) therefore there is no evidence for the genetic hypothesis.

We have already seen one major flaw in this reasoning, since MZ_A in particular and twins in general are now regarded as not being typical of the population as a whole and few people would base a genetic analysis solely upon MZ$_A$. Shields (1978) is an example of how a scientist who has devoted much time to MZ$_A$ can reach this conclusion. Furthermore, Henderson (1982, Table 1) demonstrates that a review of genetic analyses of intelligence can omit MZ$_A$ altogether.

A second point explaining why the defects in Burt's data went unnoticed for so long is that Burt's data were little different from anyone else's. Genetic analyses with or without Burt's data yield similar conclusions and intrapair differences in Burt's sets of twins are comparable with those in other MZ$_A$ studies (Fig. 6.6). If Burt had really biased his results towards a genetic argument, one would have expected that the intrapair differences would have been noticeably smaller than in the other studies where the authenticity of the data is not in question.

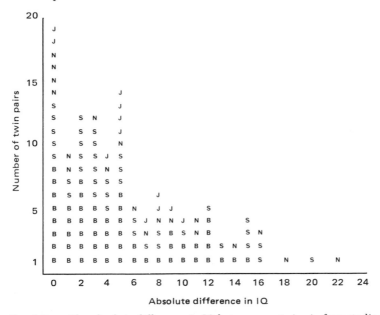

FIG. 6.6   The absolute difference in IQ between co-twins in four studies of MZ$_A$. The IQ scores have been corrected for test measurement error. N. Newman et al 1937, N = 19. J. Juel-Nielsen 1965, N = 12. S, Shields 1965, N = 38. B, Burt, 1966, N = 53. (From A. R. Jensen (1973) *Educability and Group Differences*. Methuen, London.)

Are twins representative of the population as a whole?

Because the majority of the population are not twins, we must consider whether a genetic analysis based solely on twins applies to the population as a whole. Or are there factors unique to the twin situation which do not contribute to variation among other people? There is no doubt that twins do differ from other children — on average they are slightly premature, the usual gestation length being 3–4 weeks less than in single-born children. Many obstetric complications are more frequent in twin pregnancies (MacGillivray *et al* 1975). At birth they are 30% lighter and 17% shorter than single-born children and it may not be until school-age that twins fully catch up.

There are also effects on later ability. A major study based on the routine scholastic assessment formerly given to all 11 year old children in England and Wales showed (Fig. 6.7) that twins did not do as well as single-born children on a test of verbal reasoning, and that this effect was consistent over a range of family sizes and maternal ages. The average score of twins was 95 compared with the single-born mean of 100. Twins where the co-twin had died either at or shortly after birth had an average of 99, suggesting that the causes of the decrement must be postnatal.

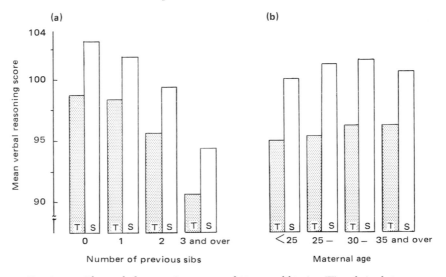

FIG. 6.7   The verbal reasoning scores of 11 year old twins (T) and singletons (S), classified (a) by birth order and (b) by maternal age. (From R.G. Record, T. McKeown and J. H. Edwards (1970) *Annals of Human Genetics* **34**, 11.)

At least two factors in the social environment distinguish twins from single-born children. Firstly, the mother's time is more fully absorbed in basic care-taking of the twins so that she has less time to interact with them in a play situation. The twins may develop a language strategy, competing with each other for parental attention. Secondly, the twins are together much of the time and may become so attuned to each other's needs that the development of adequate communication skills with other people is significantly delayed. The most extreme case occurring in an estimated 40% of twins is 'cryptophasia' or 'secret language', speech, or at least a few words, incomprehensible to anyone other than the twins.

One result is that language deficits are very common in young twins (Savic 1980). Non-verbal skills are generally more adequate although young male twins often lack concentration, leading to poorer ability on quite a variety of tasks, both verbal and non-verbal (Hay and O'Brien 1981). Although the twin deficit in verbal IQ is small (Fig. 6.7) other abilities may suffer more. For example, language delay is often symptomatic of later reading delay and recent analysis of a national survey of literacy and numeracy in Australian children has shown that only 42% of twin boys achieve adequate literacy skills by age 14 compared with 70% of single-born boys. Twin girls are only slightly behind their single-born peers. Considering reading disability which is formally defined as being at least 18 months behind in reading skills (Chapter 7) the situation is even worse. Twin boys are at least four times more likely than singleton boys to be in this category (Johnston, Prior and Hay, in press).

None of these things matter as long as all twins are affected equally. As mentioned earlier in the context of the elevated ability of adopted children, genetic analysis involves the degree of relationship and not the mean performance. But there are several examples indicating that MZ twins are exposed to different stresses from DZ twins and that the relationship between the performance of co-twins can be affected by environmental factors unique to twins. Three such examples are summarized in Table 6.5.

TABLE 6.5   Examples of MZ-DZ differences (see text for details)

		MZ_T	DZ_T
(a) *Correlations for infant mental development*			
Louisville Twin Study	Whole sample	0.85	0.62
	Retarded excluded	0.81	0.64
Collaborative Perinatal Project	Whole sample	0.84	0.55
	Retarded excluded	0.55	0.55
(b) *Mortality rates per 1000*			
Stillborn	Male	92	44
	Female	42	21
Stillborn plus infant death	Male	260	159
	Female	177	140
(c) *Correlations for height and weight in twins*			
At birth	Height	0.62	0.79
	Weight	0.63	0.68
At 24 months	Height	0.89	0.54
	Weight	0.88	0.53
(d) *Correlations for stuttering*			
Total disfluencies		0.67	−0.09

(a) From R. S. Wilson and A. P. Matheny (1976) *Behavior Genetics* **6**, 353, and P. L. Nichols and S. H. Broman (1974) *Developmental Psychology* **10**, 442. (b) From S. G. Vandenberg (1976) in *Human Behavior Genetics*, Kaplan A. R. (ed.), p. 71. Thomas, Springfield, Ill. (c) From R. S. Wilson (1979) *Annals of Human Biology* **6**, 205. (d) From P. M. Howie (1981) *Behavior Genetics* **11**, 227.

Infant mental development   Although performance on the mental development scales administered to babies does not relate highly to subsequent IQ scores (Chapter 7), it is still important to know what determines individual differences at this early age. Two major studies of infant twins have been completed, one as part of the Louisville Twin Study involving volunteer families and one as part of the US Collaborative Perinatal Project involving expectant mothers attending clinics at university medical centres. Although both studies reported higher MZ than DZ correlations (Table 6.5a) the Collaborative Project attributed the MZ–DZ difference to there being more MZ pairs where both suffered severe retardation. When these were excluded the MZ and DZ correlations were similar. No such effect was observed in the Louisville Twin Study and may reflect differences in the samples. In the Collaborative Project where many were of lower than average socioeconomic status, Blacks comprised 47% of the entire sample and 75% of those MZ pairs where both were retarded. The sibling correlation in the Collaborative Project was only 0.22, much lower than in the DZ group which has the same degree of genetic similarity. So there must be some twin-singleton difference at this age, as well as possibly an MZ–DZ difference.

Why should more MZ than DZ twins be retarded? Many more MZ twins are premature, with 53% of MZ being born before 37 weeks compared with 19% of DZ. There are also corresponding differences in mortality (Table 6.5b). Although these mortality figures are based on data obtained in 1949–50 and obstetric practices have improved since then, they do show that MZ and DZ twins represent groups with different risk factors. An additional complication shown in our second example is that there can be biological factors which not only put more MZ twins at risk, but which also create differences within MZ twin pairs in ways which cannot happen with DZ twins.

Height and weight in twins While we might expect genetic influences on physical growth and hence on the MZ and DZ correlations to remain constant during childhood, Table 6.5c shows that this is not the case. At birth MZ twins are less similar than DZ twins and only gradually does their genetic identity become apparent.

One reason for the large MZ difference at birth is the *identical twin transfusion syndrome*, something which can happen only in MZ twins as a result of their placentation. While DZ twins must have separate placentas although these may be fused together to appear to be one, there is variation among MZ twins in the nature of the placenta and the two membranes, the outer chorion and the inner amnion which surround the fetuses. This variation reflects the time after conception when the fertilized egg splits to form the two twins. The earlier the split, the more the twins are isolated from each other by having separate membranes or placentas (Fig. 6.8). In twins who are monochorionic a third circulation may develop in addition to those between the mother and each twin. This circulation through the placenta from one twin to the other is generally one-way, so that one twin (the 'donor') is shunting blood into the other (the 'recipient'). By the time of birth such twins are often very different with the recipient

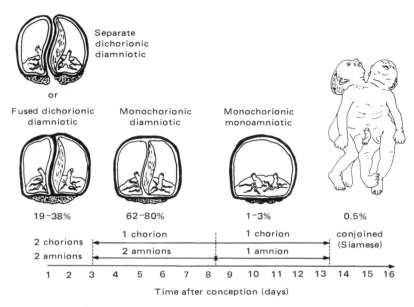

Separate
dichorionic
diamniotic

or

Fused dichorionic
diamniotic

Monochorionic
diamniotic

Monochorionic
monoamniotic

19–38%   62–80%   1–3%   0.5%

| 2 chorions | 1 chorion | 1 chorion | conjoined |
| 2 amnions | 2 amnions | 1 amnion | (Siamese) |

1 2 3 4 5 6 7 8 9 10 11 12 13 14 15 16

Time after conception (days)

FIG. 6.8   Placentation and the time of splitting of the fertilized egg to produce MZ twins. (Compiled from diagrams in I. MacGillivray, P. P. S. Nylander and G. Corney (eds) (1975) *Human Multiple Reproduction*. Saunders, London.)

weighing as much as 2000g more than the donor. There are also medical problems for both twins, the strain on the donor twin's system being such that it can lead to heart failure at birth.

While the transfusion syndrome has effects on early physical development which dissipate gradually so that the genetic resemblance of MZ twins becomes evident, what happens to behaviour? Munsinger (1977) has argued that the transfusion syndrome reduces MZ correlations for IQ and leads to an underestimate of heritability but the evidence is not clearcut. Few research programs carry out the pathology tests necessary to show if MZ twins are monochorionic (MC) or dichorionic (DC). Such tests were done in the Collaborative Perinatal Project mentioned earlier and it was found that the MC-MZ twins were actually more similar than the DC-MZ twins, partly because there may be competition for maternal resources in the DC-MZ twins who have different placentas and partly because MC-MZ twins have a very high mortality rate, with many of the severely affected MC donors dying or both being retarded. The McMaster Twin Registry results suggest that things may be even more complicated with differential effects on verbal and non-verbal IQ. Their MC-MZ and DC-MZ were equally similar on the verbal tasks, but on non-verbal tasks the DC-MZ were only as alike as DZ twins.

One must remember that the effects of chorion type may not only be biological. If one twin is weak at birth, the parents may develop different expectations of that child's abilities with consequent effects upon performance. Although many of the earlier claims about the importance of expectations by parents and teachers have been found to be overstated (Snow and Yalow in Sternberg 1982), the twin situation is a unique one because of the frequency with which comparisons between the co-twins are made by themselves and by others (Hay and O'Brien 1984).

Fɪɢ. 6.9   Placentation is an important consideration in twin development. (From N. Matterson (1982) *It's a Baby!* Marion, Greensborough, Australia.)

Stuttering in twins Stuttering in twins is an example where comparisons between the behaviour of co-twins influence the correlations. There are different methods of quantifying stuttering and the MZ and DZ correlations in Table 6.5d are based on the total number of disfluencies, the blockings, prolongations and repetitions which together characterize the stutterer. The point to note is that while the MZ correlation is fairly large, the DZ correlation is negative. That is, the higher the incidence of stuttering in one DZ twin, the lower the incidence in his co-twin ('his' co-twin, because stuttering is predominantly a male problem). The reason for this negative relationship is *competition* — if one twin stutters more than the other, he is more likely to be ridiculed by his peers and to be compared unfavourably with his co-twin by their parents. As he loses confidence, his stuttering becomes worse while his brother in contrast improves. Such competition is less likely in MZ pairs because there is not the genetic difference between the twins on which the comparisons can begin.

There are now several examples of competition in behaviour genetics, mainly from the area of personality where twins have been found to become more similar once they leave home and lead independent lives. Although they are then exposed to different environments, the possibility for competition no longer arises.

We have described three specific examples where the twin situation affects the MZ-DZ correlations in one way or another. There is also a general objection to twin studies, best summarized in the title of a paper by Lytton (1977) 'Do parents create, or respond to, differences in twins?'. The point in question is whether parents of MZ twins treat their children more similarly than do parents of DZ twins, merely because they feel that MZ twins are more alike and should be treated thus. This differential treatment would inflate the MZ correlation over the DZ value. Many twin studies have incorporated questionnaires on such parental treatment and on twin interactions (Vandenberg 1976). Although these questionnaire surveys have tended not to find a greater MZ environmental similarity, it must be remembered that surveys of most aspects of the environment especially in relation to intelligence (Urbach 1974) have generally not been very conclusive.

The best information comes from studies such as Lytton's where parent–child interactions are directly observed. He concluded that parents do treat MZ twins more similarly than DZ in some respects but

1 Only in situations where the parent is responding to some aspect of the child's behaviour. The parents do not themselves initiate more similar activities, but only respond to the more similar behaviour of MZ twins.

2 Twins are treated according to their actual and not their perceived zygosity. For various reasons parents are often mistaken about the zygosity of their twins. For example, MZ twins may be thought to be DZ because of the old and inaccurate adage, 'two placentas = non-identical, one placenta = identical' or because MZ transfusion syndrome twins initially seem so different that the parents are certain they are DZ. When accurate zygosity determination by blood-typing is performed, Lytton and others have all found that the actual MZ are treated more similarly, irrespective of whether the parents thought they were MZ or DZ.

To answer Lytton's question, parents respond to the differences in their twins and do not create them.

Model-fitting to different relationships: the solution to the problem

We have examined three major methods of human behaviour genetic analysis and found all of them wanting. Each method has specific disadvantages which may actually limit their usefulness or which critics such as Kamin (1974) can seize upon as potential objections to their validity. Subsequent to Jinks and Fulker (1970) a variety of methods have become available which combine these different types of data in the same analysis. If the same set of genetic and environmental parameters apply equally to family, adoption and twin data, then these specific objections cannot be of major consequence. We shall illustrate this approach with an analysis of intelligence test performance based on the Birmingham method. Rao and Morton (1978) provide an alternative approach with somewhat different models and assumptions.

The example is the analysis by Fulker and Eysenck (1979) of an augmented version of the data set compiled by Erlenmeyer-Kimling and Jarvik in 1963. They had gathered correlations on different familial relationships from 52 studies ranging over eight countries, many years and a wide variety of intelligence tests. Fulker and Eysenck used these data plus information from studies which emerged in the intervening 16 years, to provide the correlations for ten different relationships shown in Table 6.6.

TABLE 6.6   Analysis of familial correlations for IQ from the augmented Erlenmeyer-Kimling and Jarvik summary. The observed correlations are the medians (the middle of the range reported from the various experiments) (modified from D. W. Fulker and H. J. Eysenck (1979) in The Structure and Measurement of Intelligence Eysenck H. J., ch. 5, p. 102. Springer-Verlag, Berlin)

Relationship	Observed correlation (O)	G	E_2	Expected correlation (E)	(O–E)
Unrelated pairs apart	−0.01	0	0	0.00	−0.01
Unrelated pairs together	0.23	0	1	0.18	0.05
Foster parent-child	0.19	0	1	0.18	0.01
Sibs apart	0.34	½	0	0.35	0.01
Parent-child apart	0.32	½	0	0.35	−0.03
Sibs together	0.49	½	1	0.53	−0.04
Parent-offspring together	0.50	½	1	0.53	−0.03
DZ together	0.53	½	1	0.53	0.00
MZ apart	0.75	1	0	0.69	0.06
MZ together	0.87	1	1	0.88	−0.01

Also in Table 6.6 is the genetic and environmental model which Fulker fitted to the data. He could not use the distinction of G_1 and G_2 we introduced for the twin data, since G_1 and G_2 involve dominance which does not enter into all parts of Table 6.6, e.g. the parent–offspring relationship is purely additive. If dominance is assumed to be absent then $G_1 = G_2$ and a parameter $G = G_1 + G_2$ can be fitted, where $G = \frac{1}{2}D$. MZ twins will have a value of G of 1 since they share exactly the same genes, while DZ twins, sibs and parent–offspring relationships have $G = \frac{1}{2}$, since half their genes ($\frac{1}{4}D$ or $\frac{1}{2}V_A$ in Table 3.4) are common. With correlations having a maximum value of 1 there is no need to fit a value for E_1 since this comes by subtraction if one assumes that the only major factors contributing to the total variation are G_1, E_1 and E_2. Remember that these assumptions are testable assumptions — if dominance were important or if some other factor contributed to the variation, then the model would not fit.

Fulker then used a statistical technique called least-squares analysis to estimate the two variables G and E_2. Least-squares analysis is a more sophisticated version of the usual procedure for solving simultaneous equations such as $\frac{1}{2}x + y = 1$ and $x + y = 2$, which yields answers of $x = 2$ and $y = 0$. But far more combinations of data and variables may exist than there are variables. In Table 6.6 there are ten relationships and only two variables for which to solve. Least squares analysis selects the combination of values of G and E_2 which give the closest fit to the observed correlations, and in this case gives $G = 0.69$

and $E_2 = 0.18$. By subtraction $E_1 = 0.13$. In the last two columns of Table 6.6 are the expected correlations, given these values and the discrepancies between the observed and expected values. (The term least squares comes from the fact that the method calculates G and E to give the *least* value for the *squares* of these discrepancies.)

This final column is the test of our genetic and environmental model, since if the model is inaccurate some of these discrepancies will be very large. But the largest discrepancies in Table 6.6 are 0.06 for MZ_A and 0.05 for unrelated children reared together. Both can be explained by selective placement, of MZ twins with relatives or of unrelated children in homes rather like those provided by their biological parents. While the model assumed MZ_A would be placed at random (no E_2) and that unrelated children would have no genetic similarity (no G), the selective placement violates these assumptions.

However, the discrepancies in Table 6.6 are still small and inconsequential. A check can be made with a χ^2 test which is a statistical measure of the magnitude of the differences between observed and expected (Chapter 3). If the discrepancies turn out to be due to more than just chance, an extra parameter can be inserted, or if only one family relationship is affected it can be omitted and the analysis repeated without it. An example of the latter is given by Fulker (1974, Table 4.2). He fitted a G_1, G_2, E_2 model to Burt's data on twins, sibs and unrelated children and found the expected value of the unrelated children reared together correlation to be 0.18 less than the observed, presumably through selective placement. Eliminating this group from the analysis reduced the discrepancies and the χ^2 to an acceptable value. It is interesting that Burt's MZ_A sample, over which there has been so much dispute, fitted the model accurately.

But the discrepancies in Table 6.6 are so small that such a procedure is unnecessary. Consider what implications this has for the problems we raised with the family, adoption and twin methods. For example, the DZ, the sibs together and the parent–offspring correlations are all given the same expected values. If say, there had been anything different about the twin situation or if the age difference between the parents and children had been important, then these expected values would have been inaccurate. To take another example, the MZ correlation is reduced by 0.12 (0.87–0.75) when the twins are reared apart. In the same situation the correlation between siblings is reduced by not much more, 0.15 (0.49–0.34). If parents had really treated MZ_T very similarly, a larger difference between the MZ and sibling situations would have been expected. If the transfusion syndrome had decreased the MZ correlation or competition the DZ correlation, then discrepancies should have occurred since these factors would not enter into the non-twin situations.

Implicit in the model is a test of all such assumptions. Therefore model-fitting is a theoretically very elegant means of determining the genetic control of human behaviour given the problems which arise or are claimed to arise with the different sources of data.

One advantage of model-fitting over standard estimates of heritability is that we also learn something about the environmental factors involved and can put these in a proper perspective. In the analysis in Table 6.6 $E_1 = 0.13$, so that only 13% of the variation in IQ is attributable to variation within the family. One of the very few testable environmental hypotheses about variation in IQ has been Zajonc's confluence model based on the quality of attention a child receives from other family members. This model accurately predicts results of the sort shown in Fig. 6.7 where later-born children and twins do less well, since they receive more immature attention from their siblings or co-twins respectively. But even if the confluence effects were the sole contributor to within-family environmental differences, they can still account for no more than 13% of all the IQ variation.

While model-fitting thus puts a more realistic interpretation on the importance of specific environmental effects, this is perhaps also the chief drawback of this approach. It plays down the relevance of the factors which may be peculiar to twin, adoption or other family situations. While such factors may be immaterial to genetic analysis they may be important for other reasons. For example Zazzo argues very strongly that it is time twins ceased to be merely part of genetic analysis and instead were studied as a group which could provide unique information on behavioural development

According to the classic approach, data derived from twin studies have no general value unless the peculiarities of twins are denied. In our new perspective, it is precisely their peculiarities that give us information on the most complex processes of our personality. We may thus start, at the same time, with a twin psychology and our own psychology. (From R. Zazzo (1978) in *Twin Research: Progress in Clinical and Biological Research 24A*. Nance W. E. (ed.), p. 11. A. R. Liss, New York.)

New perspectives on the genetics of intelligence

There can never be a definitive value for the heritability of intelligence. The values entering into Table 6.6 and similar compilations will change as results become available from newer, more standardized and much larger projects such as the Hawaii Family Study and the Texas Adoption Study. Figure 6.10 is from a summary by Plomin and DeFries of the studies of intelligence published since 1976. To give a complementary perspective to Table 6.6, the results are expressed instead as the average IQ differences ($\bar{Z}$) in the various relationships rather than as correlations. The two are related by the formula $\bar{Z} = 1.13 \times SD \times 1 - r$, where $SD = 15$ in the case of IQ tests. The major changes from Table 6.6 are that the DZ correlation is 0.09 higher at 0.62 and that the sibs together and the parent–offspring together correlations both drop by 0.15 to 0.34 and 0.35 respectively. Henderson (1982) discusses some of the differences between recent and older studies.

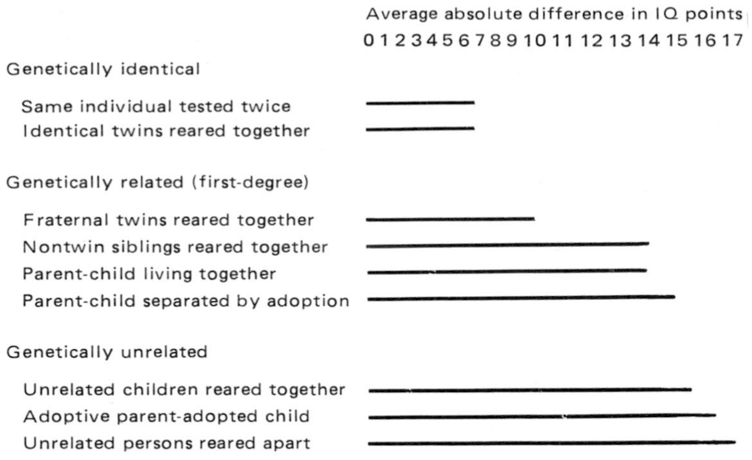

Average absolute difference in IQ points

0 1 2 3 4 5 6 7 8 9 10 11 12 13 14 15 16 17

Genetically identical

  Same individual tested twice

  Identical twins reared together

Genetically related (first-degree)

  Fraternal twins reared together

  Nontwin siblings reared together

  Parent-child living together

  Parent-child separated by adoption

Genetically unrelated

  Unrelated children reared together

  Adoptive parent-adopted child

  Unrelated persons reared apart

FIG. 6.10   Average absolute differences in IQ for individuals of different relationships. (From R. Plomin and J. C. DeFries (1980) *Intelligence* 4, 15.)

Apart from indicating a twin-environment effect since the DZ correlation is now appreciably larger than the sib correlation, what of the overall genetic architecture? Unfortunately Plomin and DeFries consider the twin, adoption and family data independently of each other, but they do find that each type of data is consistent with a heritability of 0.5 or 0.19 less than that found in Table 6.6. Whether their approach is more convincing than the model-fitting one is for the reader to decide. One difficult observation they do not try to explain is why the parent–offspring together correlation at 0.35 is little larger than the 0.31 correlation between biological parent and offspring reared apart. Surely the E_2 effect of being reared by one's own parents should affect the resemblance in IQ more than this? Caruso (1983) suggests that low correlations may have been caused by recent studies sampling people with a limited range of scores. Correcting for this and for attenuation raised the parent–offspring correlation considerably and the sibling–sibling correlations to some extent.

As more and better studies are completed the pattern of relationships and hence the genetic architecture will continue to change. What considerations need to be taken into account in planning, analysing and interpreting the results?

Considerations in human genetic analysis

The model-fitting approach to human behaviour genetics is a product of the last decade and still has some way to go before it is a comprehensive and generally-accepted approach. The limitations on it at present include the following three issues.

Sample size and number of familial relationships   There are very few sets of data on human behaviour which have the numbers and range of relationships needed for genetic analysis. Mittler (1971) exemplifies the data used in the past when 20–30 sets of twins were often considered sufficient to estimate 'heritability' with the biases which these methods of calculation involved. Now we need far larger samples and a range of relationships with which to estimate and test our model.

But how big a sample and which relationships? Computers can be used to simulate different designs and sample sizes in order to determine just which relationships in which proportions are needed to answer specific questions about the genetic architecture. Although the results of such simulations are complex to explain, one conclusion is particularly noteworthy:

From the point of view of cost-effectiveness, therefore, statistical and practical considerations coincide in favour of designing experiments in human psychogenetics which incorporate foster-siblings rather than monozygotic twins reared apart. The simulations confirm that much of the money and effort at present directed towards the collection of twin data might profitably be diverted to the collection of data on foster-siblings. The latter are certainly easier to ascertain and, except in cases of doubtful paternity, they do not require the expensive and time consuming procedure of zygosity diagnosis which is essential for twin studies. In any case it has been remarked that the use of foster-siblings substantially reduces the number of twins required to give results of comparable efficiency. (From L. J. Eaves (1969) *British Journal of Mathematical and Statistical Psychology* **22**, 142.)

So several years before the scandal over Burt's MZ_A data, Eaves had shown MZ_A were not particularly helpful from the viewpoint of adequate genetic analysis. However, if the problems of selective placement and the other biases mentioned earlier can be taken into consideration, MZ_A provide a useful additional relationship with which to verify the genetic architecture. If only twins were to be used, Eaves found that at least 250 pairs are needed merely to detect genetic variation when the broad heritability is less than 0.5 and as many as 2500 pairs are needed for a full assessment of the genetic architecture.

The other thing we must have apart from the number of individuals is the number of relationships. There must be more relationships than parameters being fitted if we are to be able to test the model. In Table 6.6 only two parameters were fitted to ten relationships, giving many opportunities to test the adequacy of this model. There have been attempts in the past such as Cattell's MAVA method (Multiple Abstract Variance Analysis) which specified numerous parameters of potential importance and tried to fit many of these to inadequate data. Jinks and Fulker (1970) note it is unwise to fit a complex model to such data until the failure of a simpler model indicates this is either necessary or appropriate.

There is no limit to the variety of relationships that can be used. Eaves (1973) analysed IQ results on 3558 individuals from the Reed and Reed family study of mental retardation discussed in Chapter 2. These people were all descendants of 53 pairs of great-great-grandparents, one of whose grandchildren was institutionalized for mental retardation in the period 1911–18. The genetic model is different in form but similar in principle to that in Table 6.6, in that the genetic and environmental components specify variation within and between groups of people with the same parents or grandparents or great-grandparents etc. Dominance could not be adequately estimated but the narrow heritability was 0.60, which is not appreciably different from the 0.66 value estimated by Jinks and Fulker (1970) from Burt's data. These data had another parallel with Burt's data in that both showed a high degree of assortative mating (a correlation of 0.4 or more between spouses' abilities) had to be an integral part of the genetic model, in contrast to the Hawaii Family Study data in Table 6.1.

Assortative mating   One problem for human genetic analysis is that people do not choose their mates at random. We introduced assortative mating for IQ in the context of the Hawaii Family Study and it constitutes one difficulty for the biometrical models we have been describing. Recall our genetic parameters within and between families, $G_1 = \frac{1}{4}D + \frac{3}{16}H$ and $G_2 = \frac{1}{4}D + \frac{1}{16}H$. If there is no dominance $G_1 = G_2$ and if there is dominance, $G_1 > G_2$. But the effect of positive assortative mating is to increase G_2, the variation between families, while leaving G_1 unaltered. Therefore, G_1 could equal G_2 because one is increased by dominance, the other by assortative mating.

This is more than a statistical technicality. In Table 6.6 we assumed dominance to be absent and were still able to fit an adequate model to the data. But shortly we consider a situation demonstrating pronounced dominance for high IQ. Assortative mating is needed to explain the discrepancy. With enough familial relationships involved, more complex models can be fitted including both dominance and assortative mating. Examples are to be found in Fulker and Eysenck (1979).

The main differences (Loehlin 1978) between the Birmingham and the Honolulu approaches to analysis are that Birmingham methods allow for dominance and assume that the choice in assortative mating is based on characteristics of the spouse. The Honolulu model assumes that dominance is absent and that assortative mating reflects social homogamy, the family background. A more realistic approach may be that assortative mating occurs for different reasons for different behaviours. For example non-verbal IQ generally shows more assortative mating than does verbal IQ, the opposite of what one might have expected. One suggested possibility for this difference has been that assortative mating for verbal IQ is more on the basis of the family background, whereas mating for non-verbal IQ is determined by the individual's own phenotype and less by the social situation in which that person developed.

Assortative mating is assuming greater importance in behaviour genetics for various reasons. It can best be studied through the recently developed twin-family method (Fig. 7.8), it is an important consideration in Eckland's view of human behavioural evolution (Chapter 8) and it creates problems for attempts to study the genetic and environmental correlations between traits. To take a simple example, people mate assortatively for both height and IQ so any genetic analysis relating physical and intellectual traits has to consider this potential source of covariation between the traits.

Genotype–environment interaction and correlation Some critics of genetic analysis do not deny that genetic variation for behaviour may exist but then argue that the genetic and environmental factors are so intertwined that they can never be successfully separated. While this may be true within the one individual, the ironic thing is that if it were true for the population and for individual differences within the population, it would have serious implications for any attempts to specify environmental factors which could improve behaviour. The role of unsystematic g–e interaction was demonstrated in Fig. 3.14 for the effects on mice of drug administration. If these had been human data with some environmental treatment replacing the drug, then the unacceptable situation arises where the treatment improves some individuals but disadvantages others.

As distinct from interaction where genotypes respond differently to some feature of the environment, g–e correlation concerns the frequency with which certain genotypes and environments occur together. In a major discussion of this topic, Plomin, DeFries and Loehlin (1977) recognize three types of g–e correlations:

1 Passive where the individual receives both the genotype and the environment without doing anything to influence this process, e.g. a child receiving both the genetic potential for high ability and an environment in which it can thrive.

2 Reactive where people react differently to individuals with different genotypes, e.g. a teacher encouraging students who are doing particularly well.

3 Active where people of different genotypes seek out a particular environment, e.g. the gifted child seeking playmates with similar abilities.

Such correlations may be positive, enhancing individual differences or negative, making people more alike, a phenomenon which has been termed 'coercion to the biosocial norm'. Peer-group pressures against the very able schoolchild would be an example of this latter situation. Although we can make such theoretical distinctions among the different types of g–e correlation, it is another matter deciding if such correlations should be distinguished from the genetic variance. Situations, for example where the gifted child will seek further stimulation, may be inevitable consequences of the genotype in the same way that parents' more similar treatment of MZ twins resulted from the children's similar behaviour. Such activities stem from the genotype and cannot realistically be separated from it.

How do we go about distinguishing g–e interaction and correlation? Jinks and Fulker (1970) proposed specific tests which they undertook before fitting models. For g–e interaction they compared the sum and the difference of co-twins' scores. MZ twins differ only because of the environmental effects upon them, while their sum or the average of their scores also reflects their genotype.

Interaction will emerge if twin pairs with different average scores do not show the same differences between co-twins. For example on a measure of neuroticism, one aspect of personality, twins with mean scores near the population average showed larger differences from their co-twin than those with means at either extreme. That is, people near the middle on this measure are more susceptible to environmental influences.

For g–e correlation, they compared the total variance for individuals of the same familial relationship reared together and apart e.g. MZ_T with MZ_A. Being reared apart should at least diminish if not eliminate passive g–e correlation and reduce the total variance. By this method, g–e correlation was detected in performance on a test of educational attainment at the primary school level, whereas it was far less obvious with IQ test performance. The reason is probably that parents are more likely to assess the school performance they expect from their child and to encourage the child to reach this level, whereas IQ lends itself less to such a procedure.

An extension of this approach has been to add parameters for g–e interaction and correlation in the usual model-fitting approach. Models can now be specified which include such parameters as the influence of siblings on each other and cultural transmission where parents pass environment as well as genes to their children. Plomin et al (1977) advocate a different approach based on the adoption situation where data on the biological parents provide some measure of genotype and features of the adoptive home indicate the

TABLE 6.7   Test for g–e interaction in the Skodak and Skeels study* (from R. Plomin, J. C. De Fries and J. C. Loehlin (1977) *Psychological Bulletin* **84**, 309)

		Environment (adoptive parents)		
		Low (<12.1 years schooling)	High (>12.1 years schooling)	Row mean
Genotype (biological parents)	Low (<9.9 years schooling)	111.8	106.3	109.4
	High (>9.9 years schooling) 122.5		121.4	121.9
	Column mean 116.6	115.4		

*Figures are the IQ scores at age 13.

environment. Table 6.7 shows how Skodak and Skeels' data can be used to test for g–e interaction in this way. Although no interaction effects were found, one may argue that years of parental education are hardly comprehensive measures of either genotype or environment. Using the same measure of genotype, g–e correlation can be tested by relating genotype to features of the adoptive child's physical, educational and social environment. While these adoption methods are a lot easier to comprehend than the elaborate biometrical methods, reliance only on adoption data with their potential difficulties rather than on a wider spectrum of relationships may limit their usefulness.

Alternatives to the analysis of resemblance between relatives

We have approached human genetic analysis only through the resemblance between relatives. But there are at least two situations where it is also possible to use group means in human genetic analysis:

Regression to the mean Although this topic was mentioned in Chapter 3, Table 6.6 provides data on the genetic architecture of IQ which allow more quantitative predictions. Table 6.8 reintroduces

TABLE 6.8   Average IQ scores for retarded patients and their relatives (data from L. S. Penrose (1963) *The Biology of Mental Defect* (3rd edn). Sidgwick and Jackson, London)

	Type of relationship to patient	Patients' IQ	Relatives' IQ	
			Observed	Expected
Patients with an IQ > 50	Sib	65.8	84.9	85.1
	Half-sib, nephew or niece	63.2	89.5	92.0
Patients with an IQ < 50	Sib	24.2	87.4	67.0
	Half-sib, nephew, or niece	33.3	95.1	85.5

the topic of familial mental retardation discussed in Chapter 2. If genetic and environmental effects common to family members comprise 87% of the variation in IQ (Table 6.6), then the best prediction on the polygenic model is that sibs of the retarded should have half those genetic and environmental effects which contributed to the retarded person's low IQ. The other 13% comprises effects unique to each family member (E_1). The score of sibs would therefore be expected to be reduced below the population mean of 100 by ($\frac{1}{2} \times 0.87 \times$ the difference of the retarded's IQ from 100). On the same basis, using the values in Table 3.4, the performance of half-sibs and similar relatives should be reduced by ($\frac{1}{4} \times 0.87 \times$ the difference in IQ).

Table 6.8 shows that these predictions hold for families where the retarded person has an IQ greater than 50. The IQs of relatives of the more severely retarded are underestimated for the reason discussed in Chapter 2, that severe retardation is more often due to specific genetic or environmental defects unique to that retarded individual rather than to polygenic causes affecting all relatives to some extent.

Inbreeding depression Families with a history of inbreeding provide an unusual source of evidence for genetic influences on behaviour. Until recently there had been few adequate studies of inbreeding partly because inbreeding is relatively rare. Also effects on intelligence were confounded with socioeconomic status (SES), since inbreeding was more common in lower SES groups where it served in various communities as one means of keeping dowries and other limited resources within the family.

The Arab communities in Israel encourage consanguineous marriages, so that across the whole spectrum of SES 34% of their marriages involve either first cousins or double-first cousins who are even more closely related in that both sets of parents come from the same two families. Inbreeding of this sort where related individuals mate must be distinguished both from incest (inbreeding where the individuals are so closely related e.g. brother and sister, that the law prohibits it) and from assortative mating (where people choose mates on the basis of phenotype and not necessarily genotype).

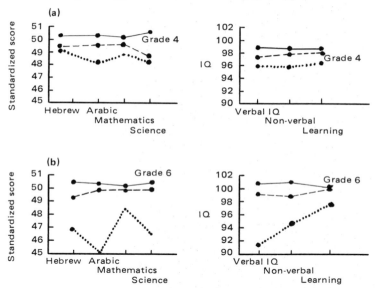

FIG. 6.11   The performance of Arab children of differing degrees of inbreeding on (a) school achievement tests and (b) ability tests. ●—● = no inbreeding; ●- - -● = from first cousin marriages; ●...● = from double first cousin marriages. (Achievement tests scores are standardized to a mean of 50, ability tests to one of 100.) (From J. Bashi (1977) *Nature* **266**, 440.)

The results of testing a representative sample of Arab children of Grades 4 and 6 (10 and 12 years approximately) are presented in Fig. 6.11. Three ability tests measuring verbal and non-verbal IQ and associative learning were used as well as achievement tests based on classroom knowledge, since ability tests may be less dependent on specific experiences and hence have a higher heritability. However, apart from associative learning which is often thought to be fairly unrelated to IQ (Jensen 1980), the results on all tests were in decreasing order, unrelated, first cousins, double-first cousins.

There are two genetic explanations for these results. The polygenic one is that there is potence (or directional dominance) for high ability. A simple single locus example is enough to show that in going from an F_2 of $\frac{1}{4}AA$, $\frac{1}{2}Aa$, $\frac{1}{4}aa$ with means $m + d, m + h, m - d$ respectively and overall mean $m + \frac{1}{2}h$ to a totally inbred situation with two equally common genotypes, AA $(m + d)$ and aa $(m - d)$ and overall mean m, the only difference in generation means is due to h. That is, in the process of inbreeding overall performance declines only if there is potence, akin to the plant example in Fig. 3.10.

The alternative explanation is that inbreeding enhances the chances of an individual being homozygous for an autosomal recessive disorder. Most single gene disorders are so severe that the children would be unlikely to be in normal classes, but there was an increased variability in performance among the double first cousins which could be attributed to some such single genes segregating.

Although the decline in performance in Fig. 6.11 may not appear to be very great, first cousins only represent one-sixteenth of the possible inbreeding and double first cousins one-eighth. The consequences of even this small depression in IQ are explained by Fulker and Eysenck (1979, p. 124). While Kamin (1980) has criticized this and other inbreeding studies for the small and often statistically insignificant effect, Daniels et al (1982) have pointed out that the effect of inbreeding is remarkably consistent across the different studies. It is also very much in line with that predicted from those biometrical studies where dominance was estimated.

Conclusion

This chapter reflects the three recent developments in genetic analysis — complexity, conflict and confusion. In 1971 it was possible for Mittler to present a straightforward account of human behaviour genetics using twins and simple statistics, but since then the situation has changed considerably. Behaviour genetic research especially on intelligence has been scrutinized both for the adequacy of the data and for alternative non-genetic explanations. While some criticisms such as the nature of Burt's data have been constructive, other criticisms have been less helpful to scientific progress and in certain cases have been quite misleading. One example would be Kamin's attempt to explain the similarity of twins' scores on the basis of their being the same age (Kamin 1974 and review by Fulker). The fact that MZ twins are more similar than DZs (Table 6.6 and Fig. 6.10) despite the same degree of age-matching in both groups immediately rules out this explanation.

At the same time, behaviour genetic analysis has changed as people seek more information on the genetic architecture than is available from a single measure of heritability. The rival schools of thought as to the methods of analysis add to the confusion of people not fully conversant with the assumptions and statistical techniques underlying the different approaches.

Many arguments within behaviour genetics and between genetic and environmental explanations are so highly technical that it is easy to forget they concern behaviour and not merely statistical methods. Yet the different genetic architectures observed with different behaviours make a particularly important point. One strength of the Birmingham methods is that they have been applied to a wide variety of behaviours and not just to IQ, e.g. Jinks and Fulker (1970), Fulker (1974). While it is easy to make sweeping criticisms of such topics as the difficulty of detecting g–e interaction or the unrepresentative nature of twins, it is quite another matter to explain why, if the methods are so bad, they detect g–e interaction in some situations but not in others or why twin data show variation in one behaviour to be genetic, in another solely environmental. So much emphasis is still placed on intelligence in texts such as Eysenck versus Kamin (1981) that it is easy to overlook the existence of other behaviours amenable to genetic analysis.

It is also easy to forget in the course of genetic analysis that we are concerned with behaviour and not with more straightforward aspects of the phenotype. Gourlay (1979) points out that the principles of genetic analysis were largely developed for characteristics of mature plants and animals, e.g. crop size or yield of meat, and not for something like intelligence measured at different ages. Genetic analyses of intelligence reflect only the psychometric approach to intellectual development and not alternative conceptions such as the information processing and Piagetian models (Siegler and Richards in Sternberg 1982). Consider the work of Piaget on the stages of thinking through which the developing child passes (Chapter 8). This approach suggests that children's abilities differ qualitatively as well as quantitatively from those of their parents. Thus parent–offspring and adoption studies become questionable and even the use of data from sets of twins of different ages is open to doubt. While Gourlay proposes that genetic models incorporate a parameter for age effects, the data on IQ discussed here suggest that intelligence test performance is remarkably robust across ages. But Chapter 7 provides some examples of behaviour where there are very marked age changes in the genetic determinants.

To carry out human behaviour genetic research, it is not necessary to become involved in the various controversies in genetics or psychology. The most important thing is to have data sets of sufficient size to provide reliable conclusions about the genetic architecture. Already we have mentioned in Chapter 2 the Reed and Reed pedigree study with over 83 000 individuals and in this chapter the Hawaii Family Study with some 6 000 individuals, the Colorado Adoption Project, the Texas Adoption Project and the Minnesota Adoption Studies with 152, 300 and 216 adoptive families respectively, the Louisville Twin Study, the La Trobe Twin Study and the Collaborative Perinatal Project with 494, 470 and 350 pairs of twins respectively and Bashi's inbreeding study with 3 200 children. Finland, Norway and Sweden each have twin registries based on the national records and involving all twins in the country, which have been invaluable in psychiatric research (Chapter 7).

The discussion of recent data summarized by Plomin and DeFries (Fig. 6.10) introduces one danger in very large studies — how representative are they? For example the new DZ correlation of 0.62 comes mainly from the National Merit Scholarship Qualifying Test (NMSQT) administered to US high school students and including measures of academic as well as general intellectual ability. Only a selected sample of students sit for this examination — 23% of all people of that age and only 10% of US same-sex twin pairs of that age — so that the generality of these data are in doubt. The parent-offspring correlation comes from the Hawaii Family Study where again the tests involve much more than intelligence. Before substituting these results for the older studies which together covered a wider spectrum of people and abilities, more consideration needs to be given to who has been tested, in what situation and with what tests.

A summary by Bouchard and McGue (1981) of 111 studies highlights the variability between studies and some of the inconsistencies. Recall the parent–offspring correlations of 0.35 and 0.31 for families reared together and apart discussed in connection with Fig. 6.10. The comparable figures from Bouchard and McGue are 0.42 and 0.22, giving a very different picture of the effects of being reared with one's biological parents.

There is still the issue of comparability between studies as raised by Caruso (1983) with his discussions of the limited sampling in some newer work. Recently Bouchard proposed that there be a 'benchmark', a reference set of tests used in every study. Once one knew that studies corresponded on the benchmark tests, they could be compared in other respects with more confidence. Thus in a recent Danish study of males adopted into the same or different families, Teasdale and Owen (1984) analysed the genetics of height along with IQ and educational level. While the two behavioural measures correlated highly, educational level had a much larger E_2 effect and a lower heritability (see Chs 7 & 8) than height or IQ.

Although the emphases have been on alternative explanations and analyses, it is more important that adequate data be collected in the first place. It is worth recalling that data sets like the Erlenmeyer-Kimling and Jarvik compilation of 1963 which have been invaluable in the development of analytical methods were created long before the current round of arguments developed or that the NMSQT twin data published in 1976 have already been the subject of many analyses from different perspectives. Given the rate of development since the early 1970s it is likely that quite different methods of analysis will be available in the next decade. Recently, methods of pedigree analysis have been introduced which rely on massive computing resources to analyse each family member separately rather than through covariances and the other population statistics we have discussed. But there is no point in having such analyses if there are not the data on which to use them. Can the enormous effort and resources required to collect adequate data be justified? Since people have and always will query the source of individual differences in behaviour, it seems better that judgements be based on fact and not ignorance. To put it more forcefully:

It is of pressing scientific interest to resolve competing claims about the basis of behavioural variation in man and to delineate those areas of human behaviour where social and personal factors are paramount. In our view the issue should be decided on the basis of data and not on philosophy, politics and prejudice. The model-fitting approach will not avoid human error nor preclude controversy, but provides a scientific basis for resolving competing claims and for quantifying our degree of ignorance. (From L. J. Eaves, K. A. Last, P. A. Young and N. G. Martin (1978) *Heredity* **41**, 315.)

Discussion topics

1 R. C. Lewontin (*Annual Review of Genetics* (1975) **9**, 387) lists six criteria which any research on genetic variation and IQ must meet "to qualify as a proper experiment whose results are worthy of serious consideration". J. C. DeFries and R. Plomin (*Annual Review of Psychology* (1978) **29**, 473) object to two of Lewontin's criteria but add a further five criteria of their own. Are all these criteria really needed for useful research on the genetics of IQ and would such ideal data add much to the existing knowledge based on far less perfect experiments?

2 "Although the environmentalist programme can explain successfully some of the correlation data, taken as a whole these data constitute much greater empirical progress for the rival hereditarian programme. First, while the environmentalist theories only account for the *relative values* of some of the observed correlations, the hereditarian programme predicts precise quantitative values for each of the correlations ..." (P. Urbach (1974) *British Journal for the Philosophy of Science* **25**, 119).

Using the data in Table 6.6 or Fig. 6.10, can you suggest a purely environmental model which would accurately predict the values for the different relationships?

3 Having read this chapter what is your feeling about the quote from Henderson in the Introduction? How much criticism is useful and how much misleading? (Eysenck versus Kamin (1981) provide a good guide to the range of possible criticisms and countercriticisms.)

References

Annotated bibliography

Fulker D. W. (1974) Applications of biometrical genetics to human behaviour. In van Abeelen J. H. F. (ed.) p. 9. *The Genetics of Behaviour* North Holland, Amsterdam. (Although more technical than the present chapter this is one of the few discussions of model-fitting which does not assume detailed knowledge of biometrical genetics. The comparative analyses of Burt's and Jencks' data in Table 4.3 is particularly noteworthy.)

Henderson N. D. (1982) Human behavior genetics. *Annual Review of Psychology* **33**, 403. (While using a different terminology to this Chapter,

Tables 1 and 2 and the accompanying discussion are a useful example of combining data from different relationships.)

Jinks J. L. and Fulker D. W. (1970) Comparison of the biometrical genetical, MAVA and classical approaches to the analysis of human behaviour *Psychological Bulletin* **73**, 311. (More technical than Fulker (1974) but particularly important for demonstrating the range of results which emerge from biometrical analyses of different behaviours.)

Kamin L. J. (1974) *The Science and Politics of IQ*. Erlbaum, Potomac, Md. (and review by D. W. Fulker (1975) *American Psychologist* **88**, 505). (Kamin's book remains the most widely quoted critique of genetic analyses of intelligence. Apart from predating or ignoring most large-scale studies, there are also other faults which Fulker discusses.)

Mittler P. (1971) *The Study of Twins*. Penguin, Harmondsworth. (An example of what twin research used to be like before Kamin and before the development of more complex analytic methods.)

Plomin R., DeFries J. C. and Loehlin J. C. (1977) Genotype-environment interaction and correlation in the analysis of human behavior *Psychological Bulletin* **84**, 309. (A largely non-technical summary of these two complications in genetic analysis. They emphasise adoption data, in contrast to the biometrical approach summarized in L. J. Eaves, K. Last, N. G. Martin and J. L. Jinks (1977) *British Journal of Mathematical and Statistical Psychology* **30**, 1)

Shields J. (1978) MZA twins: their use and abuse. In *Twin Research: Progress in Clinical and Biological Research* Nance W. F. (ed.), **24A**, 79. (A realistic account of the practical issues involved in assessing MZ_A and a reply to Kamin's criticisms of some of this research.)

Sternberg R. J. (ed.) (1982) *Handbook of Human Intelligence*, Cambridge University Press. (The most complete summary of all aspects of intelligence testing. Apart from Scarr and Carter-Saltzmann's useful review of family, adoption and twin data on intelligence, the chapter by Carroll on intelligence testing is relevant to Chapter 1 and the Chapters on education by Snow and Yalow and on social policy by Zigler and Seitz to Chapter 8.)

Urbach P. (1974) Progress and degeneration in the 'IQ Debate'. *British Journal of the Philosophy of Science* **25**, 99. (An account of the genetic and environmental approaches to IQ from a philosophical perspective, which seeks to distinguish constructive or 'progressive' criticism from *post hoc* 'degenerative' criticism. It helps make sense of some of the claims and counterclaims discussed in this chapter.)

Vandenberg S. G. (1976) Twin studies In *Human Behaviour Genetics*, Kaplin A. R. (ed.), p .71. Thomas, Springfield Ill. (An illustration that there is more to twins and the twin situation than just providing statistics for genetic analyses.)

Additional references

Bouchard T. J. and McGue M. (1981) Familial studies of intelligence: a review. *Science* **212**, 1055.

Caruso D. R. (1983). Sample differences in genetics and intelligence data: sibling and parent-offspring studies. *Behavior Genetics* **13**, 453.

Daniels D., Plomin R., McClearn G. and Johnson R. C. (1982) "Fitness" behavior and anthropometric characters for offspring of first-cousin matings. *Behavior Genetics* **12**, 527.

DeFries J. C. and Plomin R. (1978) Behavioral genetics. *Annual Review of Psychology* **29**, 473.

Eaves L. J. (1973) Assortative mating and intelligence : an analysis of pedigree data. *Heredity* **30**, 199.

Eysenck H. J. versus Kamin L. (1981) *Intelligence: the battle for the mind.* Pan, London.

Fulker D. W. and Eysenck H. J. (1979) Nature and nurture : Heredity. In *The Structure and Measurement of Intelligence* Eysenck H. J. (ed.), ch. 5, Springer-Verlag, Berlin.

Gourlay N. (1979) Heredity versus environment — an integrative analysis. *Psychological Bulletin* **86**, 596-615

Hay D. A. and O'Brien P. J. (1981) The interaction of family attitudes and cognitive abilities in the La Trobe Twin Study of Behavioural and Biological Development. In *Twin Research: Progress in Clinical and Biological Research* **69B**, 235, Gedda L., Parisi P. and Nance W. E. (eds). A. R. Liss, New York.

Hay D. A. and O'Brien P. J. (1984) The role of parental attitudes in the development of temperament in twins at home, school and in test situations. *Acta Genetica Medicae et Gemellologiae* **33**, 191.

Hearnshaw L. S. (1979) *Cyril Burt: Psychologist.* Hodder and Stoughton, London.

Horn J. M., Loehlin J. C. and Willerman L. (1982) Aspects of the inheritance of intellectual abilities. *Behavior Genetics* **12**, 479.

Kamin L. J. (1980) Inbreeding depression and IQ. *Psychological Bulletin* **87**, 469.

Jensen A. R (1980). *Bias in Mental Testing.* Free Press, New York.

Johnston C., Prior M. and Hay D. A. (in press). Prediction of reading disability in twin boys. *Developmental Medicine and Child Neurology.*

Loehlin J. C. (1978) Identical twins reared apart and other routes to the same destination. In *Twin Research: Progress in Clinical and Biological Research* **24A**, 69, Nance W. E. (ed.). A. R. Liss, New York.

Lytton H. (1977) Do parents create, or respond to, differences in twins? *Developmental Psychology* **13**, 456.

MacGillivray I., Nylander P. P. S., and Corney G. (eds) (1975) *Human Multiple Reproduction.* Saunders, London.

McAskie M. and Clarke A. M. (1976) Parent–offspring resemblances in intelligence : theories and evidence. *British Journal of Psychology* **67**, 243.

Munsinger H. (1975) The adopted child's IQ : a critical review. *Psychological Bulletin* **82**, 623.

Munsinger H. (1977) The identical-twin transfusion syndrome : a source of error in estimating IQ resemblance and heritability. *Annals of Human Genetics* **40**, 307. (See also reply by Kamin L.J. (1978) *Annals of Human Genetics* **42**, 161.)

Rao D. C. and Morton N. E. (1978) IQ as a paradigm in genetic epidemiology. In *Genetic Epidemiology* Morton N. E. and Chung C. D. (eds), p. 145. Academic Press, New York.

Savic S. (1980) *How Twins Learn to Talk: a study of the speech development of twins from 1 to 3.* Academic Press, London.

Teasdale T. W. and Owen D. R. (1984) Heredity and familial environment in intelligence and educational level — a sibling study. *Nature* **309**, 620.

7 Genetics and human behaviour

Topics of this chapter

1 Three common misconceptions about the implications of genetic variation in human behaviour.
2 The ways in which the genetic analysis of cognitive abilities has gone beyond IQ scores to consider specific skills and disabilities and the process of development.
3 The genetics of personality in children and adults, why they differ and what the implications are for personality theory.
4 Genetic approaches to psychopathology using adoption and single gene marker studies, with emphasis on the possible heterogeneity within disorders.

Three common misconceptions

In this chapter we consider various applications of the techniques of human behaviour-genetic analysis introduced in Chapter 6. But first it is important to realize what the outcome of any genetic analysis does *not* imply:
1 Evidence for genetic involvement in one behaviour does not mean that genes are implicated in all behaviours. Popular questions such as 'Is our behaviour inherited — nature or nurture?' or even the slightly more accurate 'To what extent is our behaviour determined by our genes?' are not scientific. The answer depends on precisely what behaviour is measured in what situation, a point introduced in Chapter 1 with the discussion of the differences between schizophrenia and the neuroses (Fig. 1.2). With the techniques of Chapter 6 we can be even more exact and distinguish behaviours not just in terms of their 'heritability', but in terms of the types of gene action (additive, dominance) and of environmental effects (within or between families) as well as in the degree of assortative mating and genotype–environment correlation and interaction. The differences between genetic analyses of intelligence and personality measures will make such a distinction clearer.
2 Genetic determination is not an argument for the maintenance of the *status quo*. Just because a particular trait is largely genetic does not mean people are bound by their genes to a particular social niche. Figure 7.1 illustrates the difference between this erroneous concept of genetics as 'like begats like' (Fig. 7.1a) and the real situation (Fig. 7.1b).

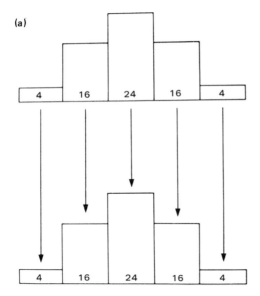

(a)

| 4 | 16 | 24 | 16 | 4 |

| 4 | 16 | 24 | 16 | 4 |

Fig. 7.1   The distribution of performance in one generation and the next. (a) The wrong concept of genetic determinants of human variation. Only an inflexible system of cultural transmission can produce such a fixed relation between parents and offspring.

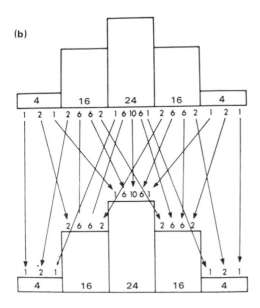

(b)

Fig. 7.1   (b) The reassortment of individuals in a two locus, random mating population with no dominance (see Fig. 3.7). The mean and variance remain the same from one generation to another, but offspring are not necessarily in the same class as their parents.

(c)

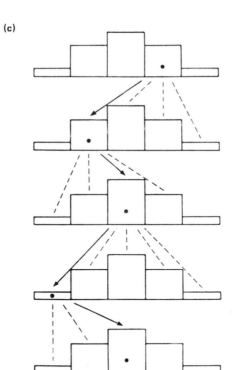

FIG 7.1   (c) A hypothetical example of how the performance of members of the one family can vary between generations (From C. C. Li (1971) In *Intelligence: Genetic and environmental influences* Cancro R. (ed.), p. 162. Grune and Stratton, New York.)

As a result of segregation and recombination children are not exactly like their parents, but some will do better and some worse. Figure 7.1c is a hypothetical example of how much shift can occur from one generation to the next. That social mobility is partly a consequence of meiosis and not of environmental opportunity alone is particularly important for intelligence test performance where research on tests and on their genetic determination has been condemned as a means of justifying and preserving socioeconomic group differences. This point is discussed further in Chapter 8.

3   Genetic does not mean unchangeable. Significant heritability does not imply that a behaviour cannot be changed by suitable environmental manipulations although the genotype may set limits on the effectiveness of such manipulations. Gottesman used the term 'reaction range' to illustrate this concept (Fig. 7.2). The idea initially came from studies of the height of Japanese children growing up in three dietary situations: in postwar occupied Japan (restricted, low-protein diet), in contemporary Japan (natural diet) or where parents had emigrated to California (enriched, high-protein diet), but can be extended to intelligence or any other trait. While the size of the reaction range is an empirical question for that particular trait, Fig. 7.2 suggests that the more able have the greatest scope for being

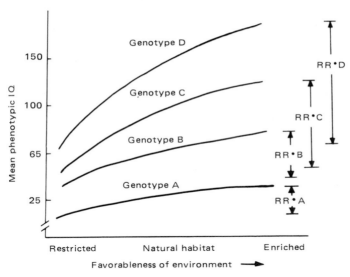

FIG. 7.2   The hypothesized reaction ranges (RR) of four genotypes, A is single-gene disorder and B,C and D, represent three points in the range of polygenic variation. (From I. I. Gottesman (1963) in *Handbook of Mental Deficiency: Psychological theory and research* Ellis N. (ed.), p. 253. McGraw-Hill, New York.)

influenced by their environment and that an enriched environment will not necessarily enable the less able (genotype B) to catch up with the most able (genotype D). Chapter 8 examines some preschool environmental enrichment programs where this is the general result.

From the start behaviour genetics has accepted the role of environmental influence. In 1928 Burks reported a study of foster children which incorporated quite detailed measures of the home environment. While her genetic analyses have been criticized, she did appreciate the role of the environment:

Assuming the best possible environments to be three standard deviations above the mean of the population (which, if 'environments' are distributed approximately according to the normal law, would only occur about once in a thousand cases) the excess in such a situation of a child's IQ over his inherited level would be between 9 and 27 points — or less if the relation of culture to IQ is curvilinear on the upper levels, as it well may be. (From B. S. Burks (1928) *Yearbook of the National Society for the Study of Education* **27**, 219.)

Her data are based on fostered children and may not have general application but the point remains. Within the usual range of environments experienced by children, very few are sufficiently extreme as to greatly affect IQ. A widely quoted example in this context is that of the MZ twins known in the literature as Gladys and Helen who were reared in exceedingly different situations and ended up differing by 24 IQ points. While it is an indication of how large environmental effects can be, it is also consistent with the predictions of genetic analysis as the quote from Burks shows.

A more awkward question is to specify just which aspects of the environment are important for cognitive development (Wachs in Plomin 1983). In an article appropriately entitled 'In search of the missing environmental variance in cognitive ability', Vandenberg and Kuse (in Gedda *et al* 1981) examined the contributions of 44 environmental variables to performance in the Hawaii Family Study of Cognition mentioned in Chapter 6. These environmental variables covered many aspects of the family situation but explained no more than 20% of the total variation. The highest correlations with cognition involved such factors as family income and knowledge of foreign languages which are partly dependent on IQ and hence may reflect genetic as well as environmental influences.

While behaviour genetics does recognize the role of the environment, one cannot modify behaviour unless one can identify influential environmental variables, a task which has proved particularly difficult. Vandenberg and Kuse focussed on environmental factors *common* to family members, but McCall (in Plomin 1983) suggests that behaviour genetics, like much of psychology, has paid insufficient evidence to environmental differences *within* the family. The analyses of intelligence in Chapter 6 indicated the role of such environmental factors and we see shortly that they appear to be even more important for personality.

Beyond IQ

Figure 7.3 introduces our discussion of two ways in which behaviour genetics can do more with cognition than simply analyse IQ. The first question is whether the same genes and environmental factors influence ability during the course of development. The hypothetical model in Fig. 7.3a is one where additional genes are 'switched on' at different stages of development (akin to the regulatory genes

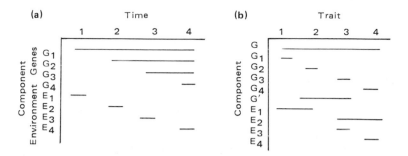

FIG. 7.3   Two hypothetical examples of (a) how gene expression may change with time and (b) how genetic and environmental effects may contribute to four correlated measures. G refers to genes having a general effect over all traits, G1 . . . G4 to time- or trait-specific genes and G' to genes affecting a pair of traits. (From L. J. Eaves (1982) In *Genetic Basis of the Epilepsies* Anderson V. E., Hauser W. A., Perry J. K. and Sing G. F. (eds) p. 249. Raven Press, New York.)

mentioned in Chapter 1) while the environmental factors are occasion-specific rather than operating throughout development.

Any other combination of timing of genetic and environmental factors may also be considered, e.g. where there is an early environmental influence with long-term consequences. Second is the question of what determines the different types of ability which make up intelligence. The illustrative model in Fig. 7.3b, again only one of many possible, has one common set of genetic influences plus specific ones for each ability or pair of abilities, with a variety of environmental factors some influencing more than one ability.

Genetics of cognitive development

The first problem in studying cognitive development is that of measurement. The same difficulty arises with testing infants and young children as happened with rodent learning (Chapter 5). The various tests suitable for youngsters may often make quite specific sensory or motor demands on the child, limiting interrelationships between tests and as shown in Fig. 6.1, predictability to later performance (see Lewis 1976 for a review of this area). Even the same test given at different ages may tap somewhat different skills. For example the Wechsler Intelligence Scale for Children-Revised (WISC-R) can be given to 7–16 year olds, but the speed at which the children do particular tasks usually only counts towards their scores after the age of 10.

Spurts and lags in development — the Louisville Twin Study The problems of comparing tests across different ages make the results of the Louisville Twin Study (Wilson in Plomin 1983) all the more

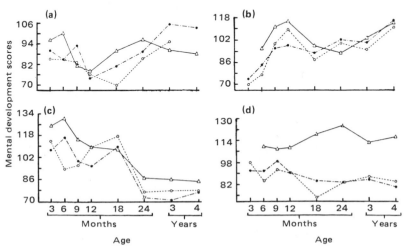

FIG. 7.4   Typical developmental curves for each member of MZ or DZ twin pairs (o- -o, ●- · -●) and their younger siblings (△—△) between 3 months and 4 years. (Derived from R. S. Wilson (1978) *Science* **202**, 939, 1978 by the AAAS.)

impressive. Started in 1957 this project has been following twins and their younger siblings on a series of tests from birth through to 9 years with a test again at 15 years. At present 494 sets of twins are actively involved. The children fluctuate in performance from year to year as the examples in Fig. 7.4 show, but the fluctuations are systematic for each family. The twins and their sibling in A decline and then improve relative to the mean of 100, whereas those in C decline throughout the 4 years of testing.

TABLE 7.1   Correlations among twins for overall level of performance and for age-to-age changes (derived from R. S. Wilson (1978) *Science*, **202**, 939)

| | Overall level | | Age-to-age change | |
	MZ	DZ	MZ	DZ
3–12 months	0.84	0.78	0.40	0.15
12 months–3 years	0.89	0.79	0.67	0.42
3–6 years	0.90	0.75	0.47	0.30

The resemblance of MZ and DZ pairs on the age-to-age fluctuations in ability can be compared just like their overall similarity with the results shown in Table 7.1. Initially MZ twins are only a little more alike than DZ's in their overall and age-to-age similarity, but the MZ-DZ distinction becomes clearer as the children get older and the tests tap more than just sensori-motor skills. The fluctuations appear to be partly genetically determined but to a lesser extent than the overall level.

What do these results mean? Wilson (in Gedda *et al*, 1981, p. 208) concludes:

These results strongly suggest that developmental processes are initiated and guided by timed gene-action systems which are activated in sequential fashion, and on a schedule largely determined by the genotype. In addition to the profound species-wide programming of developmental processes, there are distinctive variations in rate and schedule superimposed upon the main trends, and these furnish the dispersion of individual differences in the population. Each zygote contains a preprogrammed set of instructions that constantly propels the developmental processes along predetermined pathways, and maintains the directional focus in the face of deflecting agents.

But these spurts and lags are specific to each set of twins and it is not as if there are particular ages or stages of development when all children are likely to change. These results must be distinguished from the Fels Longitudinal Study which continued over such a time period that children followed from 2.5 to 17 years of age could subsequently have their own children followed in the same way. The average changes in IQ within each child in their 14 year test period was 28.5 IQ points with one in seven children shifting more than 40 points. These changes were not random but generally fell into five different categories (Fig. 7.5), with the most likely times for change being 6 and 10 years.

The Fels Study is concerned with fluctuations in cognitive development which are common to all families rather than specific to each family as in the Louisville Study. However, analyses of Fels

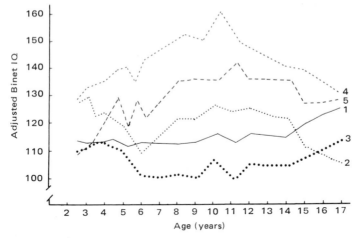

FIG. 7.5   Five profiles of cognitive development in the Fels Longitudinal Study. (From R. B. McCall, M. I. Appelbaum and P. S. Hogarty (1973) *Monographs of the Society for Research in Child Development* **38**, No. 3, 1.)

sib–sib and parent–offspring data (which are far more limited than the Louisville data) do not support the idea that these fluctuations are genetically determined. Rather the different profiles have been related to parental differences in attempting to accelerate their child's development and in how they exercise control over their children.

It is unclear why the Fels and Louisville studies reach such different conclusions but one simple explanation can be dismissed. With the Louisville twins being the same age, the fluctuations may reflect the particular changes or stresses their family underwent at that time. That is, there may be genetic variation for the reaction to family stress rather than for fluctuations in cognitive ability. What makes this unlikely is that a sample of younger siblings tested like the twins but who would have experienced such stresses at a different age, gave twin–sib correlations for age-to-age changes very similar to the DZ correlations in Table 7.1.

Whatever the reason, it is clear that fluctuations in IQ scores within the one child are not random (a frequent criticism of the IQ test) but are systematic and open to investigation with twin or family studies.

Changes in genetic determination with age   The Louisville data in Table 7.1 showed that in overall IQ MZ twins became more similar relative to DZ by the age of 2 or 3. While this probably relates to the twins outgrowing their birth stresses, there are various other mechanisms by which MZ-DZ differences may change during childhood. Fischbein (in Gedda *et al* 1981) hypothesizes a divergence effect where MZ react similarly to environmental events while DZ experience different events and react differently to them, so that the

DZ correlation declines with age. This would happen in a permissive or enriched environment while in contrast there may be a convergence effect in a restricted environment, where DZ twins progressively become more similar.

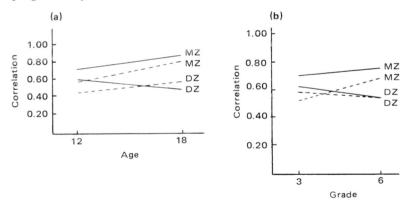

FIG. 7.6   (a) Correlations for verbal (solid line) and inductive reasoning (broken line) tests in Swedish male twins. (b) Correlations on a mathematical achievement test at Grades 3 and 6 (approximately 9 and 12 years of age respectively). Solid line, boys. Broken line, girls. (From S. Fischbein (1981) in *Twin Research* **3** (Part B): *Progress in Clinical and Biological Research* **69** Gedda L., Parisi P. and Nance W. E. (eds) 211.)

Figure 7.6 presents data from the Swedish longitudinal twin study, the SLU project. Between 12 and 18 years male MZ and DZ twins diverge for verbal ability but not for inductive ability (as measured by questions of logic such as 'If Stockholm is situated north of Gothenburg, mark the circle to the right of the left triangle.') Fischbein interprets this difference as there being more scope with verbal ability for DZ twins to react differently to home and school environment. The tests at 18 years were at military enrolment and no data for females are available. But a mathematical achievement test shows a divergence for both boys and girls between Grades 3 and 6 (Fig. 7.6b). This test reflects the children's response to the teaching given in school where according to Fischbein the free and permissive Swedish educational environment should favour such divergence.

A similar divergence at adolescence emerges from US studies of siblings and adopted children. Instead of possibly passive genotype-environment correlation of young children exposed to particular home situations (Chapter 6). Scarr and McCartney (in Plomin 1983) postulate that at adolescence there is an active g–e correlation where individuals begin to select environments which fit their capabilities and interests. Such selection still reflects genetically determined capabilities and so in the adopted children depends more on their biological parents than on any features of their adoptive home. Thus in the Minnesota Adoption Study, they found that the correlations in mental development between unrelated infants adopted into the same home dropped almost to zero when they reached adolescence. At this age the correlation of the adopted child's IQ was only 0.09 and

0.11 with the adoptive mother's and father's education respectively, compared with 0.28 and 0.43 with the biological mother and father with whom they had had little or no contact for many years.

What has not yet been considered is the issue posed at the start of this section. There has been no investigation of the possibility that the divergence between MZ and DZ twins or the decreased similarity of the adopted children to their adoptive parents is a result of additional genes being switched on at puberty, in addition to those active earlier in development.

Discussion

While behaviour genetics is coming to grips with the fact that cognitive ability is not fixed, its methodology is limited and often falls back on the traditional twin study whose limitations were discussed in Chapter 6. The reason is the need for genetic analyses carried out on children in the same age-range, whether it be to study fluctuations with age or processes of thinking which are restricted to children. An example of the latter is Piaget's concept of concrete operations, the stage reached between 6 and 11 years when children come to realize say, that a squat, broad vessel can contain the same volume of liquid as a tall, thin one (the 'conservation of volume'). Clearly it would be impractical to compare children with their biological or adoptive parents, since the parents would have proceeded to Piaget's next stage of formal operations. The one exception would be studies such as the Fels project where the parents had also been tested as children, but such projects are naturally rare and are limited by the cohort differences between generations growing up many years apart.

Even the best adoption studies produce ambiguous answers with young children. The Colorado Adoption Project (Plomin and DeFries in Plomin 1983) is a very comprehensive longitudinal study where the adopted children have so far only reached the age of 2–3 years. Most of the correlations with parental measures are less than 0.2. Does this mean that parental genes have little influence on behaviour at this age or just that the measures made on these young children are not tapping the same skills which are being measured in the parents, as indicated in Fig. 6.1?

But can we rely on twins to study processes specific to childhood or is their course of development different? The few twins and triplets involved in the Fels project were less variable than the singletons, leading to the suggestion that 'the twin situation acts as a buffer to environmental inputs in general'. A more major problem is that twins do start at an intellectual disadvantage to singletons and gradually catch up, although Fig. 6.7 showed that the recovery may not be complete. If twins are developing faster in order to catch up with singletons, it may be inappropriate to base genetic analyses of development just on twins.

To examine this problem the Louisville Twin Study involves siblings and the La Trobe Twin Study (Hay and O'Brien in Plomin 1983) has gone further and used the cousins as well, as an additional group of relatives growing up without twins in the family. One example illustrates the importance of such additional data. Wilson (in Plomin 1983) compared DZ and twin–sibling correlations for intelligence and found the former was consistently higher until age 15, even though the two groups are comparable genetically. He proposes that the additional similarity in IQ resulting from the closeness of child twins takes until adolescence to dissipate. This contrasts with Fischbein's theory of divergence in adolescent DZ twins (Fig. 7.6) and is supported by Wilson's twin–sibling correlations which do not show the change at adolescence predicted by the divergence theory.

Genetics of specific abilities

An intelligence test score is a composite of many different abilities so that people may achieve the same overall IQ score in different ways. Figure 7.7 shows the performance of three MZ twin pairs on the ten WPPSI subtests. The twins in pair C have very similar verbal IQs even though their profiles across the sub-tests are quite different, one doing well on vocabulary and poorly on arithmetic and the other the reverse. With this exception, the profiles across the ten sub-tests within each twin pair lead us to the question of whether there is a common, partly inherited, ability underlying the ten sub-tests, in which case the fluctuations of pairs from one sub-test to another are environmentally induced. Alternatively are these different abilities inherited to different extents, which would help explain why one pair, or one of a pair, does well on one sub-test, one on another? Many of the early studies of cognition quoted heritabilities for different abilities even though their estimates were really too imprecise to warrant such a distinction. Thus Mittler (1971) evaluates the work of Blewett who reported a higher heritability for verbal and reasoning than for numerical and spatial ability and speculated that perhaps 'good art students are born, whereas mathematicians, engineers and statisticians are made by training'.

In a review of many studies along these lines, DeFries et al (1976) indicate that verbal skills generally do show more genetic variation. In the Hawaii Family Study, the four factors of verbal and spatial abilities, perceptual speed and visual memory summarize the information on 15 different cognitive tests. Remembering that a family study only gives an upper estimate of heritability, the regressions of offspring on midparent were 0.65, 0.61, 0.46 and 0.44 respectively for the four factors. That is for cultural and/or genetic reasons, children resembled their parents most on verbal ability.

A more precise question for cognition is whether the same genes and environmental factors are involved in the different abilities. Only very recently have adequate methods become available to answer this question. In the same way that genetic models are specified in advance (Chapter 6), these methods require also hypotheses about the relationships between the abilities to be specified as in Fig. 7.3b. The analysis is then a question of testing and modifying hypotheses about

(a)

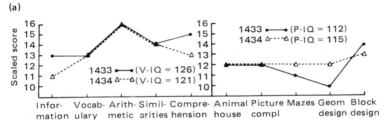

(b)

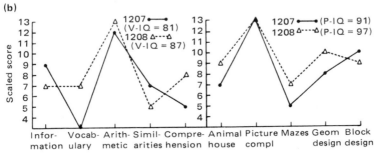

(c)

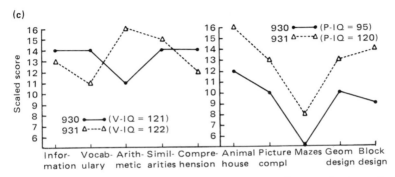

FIG. 7.7   The profiles of three MZ pairs in the Louisville Twin Study on the ten subtests comprising the WPPSI. V-IQ is verbal IQ, P-IQ is performance IQ. (From R. S. Wilson (1975) *Developmental Psychology* **11**, 126. Copyright 1975 by the American Psychological Association. Reprinted by permission of the author.)

both the genetics and the relationships between abilities. Table 7.2 summmarizes the results of such an analysis of MZ and DZ twins on Thurstone's Primary Mental Abilities test. Since Thurstone did not

TABLE 7.2   General and specific components of variation in the five primary mental abilities* (From N. G. Martin and L. J. Eaves (1977) *Heredity* **38**, 79.)

Ability	E_1 General	E_1 Specific	E_2 General	E_2 Specific	½D General	½D Specific
Numerical ability	—	0.165	0.139	—	0.696	—
Verbal comprehension	0.096	0.046	0.606	—	0.127	0.125
Spatial ability	—	0.177	0.066	0.294	0.164	0.299
Word fluency	0.058	0.224	0.215	—	0.140	0.363
Reasoning	0.071	0.142	0.379	0.231	0.177	—

*For ease of comparison across abilities, these values are scaled to make the total variance 1.0 for each ability.

incorporate a general factor in his views of ability (Chapter 1) and designed this test to tap discrete skills, it is a particularly good one on which to see if any common factor does emerge or if the effects are specific to each of his five abilities.

As might be expected nearly all the E_1 variance is specific, except for a small general E_1 factor on the three verbal traits. That is, the environmental factors unique to individuals within a family are also unique to particular abilities. E_2 is almost the opposite with differences between families having a general influence on all abilities but particularly verbal skills. This makes one ask whether the high heritabilities previously reported for verbal ability were actually high 'familialities', to use the term coined for the Hawaii Family Study to describe situations where genetic and between family environmental influences are confounded. The genetic estimates do not totally support Thurstone's view of independent primary abilities. There is a general additive genetic component (a dominance parameter was not needed) underlying all five abilities and for only three abilities is it necessary to invoke specific genetic effects.

Other approaches to specific abilities

Other lines of behaviour-genetic research distinguish particular abilities and their determinants.

Maternal effects and the twin-family design   The twin-family design (Fig. 7.8) incorporating adult twins, their spouses and offspring represents a powerful technique for genetic analysis without the specific problems which can arise in the twin and adoption situations. The children of MZ twin pairs are in genetic terms half-sibs rather than just cousins. The variation in such extended sibships can be partitioned into: I — within sibships, II — between sibships within families and III — variation among families. The role of maternal effects can be assessed by comparing these variances. The offspring of male MZs are born to genetically unrelated mothers and any variation

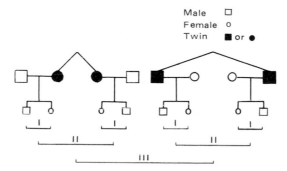

FIG. 7.8   The three types of variation among the offspring in kinships of MZ twin parents. I, within sibships. II, between sibships within families. III, among families. (From R. J. Rose *et al* (1980) *Nature* **283**, 375.)

in maternal effects between these women will inflate variance II, relative to the comparable estimate from female MZs where it is the fathers who are genetically unrelated. Conversely any similarity of maternal effects in these female MZs will inflate the variance III among families relative to the estimate from male twins.

The relevance to specific abilities is that the twin-family design reveals maternal effects on two Wechsler verbal subtests, Information and Vocabulary but not on one performance subtest, Block Design. This is in spite of Block Design being a widely-used measure of general intelligence which correlates highly with the verbal subtests. (The verbal-performance distinction refers to the fact that the performance tests depend to a much lesser extent on verbal skills e.g. the Block Design sub-test requires blocks to be fitted together to copy a pattern.)

What could be the basis for such a distinction between the tests? After excluding on empirical grounds such explanations as twin sisters spending more time together as children (which would make their verbal IQs more similar) or as adults (where their children may share more IQ-relevant information), it was concluded that either the twin mothers had provided more similar intrauterine environments or more similar patterns of child rearing. Such pre- and postnatal effects could not be distinguished with these data. The maternal effect does not mean that the considerable heritability for verbal subtests can be dismissed but rather goes some way to identifying the source of the large specific E_2 component for verbal ability in Table 7.2.

Spatial ability   One of the most widely discussed sex differences has been the greater ability of males than females on a wide range of spatial tests of the sort shown in Fig. 7.9. The common feature of these tests is recognizing an object after the rotation of a two-dimensional representation. While this sex difference is generally accepted, both biological and cultural explanations are contested (Vandenberg and Kuse 1979). Genetics is not much help in that sex differences like ethnic group differences are not suitable for most genetic analyses — one cannot have MZ twins, one male and one female.

What has developed is the concept of an X-linked gene determining spatial ability. If the recessive allele enhances spatial ability then the mean of the female population should be below that of males who only have one X chromosome. Vandenberg and Kuse point out that this hypothesis is inconsistent with the impaired spatial ability of Turner's syndrome females. Table 7.2 showed that there was considerable specific genetic variance for spatial ability, but to prove X-linkage requires detailed analyses of particular family relationships. The features of X-linked inheritance (Fig. 2.19b) is that sons get only a Y chromosome from their father, so that for X-linked characters the father–son correlation should be zero, the mother–son correlation should be greater than the mother–daughter correlation and so on.

Although earlier data lent some support to this hypothesis, Vandenberg and Kuse review five recent studies of parent–offspring relations and find no support for the X-linkage hypothesis. The

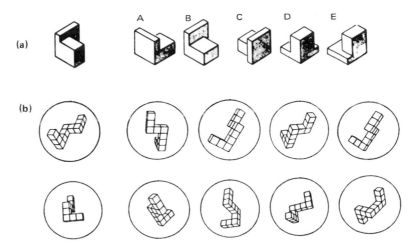

FIG. 7.9   Examples of three-dimensional spatial tests. All involve matching the object on the left with some aspects of one or more of the alternatives on the right. (a) Stafford identical blocks. (From S. G. Vandenberg (1977) In *Genetics, Environment and Intelligence* Oliverio A. (ed.), p. 285. North Holland, Amsterdam.) (b) Two sample items from the Mental Rotations test. (From S. G. Vandenberg and A. R. Kuse (1978) *Perceptual and Motor Skills* **47**, 599.)

difficulty may lie partly in the age differences between parents and young children — sex differences in spatial ability first appear in early adolescence and remain through adulthood even though the scores of both males and females start to decline in the thirties. Analyses of sibling–sibling correlations get round this age problem but give no definitive support for the X-linkage hypothesis. Even the very different technique of searching for linkage to known biochemical marker loci on the X chromosome has not been successful.

These results do not exclude biological bases to the sex difference in spatial ability, but only mean that this particular biological hypothesis is not appropriate. Since spatial ability is correlated with success in a wide variety of technical, vocational and occupational activities (McGee 1979), the problem of sex differences cannot be ignored.

Reading disability   By definition, reading disability is an example of a specific ability independent of general intelligence. To distinguish them from children unable to read because of general educational subnormality, the diagnosis of reading disability (sometimes called specific developmental dyslexia) only applies to children whose intelligence is in the normal range but whose reading skills are at least 18 or 24 months below those expected for their age.

Figure 7.10 comes from a study in Melbourne of reading disabled children (technically called the probands, the individuals from whom the families were identified) and their siblings. All children are above average on Block Design and the Vocabulary test, so that their non-

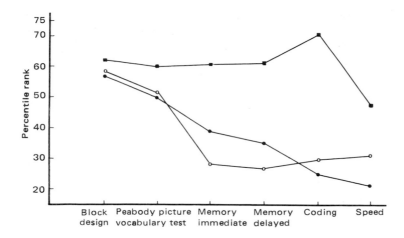

FIG. 7.10   The profiles of reading-disabled children (the probands) (●—●) and their brothers (o—o) and sisters (■—■). The scores are in percentile ranks where 50 is the average (the score below which 50% of the population lies). (From R. Rosewarne, unpublished data, La Trobe University, Melbourne.)

verbal and verbal IQs are adequate. However all boys in the families (and reading disability is 3 to 5 times more common in males) are very poor at memory tasks, both of short and long durations. They lack the ability to handle information quickly which is tapped in the coding and speed tests, where items have to be sorted rapidly by particular criteria. Other family studies such as the Colorado Family Reading Study (DeFries and Baker in Plomin 1983) confirm the familial nature of reading disability, in that parents as well as siblings of the probands show specific disabilities particularly in these tests of coding and speed.

The family method is complicated by the sex difference in reading skills with its implications for the correlations between same- and opposite-sex relatives. Females employ phenological coding in reading more than males. For example, Coltheart et al. (1975) required students to go through the alphabet mentally, counting the number of letters either with the sound 'ee' or where the upper-case form had a curve in it. No visual information could help in the former task where females did better and made fewer mistakes, while the males did better on the latter task where sound information was of no help.

But this is only one problem for family studies of reading. Apart from the question of whether reading ability is heterogeneous, so that one model of genetic transmission is inappropriate for all families (Finucci et al. 1976), the major issue is that family members share environments as well as genes. Parents of probands in the Colorado Study reported fewer attempts to teach their children to read before starting school and family members generally read less and watched more television. The family method cannot determine if these are a consequence or a cause of reading disability. This problem confounds the frequent report that reading disability fits an autosomal dominant mode of inheritance. If there is one parent who did not learn to read properly and who discourages his or her children from doing so, this will appear as autosomal dominant inheritance. DeFries and Baker (in Plomin 1983) examined families longitudinally and detected a parental influence on the age-to-age continuity of reading disability which was absent in families without a history of reading disability. This parental influence was not found for the coding and speed tests, suggesting it may reflect specifically the family's lack of encouragement of reading skills. Thus attempts to go further and locate a dominant gene for one particular subtype of reading disability on chromosome 15 through association with particular chromosome bands (Pennington and Smith in Plomin 1983) must be viewed with caution especially as the only families included were those where 'the pedigree was consistent with autosomal dominant inheritance' (p. 375). The one line of evidence which justifies such an approach is that reading disabled children may have normal siblings, which is more consistent with segregation than with a family environment effect.

Twin studies have not helped greatly to clarify the situation (Herschel 1978). Apart from reading disability being more common in twins (Chapter 6) and possibly arising for different reasons than in singletons, there are two further problems. Firstly there may be more than one kind of reading disability so that the diagnosis of reading disability in twins may not mean that co-twins have the same disorder, confounding those genetic analyses which simply compare 'reading disability' in MZ and DZ twins. Herschel reviews the various classifications of reading disability, including Naidoo's distinction of five types on the bases of differences in test performance and family history. Several of these groups but not all are characterized by lateralization differences, such as an excess of left-handedness or a lack of correspondence for eye, foot and hand dominance. Pennington and Smith (in Plomin 1983) also discuss subtypes of reading disability and draw some interesting parallels with X chromosome anomalies and their hemispheric-specific deficits mentioned in Chapter 2.

Secondly, twins and specifically MC-MZ twins who split later than DC-MZ when the cell mass is more developed (Fig. 6.8), show unusual lateralization. This could either result from the phenomenon of mirror-imaging (where the symmetrical cell mass splits into a right and left half so that one twin is right-handed one left handed) or through some other disturbance of symmetry (Boklage et al 1979). Boklage et al point out that twin data are not consistent with data from other relationships on the genetics of lateralization and figure prominently in arguments against any genetic basis to handedness. Therefore where lateralization effects have been implicated in certain types of reading disability, twin studies may not help to establish the causes of reading disability in the general population.

Where twins may contribute is in situations where one MZ twin but not the other has reading disability or where they both have a disability but of different types. Such situations are a unique opportunity for understanding the environmental or biological events leading to differences in ability between two genetically-identical individuals.

Discussion

Behaviour genetics is now capable of going far beyond the genetic analysis of one behaviour at one stage of the lifespan. We have concentrated on development in children but there is a growing interest in the genetic analysis of the decline in abilities late in life (Royce in Royce and Mos 1979). Twin and adoption studies are providing unique perspectives on aspects of development. While the twin studies are best suited to analysing the time-course of development, longitudinal adoption studies (Plomin and DeFries in Plomin 1983) have particular advantages for the analysis of g–e interaction (Table 6.7).

The genetic analyses of the components of intelligence tests have largely justified the use of the IQ score in that there clearly is a general ability. In other ways behaviour geneticists may be criticized for not keeping up with such developments in cognitive psychology as employing Piagetian tests or analysing the strategies used in solving test problems (Eysenck 1979; Sternberg 1982) or trying to measure the physiological processes in the brain which underlie intelligence (Eysenck 1982). While studies of this sort are underway and are reviewed by Henderson (1982), the investment of resources required for a genetic study leads to a justifiable reluctance to use new tests until their value is clearly established.

Analyses of the components of IQ provide a model for other behaviours. The most obvious link is with scholastic achievement — is there a common factor underlying performance in English, mathematics and the other skills and how does this relate to IQ? Two recent studies of twins in national testing programs, the American College Test (ACT) and the National Merit Scholarship Qualifying Test (NMSQT), reveal a single factor accounting for 70–80% of the genetic variance and a single environmental factor accounting for almost the same amount of environmental variance. Henderson (1982) points out that this simple pattern emerges despite the tests comprising both intellectual ability and achievement tests, with an additional reading comprehension component.

But the genetic determinants of scholastic achievement are not exactly like those of intelligence (Fulker and Eysenck, Ch. 5 in Eysenck 1979). The environmental variance between families is larger (30–40%) on achievement tests. In contrast to the data from late adolescents taking tests such as the NMSQT, analyses of achievement tests in primary school children reveal a genotype–environment correlation as mentioned in Chapter 6. It is easy to appreciate how this difference arises. Parents' own educational status or the home environment they provide could foster or impede primary school performance, since parents will have expectations of how well children should perform in the basic skills acquired at this stage. With the more specific skills in foreign languages, in physics, in English literature etc. taught to older students, parents may have no experience on which to base such expectations. Hence, g–e correlation is not found.

It is unwise to ignore such differences between scholastic achievement and intelligence. To quote '... this means that there is much more that we can do to improve school performance through environmental means than we can do to change intelligence *per se*' (A. R. Jensen (1979) *Harvard Educational Review* **39**, 59). Or, in terms of Fig. 7.2, achievement has a much larger reaction range than ability.

Personality and temperament

Apart from intelligence, personality is the one aspect of human behaviour which most lends itself to genetic analysis. There is a wide variety of personality questionnaires which individuals can complete or which parents can complete to evaluate their children. We shall adopt a fairly neutral definition of personality:

Personality is 'that which characterizes an individual and determines his unique adaptation to the environment'. More colloquially, personality means — what sort of a person is so-and-so, what is he like? At the same time we usually restrict the term to the relatively permanent emotional qualities underlying the person's behaviour, his drives and needs, attitudes and interests, and distinguish it from his intellectual and bodily skills and cognitive characteristics. (P. E. Vernon (1963) *Personality Assessment: A critical survey*, p. 6. Methuen, London.)

However, personality theorists are not in agreement as to what their paper-and-pencil personality tests measure. Eysenck in his test the Eysenck Personality Questionnaire (EPQ) identifies three dimensions, introversion–extraversion, neuroticism–stability, and psychoticism, while Cattell by the very name of his test 16 PF (16 Personality Factors) identifies far more. Figure 7.11 illustrates

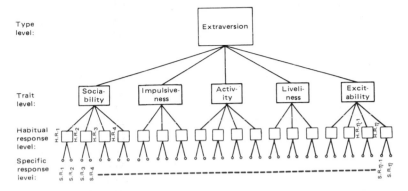

FIG. 7.11   Eysenck's view of the hierarchical nature of one dimension of personality. (From H. J. Eysenck (1976) in *Human Behavior Genetics* Kaplan A. R. (ed.), p. 199. C. C. Thomas, Springfield, Illinois.)

Eysenck's resolution of this difference. His dimensions or *types* are broader concepts which correlate with several of the more specific *traits* such as sociability. Traits in turn are defined on the basis of the observed correlations between reported behaviours in even more specific situations, e.g. sociability is based on the statistical result that individuals who like to go to parties, also like to have many friends, like to talk to strangers, dislike reading in solitude, etc. Cattell (1982) describes his somewhat different conception of personality and its genetic analysis by the MAVA method (Chapter 6).

Not all personality theorists agree with such tests, arguing that an individual's personality cannot be reduced to a few numbers. Others criticize on the grounds that one needs objective measures of what a person actually does in a particular situation, rather than an overview based on self-report.

Adult personality

Although we shall see this trait–situation dichotomy is important in children, a surprising consistency emerges in the questionnaire measures of adult personality. All personality dimensions irrespective of the questionnaire from which they are derived show that about half the variation is due to additive genetic effects (more if test unreliability is taken into account) and the remainder to specific within-family environmental variance. In the case of twins, this means that MZ twins typically have a correlation of about 0.5 compared with 0.3 for DZ twins. In the Minnesota and Texas projects mentioned earlier, children adopted into the same family are little more alike than any two individuals picked at random from the population, reflecting the absence of common environmental influences in children growing up in the same family.

Most exact biometrical analyses (Eaves and Young 1981) confirm that there are no common environmental effects, no dominance, no epistasis, no genotype–environment correlation or interaction , no sex-linkage or sex-limitation and no assortative mating. In practical terms, we have a situation where parents exert no general effects over personality development in their children.

Henderson (1982) does not completely accept this result, arguing that between family environmental variation may be reduced either by Cattell's 'sociopsychological law of environmental coercion toward the mean', where extreme phenotypes are forced to conform or by competition between siblings of the sort noted for stuttering in Table 6.5d. Competition is detectable for introversion–extraversion, but it is only of any magnitude in child and adolescent data not in adults. Goldsmith (in Plomin 1983, p. 350) has other reservations, based on the difficulty of distinguishing between family effects from the other sources of variation.

One reason for such a simple picture of personality may lie in the measures. Figure 7.11 showed that types such as extraversion are the compound of many questionnaire responses to specific situations. Table 7.3 comes from the analysis of 11 questionnaire items dealing

TABLE 7.3   Genetic and environmental components of variance for neuroticism and for its profile- and occasion-specific effects (from L. J. Eaves and P. A. Young (1981) *Dimensions of Personality: Papers in honour of H. J. Eysenck* Lynn R. (ed.), p. 129, Pergamon, Oxford)

| Source | Neuroticism | Specific effects | |
		Profile	Occasions
Genetic ($\frac{1}{2}D = V_A$)	11.4	14.0	0
Environmental (E_1)	8.6	15.5	3.4
Total variance	20.0	29.5	3.4
'Heritability'	0.57	0.47	0

with neuroticism measured in the same twins on two occasions, 2 years apart. While the differences between what was measured on the two occasions were due solely to environmental effects, there was almost as much genetic variance for the individual differences across the profile of responses to the 11 items after the neuroticism dimension had been taken account as there was for neuroticism itself. Eaves and Young (1981) conclude that we should not just think of genes operating at the all-inclusive type or trait level, but should also consider their action in specific item responses. Note that this does *not* mean there is a separate locus for each questionnaire item (an idea that was discussed in early behaviour genetics) but rather that each item taps a somewhat different facet of personality with a slightly different combination of genetic and environmental factors.

Analyses at an intermediate level of the factors such as 'depression', 'worry', 'insomnia' and 'paranoia' which derive from individual responses and go to make up the neuroticism dimension also suggest a more complex picture of personality (Eaves and Young 1981). For some factors such as 'depression' and 'paranoia' *sex limitation* is important where genes are expressed to different extents in the two sexes — in practical terms this means the DZ correlation from opposite-sex twins is much less than that for same-sex pairs. For other factors such as 'shyness' and 'psychosomatic', competition improves the fit of the genetic and environmental model.

Such results suggest an important distinction between types and traits. The global nature of types such as extraversion and neuroticism may mean that they are inherited to similar extents, because they are each the average of traits differing in their genetic determinants. This is not simply a feature of the Eysenck Personality Questionnaire. Loehlin found the same with the factors comprising extraversion and neuroticism in the California Psychological Inventory (CPI) given to the NMSQT twins, as did Carey and Rice in their analysis of the Differential Personality Questionnaire (DPQ). While they found extraversion to show the typical simple additive genetic determinants, this was not the case for the three traits which together make up extraversion:

Assortative mating was important for one trait, dominance and/or special MZ environments were for two, and sex differences were for all three, although the nature of the sex differences varied among measures. Age effects were equivocal but may play some role in one trait . . . . (G. Carey and J. Rice (1983) *Behavior Genetics* **13**, 54.)

There may be further heterogeneity between traits, masked by the fact that the same questionnaire items contribute to different traits. When such an overlap is taken into account, differential heritability does appear on the CPI with measures of sociability and compulsivity being the most heritable.

Analysis at these more detailed levels suggests that the genetic analysis can say something about how personality is structured. Rather than blindly accepting the model of behaviour provided by personality theorists, genetic analysis can indicate where the model is inadequate. For example Eysenck's model of personality is a broad one predicting individual differences in many other aspects of behaviour. We can ask to what extent the genetic determinants of his main personality dimensions influence variance on other questionnaires dealing with behaviour such as impulsiveness (comprising dimensions of 'impulsiveness', 'risk-taking', 'non-planning' and 'liveliness') and sensation-seeking (comprising 'disinhibitions', 'thrill-seeking', 'experience-seeking' and 'boredom susceptibility') which Eysenck considers to be related to personality. It turns out that only about 40% of the genetic variance on impulsiveness and sensation-seeking can be explained by the genetic variation on the personality dimensions, leaving a lot of the individual differences unaccounted for. The genetic model is not the sequel to the behavioural model but rather suggests where the behavioural formulation needs improvement. Separate genetic analyses of impulsiveness and sensation-seeking indicate sex-limitation effects, accounting for up to 30% of the genetic variance. Such sex differences will have to be incorporated in subsequent attempts to link these variables to the personality dimensions.

One argument against the idea that personality develops without family influences is that this result is an artefact of the measurement. People are asked to report how they feel about particular situations and all that can emerge from such an imprecise measure is a picture of fairly random genetic and environmental effects. That most dimensions of personality from the different questionnaires produce the same genetic results would support the criticism that one is studying the genetics of filling-in questionnaires not of personality.

Fortunately other self-report questionnaires do not yield the same result. Eysenck and his colleagues developed questionnaires of social attitudes, asking people's views on such topics as birth control, defence spending and flogging, and identified two major dimensions of tender-minded versus tough-minded and radicalism versus conservatism. Unlike personality measures there is considerable assortative mating. This contribution to differences between families together with common family environment accounts for about one-third of the total variance, with genetic and within-family environment each contributing another third. Although Eysenck has specified some dependence of social attitudes on personality, social attitudes are clearly determined much more by cultural factors and not by chance events.

Personality and development

The development of adult personality depends to a considerable extent on experiences in childhood, most of which occur in the context of relationships with parents, siblings, peers and other important figures in the child's life. The nature of those relationships are influenced by, and in turn influence, the personalities of the individuals involved. To understand the bases of adult personality we must come to terms with this dialectic between individual characteristics and relationships which starts at birth and continues throughout life. (R. A. Hinde (1981) *Nature*, **293**, 607.)

Although the analyses we have just discussed do not necessarily support Hinde's view of the influences on adult personality, there are two related reasons for studying the genetics of personality in a developmental perspective.

TABLE 7.4   Median correlations for family resemblance on measures of introversion-extraversion and neuroticism-anxiety (derived from S. Scarr, P. L. Webber, R. A. Weinberg and M. A. Wittig (1981) *Journal of Personality and Social Psychology* **40**, 885)

	Twin		Sibling		Parent–child	
	MZ	0.52	Biological	0.20	Biological	0.15
	DZ	0.25	Adoptive	0.07	Adoptive	0.04
Heritability						
Formula	$2(r_{MZ}-r_{DZ})$		$2(r_{Biol.}-r_{adopt.})$		$2(r_{Biol.}-r_{adopt.})$	
Estimate	0.54		0.26		0.22	

Firstly while twin and adoption studies of adult personality yield similar results, there are discrepancies in adolescents. The DZ, parent–child and sibling correlations should all be comparable and Table 7.4 indicates that this is generally the case. But when Scarr *et al*, used these data to calculate conventional heritabilities with all the implicit assumptions, they found that from MZ and DZ twins to be much larger. They explain this difference partly as a result of the greater postnatal environmental similarity of twin pairs and also through genotype–environment correlation where individual genotypes evoke and select different environmental responses. With MZ twins having the same genotype, their correlation will be inflated relative to the others, exaggerating the heritability.

Secondly individual differences in personality may vary during development and may contribute to the twin correlations being higher simply because twins are tested at the same stage of personality development. Recall from Chapter 6 that Kamin suggested this age-matching as a factor in cognition, but only in personality does it seem to be of major importance. To take three instances reviewed by Eaves and Young (1981):

1 Although both the juvenile and adult forms of the EPQ show

similar genetic models, the correlations between the genetic variance in the two forms are only 0.44 for extraversion and 0.32 for psychoticism, compared with 0.84 for neuroticism. In other words only 10% (0.32 $\times$ 0.32 $\times$ 100) of the juvenile genetic variation in psychoticism results from those genes which contribute to variation in adult psychoticism.

2   The 'lie' scale in the EPQ is a means of checking the validity and consistency of scores on the three personality dimensions. Its determinants change with age-juvenile MZ and DZ twins are equally similar suggesting only a common environment as the determinant, while adult MZs are more alike than DZs indicating some genetic variance.

3   Genetic variance for neuroticism increases with age in that DZ intrapair differences increase while MZ differences remain constant. In studying parent–offspring relations, neuroticism but not the other personality dimensions requires a factor specifying the age difference. Similarly in the Minnesota Adoption Study the estimate of heritability from biological and adoptive siblings was higher than from comparable parent–child relations, indicating a possible age effect.

These points mean that studies of personality in children need to be interpreted much more cautiously than those of cognition. Studies generally have to be based on twins, since many of the personality measures in young children such as the occurrence of temper tantrums have no real parallels in adults, limiting most adoption studies. Thus, the Colorado Adoption Study (Plomin in Porter and Collins 1982) found no relation of temperament at 12 months to measures of parental temperament or home environment. But twin studies introduce problems in terms of age-matching and environmental similarity as well as in the confounding effects of their unique biological factors. Wilson (in Porter and Collins 1982) found associations between birthweight and irritability in twins up to 6 months and a tendency for the larger twin to be more distressed during physical measures. But by 12 months these effects had disappeared. Also there may be limited prediction to adult behaviour. While intrapair personality differences in adult MZ twins have been related back to childhood differences, these need not imply some long-term consistency. Retrospective reporting by adults of their childhood differences may be based more on their current differences rather than on any accurate recollection of the earlier status.

As shown in a review by Goldsmith (in Plomin 1983) genetic analyses of personality in children present a far more varied picture than do those of adults, reflecting the much wider variety of behaviours which are tapped both by parental questionnaire and by observational studies. The extent of variation is such that the most appropriate question to consider is 'In what situations and at what age are genetic factors involved?'

This variability is one reason why the term 'temperament' is often preferred for children over 'personality' with its connotations of long-term consistency. As indicated by Thomas and Chess, two people who have done much to develop our understanding of this concept, it is inappropriate to define temperament as others have done in terms of its genetics:

> In this regard, we question the position of Buss and Plomin that the crucial criterion for temperament is inheritance, that "this is what distinguishes temperament from other personality attributes". Temperament, like nutrition, intelligence, perceptual organization, and other structural attributes of behavior, is best defined operationally and not tied to a theory of origin for which there is only suggestive and presumptive evidence. (A. Thomas and S. Chess (1980) *The Dynamics of Psychological Development*, p. 75. Brunner/Mazel, New York.)

TABLE 7.5   The determinants of behaviour in 13–37 month old twins when a stranger is talking to their mother (derived from R. Plomin and D. C. Rowe (1979) *Developmental Psychology* **15**, 62).

| Determinants | | Correlations | |
		MZ	DZ
'Heritable'	Positive vocalization to stranger	0.58	0.34
	Looking at stranger	0.67	0.08
	Approaching stranger	0.50	−0.05
	Proximity to stranger	0.40	−0.03
	Touching mother	0.47	0.22
Between-family environment	Positive vocalization to mother	0.56	0.46
Within-family environment	Smiling at mother	0.19	0.19
	Looking at mother	−0.01	0.11
	Approaching mother	0.14	−0.03
	Proximity to mother	0.23	0.11
	Smiling at stranger	0.08	0.25
	Touching stranger	−0.07	−0.03

Table 7.5 justifies the views of Thomas and Chess. It summarizes the behaviour of twins in one specific situation — they were also observed playing with their mother, playing with the stranger, separated from their mother, etc. Plomin and Rowe divide the determinants of the various behaviours into three categories: 'heritable' (where the MZ correlation exceeds the DZ), between-family environment (where both the MZ and DZ correlations are substantial but similar) and within-family environment (where both are small and similar). Given that some of the DZ correlations in the 'heritable' category are negative, such a simple genetic model is unlikely to be adequate.

Nevertheless two trends emerge from these results. There is more genetic variation for 'shyness', the behaviour towards the stranger than towards the mother. Rather like the adult situation, environmental determinants are more within-family than between-family. Data from twins in the Louisville Twin Study observed from 9 through 30 months support the first conclusion. MZ twins are much more similar than DZ twins in a playroom situation where the mother is absent and only an attendant present than in a more structured setting where the mother remains. While this can be interpreted as showing that there is more genetic variance in behaviour in stressful situations, it is interesting to draw comparisons with the adult data mentioned earlier which suggested that sociability has a particularly high heritability.

The longitudinal nature of the Louisville data also permit analyses of the changes with age as described earlier for the cognitive data (Fig. 7.4). Both across ages and across situations changes of the MZ twins are more in concert and hence seem to be partially regulated by genetic influences. To some extent MZ twins become relatively more similar than DZs as they develop but Fig. 7.12 shows that this is not

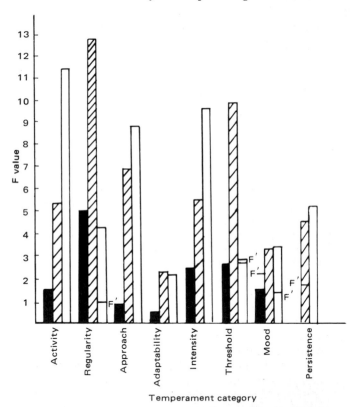

FIG. 7.12   DZ/MZ F-ratios for twin within-pair variances in temperament at 2 months (■), 9 months (◙) and 6.5 years (□). In some cases, a slightly different ratio F' is used to meet certain statistical assumptions. (From A. M. Torgersen in Gedda *et al* (1981), p. 261.)

inevitably the case. These data come from a study of twins at 2 months, 9 months and 6.5 years on Thomas and Chess's formulation of temperament as *how* a child reacts to various situations rather than *what* the child does or *why* the child does it. The data are F ratios, comparing the DZ within pair variance with the MZ one, so that if there is a smaller variance in MZ pairs, the F ratio will be greater than one. As noted in Chapter 6, this measure is a poor estimate of heritability. Activity, approach, intensity and persistence have higher F ratios at 6.5 years than at infancy. Apart from the possibilities of low reliability of the tests with young children and of genetic variance being masked by the special perinatal problems of twins, Torgersen offers an explanation of the developmental changes in terms of interaction. 'The more different the twins are in temperament at infancy, the more different will be their reactions to the environment, and consequently the more different will be their individual development.' (p. 267.)

To some extent she countered the criticism that MZ twins may have been more similar by reasons of their shared environment by finding intrapair differences to be no greater in those MZ twins who usually played apart than among those who were inseparable.

While these results show genetic variance for a range of temperaments, they do not imply that behavioural problems are partly genetic. The categories in Fig. 7.12 which have low or no significant F ratios at the age of 6, namely regularity of body function, adaptability, threshold of responsivity and mood, are those that, according to Thomas and Chess discriminate 'difficult' children from normals.

Discussion

The analyses of adult personality with their simple genetic and within-family environmental effects are distinctly different from the measures in children where almost every combination of genetic and within and between family environmental effects is possible. However, the methodological differences between the situations cannot be ignored. Not only does the observation of children in a variety of settings introduce many additional variables, but the time-consuming nature of the data-recording means that samples are much smaller and, as is obvious, the analyses much less sophisticated.

Plomin (in Gedda *et al* 1981) points out that with samples of 25 MZ and 25 DZ pairs and an MZ correlation of 0.5, the DZ correlation would have to be negative to be significantly different. Heritabilities can be calculated with smaller differences in the MZ and DZ correlations but are really meaningless, ignoring the fact that such differences are not significant and may reflect only random error. Just how much of the variation between the different categories of children's behaviour is attributable to such sampling error is unclear, but one situation where DZ correlations are frequently negative highlights an even more major difficulty for genetic analyses of children's temperament. Zero or negative DZ correlations characterize situations where parents are asked to rate their twins on what Plomin calls 'molar' features such as 'friendliness', as distinct from more specific or 'molecular' questions, e.g. 'does the child mouth objects?'

What this suggests is that parents form general impressions which exaggerate the perceived differences between their DZ twins. The less specific the question about the twins, the greater the scope for these impressions to bias the answer. In following parents of twins from before the birth, Dibble and Cohen (1980) have shown that 'subtle differences in early endowment or in parental perception of a child as vulnerable or fragile may lead to enduring patterns of interaction emphasizing children's difficulties or sensitivities' (p. 101). Such effects may differ with zygosity. For example Dibble and Cohen found that mothers of DZ twins showed more consistency in how they cared for their children if they had perceived the child to be in good health during the first month of life, while mothers of MZ twins were more consistent when the pregnancy had not been accompanied by vaginal bleeding or morning sickness.

Given that genetic analyses of early personality must often rely on twins, then more attention must be paid to parents' impressions of the twins (Dibble and Cohen 1980), to the effects twins may have on parenting practices (Lytton 1980) and to the effects being a twin may have on the children's personalities. Few data are available apart from those of Koch (1966) who identified temperament patterns characteristic of male and female MZ and same and opposite-sex DZ. Although her data can be criticized in many respects, they at least suggest specific influences on personality development unique to twins such as being together or separated in school:

At any rate, it is apparent in our data that boys, the more aggressive children, the older, and those possessed of better speech, tended to be found with greater frequency among the separated children, whereas girls, the closer, the more conforming, and the scholastically more able were found in greater numbers among the nonseparated. (p. 200.)

The analyses of adult personality show few indications of such unique twin effects. Either they disappear during development or the adult questionnaires tap different facets of personality. Whichever is the case, the fact that adult personality depends upon the 'slings and hazards of outrageous fortune' as Eaves and Young (in Gedda *et al* 1981) put it, says very little about how personality develops during childhood.

Psychopathology

The last few years have seen major developments in the genetic analysis of behavioural disorders. To some extent these are a reaction to the views predominating in the early 1970s which in different ways attacked the whole concept of psychiatric illness and its possible biological bases. For example, the behaviour modification approach rejected the concept of a 'medical model' of mental illness and any underlying biological bases — it was the symptom which had to be treated, not its cause. Laing's existentialist views of schizophrenia put the blame on the family, the diagnostician and the hospital staff for shaping any aberrant behaviour and again diverted attention from the biology.

We shall examine only three disorders, schizophrenia, manic-depressive psychosis and alcoholism, ignoring criminality (Hutchings and Mednick in Mednick *et al* 1974) and the dementias, the behavioural disorientation which can accompany old age but which can also be of earlier onset (Tsuang and Vandermey 1980). As regards other disorders, Tsuang and Vandermey aptly describe the present status:

...in the case of neuroses, personality disorders and socialopathy, the evidence for genetic influence at present is still incomplete, and the diagnostic classifications are so uncertain that any assertions that could be made would be far outweighted by hesitations and qualifications. (p. vi)

We also ignore most of the family and twin data which are adequately described in the many excellent texts on the genetics of psychopathology (Goodwin and Erickson 1979; Gottesman and Shields 1982; Mednick *et al* 1974; Mendlewicz and Shopsin 1979; Tsuang and Vandermey 1980). But one cannot neglect the famous Siamese twins of the last century, Chang and Eng Bunker where Chang drank heavily while Eng was abstemious. The reader can decide what implications this difference has for genetic and cultural determinants of alcoholism!

One must be especially careful with the interpretation of twin psychiatric data. Attention is drawn to Gottesman and Shields' discussion of possible problems with twin reseach in schizophrenia, most notably the sampling biases which may elevate and/or depress MZ concordance rates and the fact that twins, especially MZ, may be a high-risk group. While they suggest that early-onset childhood schizophrenia (sometimes called infantile autism) has no genetic basis and is quite distinct from preadolescent-onset schizophrenia, there are indications in both cases of a higher incidence in twins and an unusually high MZ concordance relative to DZ, e.g. in one study of preadolescent schizophrenia 71% for MZ and 17% for DZ. As in the discussion accompanying Table 6.5a, they hypothesize a higher MZ incidence of pregnancy and birth complications, resulting in brain damage or other abnormalities.

We concentrate on two techniques that distinguish the genetic analyses of psychiatric disorders. First is the extensive use of adoption data, mental illness being one reason why parents may offer their children for adoption. Second is the search for genetic associations or linkage. Are certain illnesses associated with a particular, definable genotype at the ABO or similarly polymorphic loci? Most attention has been paid to the main histocompatibility complex, the human leucocyte antigen (HLA) system, which has several closely linked loci each with multiple alleles inherited in a codominant fashion, so that heterozygotes can be distinguished from the homozygotes. The enormous range of possible variations (haplotypes) in this system which limits the range of successful tissue transplants (*histo*-means tissue) has led to *associations* being sought between HLA and many diseases. A more precise question than association is to determine through *linkage* studies to known markers whether there are genes contributing to the disease located on the same chromosome as the marker. Again the HLA system on chromosome 6 and various X chromosome markers such as colour-blindness are those most widely used. (The term 'marker' refers to any gene of clearcut effect whose location on a particular chromosome is known and useable in such linkage studies.)

Classification of psychiatric disorders

Table 7.6 illustrates our three examples. Schizophrenia is the most common psychosis with a population incidence of 1%. It is characterized not by the splitting of the mind into different but still functional personalities as is often thought, but by the divorce and fragmentation of different mental capacities within the one personality. The person cannot distinguish imagination from reality.

Manic-depressive psychoses are characterized by extreme swings of mood or 'affect' from normal to deep depression or to mania. The illness is classified as bipolar if the person swings to both extremes or as unipolar if the swings are from normal to either mania or depression but not both. The majority of unipolar psychotics have depression and unipolar manics are rare — some people argue they are bipolar psychotics who eventually will have periods of depression. While the incidence of unipolar and bipolar affective psychoses are both around 1% though somewhat higher in women, depression is much more common with up to 25% of the population reporting depression at some stage of their life (Tsuang and Vandermey 1980).

Table 7.6 defines alcoholism in a way that distinguishes it from heavy drinking, in that alcoholism is accompanied by a loss of control over drinking and by drinking to an extent that life is disrupted. This distinction is important for genetic analysis in that studies of alcohol usage and of alcohol metabolism in the general population may not involve the same factors that characterize alcoholism. The incidence of alcoholism is 5% in males and around 1% in females, although the incidence in females has been rising. It is unclear whether this is due to more women in the workforce who can afford to buy drink or to a greater willingness now to classify women as alcoholic rather than calling their problem by another name, such as depression.

Although the hazards of reliable diagnosis were mentioned in Chapter 1, two more problems of classification arise. One is whether each disorder is genetically heterogeneous e.g. are bipolar and unipolar psychoses due to the same genetic and environmental factors? Secondly, the disorders are not entirely distinct. The incidence of alcoholism in the fathers of manic-depressives is elevated although it is unclear whether this implies a genetic overlap or whether resorting to alcohol is one way of coping with depression.

TABLE 7.6   Examples of mental illness

(a) *Schizophrenia*
She was always afraid, insecure, introvert and lonely. For the most part she

lived in her fantasies when growing up but was successful at school. From the age of four felt compelled to count up to certain numbers and wash her hands repeatedly in order to avoid unpleasant occurrences. Later she became afraid of closed rooms, walking in open places, going by lift. She was also afraid of getting cancer. During her teens she was frightened of injuring others with sharp objects and when tired felt an inclination to steal things. By and by she came to believe that she had stolen things and that everyone was talking about it. She also worried about getting pregnant by men passing in the street and in the end did not dare to go out. From the age of 20 incapable of taking an interest in things or concentrating, she wondered if she would influence other people. She seemed indifferent to her complaints despite their being the only thing she would talk about. All her thoughts and actions had to be written down. Meaningless words formed themselves in her head and sometimes it was completely empty. Aggressive impulses sometimes gave a feeling of anxiety, sometimes satisfaction. She made accusations against her relatives and exercised terror at home. Gradually became more indolent, heard voices and felt radiation. She was looked after sometimes at home and sometimes in hospital, a total invalid. (From A. K. Nyman (1978) *Acta Psychiatrica Scandinavia* (Supplement) **272**, p. 109)

(b) *Depression*
I was seized with an unspeakable physical weariness. There was a tired feeling in the muscles unlike anything I had ever experienced. A peculiar sensation appeared to travel up my spine to my brain. I had an indescribable nervous feeling. My nerves seemed like live wires charged with electricity. My nights were sleepless. I lay with dry, staring eyes gazing into space. I had a fear that some terrible calamity was about to happen. I grew afraid to be left alone. The most trivial duty became a formidable task. Finally mental and physical exercises became impossible; the tired muscles refused to respond, my "thinking apparatus" refused to work, ambition was gone. My general feeling might be summed up in the familiar saying "What's the use." I had tried so hard to make something of myself but the struggle seemed useless. Life seemed utterly futile. From E. C. Reid (1910) *Journal of Nervous and Mental Disease* **37**, 612.

(c) *Mania*
Therapist   Well, you seem pretty happy today.
Client: Happy! Happy! You certainly are a master of understatement, you rogue! [*Shouting, literally jumping out of seat.*] Why I'm ecstatic. I'm leaving for the West Coast today, on my daughter's bicycle. Only 3100 miles. That's nothing, you know. I could probably walk, but I want to get there by next week. And along the way I plan to follow up on my inventions of the past month, you know, stopping at the big plants along the way having lunch with the executives, maybe getting to know them a bit — you know, Doc, 'know' in the biblical sense [*leering at therapist seductively*]. Oh, God, how good it feels. It's almost like a non-stop orgasm. (From G. C. Davison and J. N. Neal (1978) *Abnormal Psychology: An experimental clinical approach* Wiley, New York (2nd edn), p. 192 © 1978 Reprinted by permission of John Willey & Sons, Inc. New York)

(d) *Definition of alcoholism*
... excessive drinkers whose dependence on alcohol has attained such a degree that they show noticeable mental disturbance or an interference with their mental and bodily health, their interpersonal relations and their smooth social and economic functioning; or who show the prodromal [beginning] signs of such developments (Reprinted in N. Kessel and A. Walton (1965) *Alcoholism*, Penguin, Harmondsworth, p. 18, from WHO Technical Report Series, No. 48)

Schizophrenia and manic-depressive psychoses overlap little but have an intermediate category of schizoaffective disorders characterized by mood disorders plus schizophrenic symptoms. Schizoaffective disorders are not simply part of a continuum between the two major disorders but are themselves heterogeneous in that only some schizoaffectives respond well to lithium, a drug used with manics. The concordance rate for suicide in schizophrenic MZ twins is 3% compared with 33% in schizoaffective MZ twins, suggesting that the overlap between these disorders in small. Figure 7.13 shows that

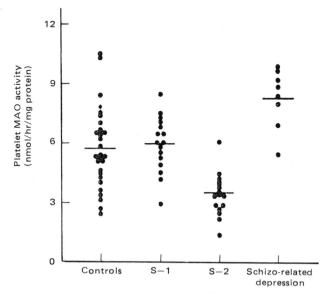

Fig. 7.13   Levels of monoamine oxidase (MAO) activity in normal individuals (control), in schizophrenics who lacked auditory hallucinations and delusions (S-1) or who had such symptoms (S-2) and in depressed patients with some schizophrenic symptoms (schizo-related depression). Points are individual values and the bars are the group means. (From J. J. Schildkraut, P. J. Orsulak, A. F. Schatzberg and J. M. Herzog (1980) *Schizophrenia Bulletin* **6**, 220.)

one group of schizoaffectives can also be distinguished biochemically by the levels of monoamine oxidase (MAO) activity. MAO is an enzyme involved in the degradation of neurotransmitters and has been implicated in many disorders including schizophrenia, manic-depressive psychoses, alcoholism and even Down's Syndrome. The evidence is complex and not consistent within any disorder as seen in Fig. 7.13 by the different levels in two schizophrenic subgroups.

One way to distinguish the disorders is on the basis of family studies. Table 7.7 suggests some genetic overlap between unipolar psychosis and schizophrenia, but shows that schizoaffective disorders are a heterogeneous group. The incidence of schizo-affective disorders in the relatives was so low that it is not included in Table 7.7, but the relatives of the schizoaffectives do have a high incidence of the other disorders — as Mendlewicz puts it 'some schizoaffective syndromes share some common genes with affective

TABLE 7.7   Risk of psychiatric disorders in first-degree relatives of probands with different disorders (derived from J. Mendlewicz (1979) in *Genetic Aspects of Affective Illness* Mendlewicz J. and Shopsin B. (eds), p. 92. Spectrum, New York)

Disorder in proband	Incidence in relatives (%)		
	Bipolar	Unipolar	Schizophrenia
Bipolar	18.7	20.6	1.8
Unipolar	2.1	27.2	3.2
Schizoaffective	13.1	22.4	10.8
Schizophrenia	1.4	7.4	16.9

disorders while others may share some with schizophrenia.' (p. 97). Even this may be an oversimplification. Bipolar affective disorders can be divided into Bipolar I which has some overlap with and a family history of schizoaffective disorders, as well as definite severe periods of mania, while Bipolar II has links with major depressive disorders and is associated with hypomania (less severe mania, not requiring hospitalization). Although these groups differ in such things as MAO level, the incidence of suicide and the age of onset, Gershon (in Kety *et al* 1983) argues that they are not genetically distinct because the various forms of manic-depressive psychoses are equally prevalent in their relatives.

Schizophrenia

Like all psychiatric disorders, schizophrenia is a threshold trait, one where behaviour must exhibit more than a particular degree of abnormality before the person is classified as 'ill'. What determines this threshold is partly societal and partly reflects the current diagnostic practices such as the UK–USA differences discussed in Chapter 1.

Gottesman and Shields (1982) present a more detailed way, the diathesis-stress theory, of viewing the various determinants of the distribution of liability, where schizophrenia forms one extreme of a continuum throughout the population (Fig. 7.14). As well as genes specific to schizophrenia, their model incorporates more general genetic and environmental assets and liabilities. The most important feature of this complex of determinants is that people with the same environmental circumstances will not necessarily both exhibit schizophrenia, nor will people with identical genotypes, i.e. MZ twins, if their environmental loads differ.

This model also predicts variation in behaviour in the non-schizophrenic part of the population which shades into schizophrenia at one extreme. Through the American studies of Heston and the Danish studies of Kety and Rosenthal on the adopted children of schizophrenic mothers has developed the concept of the schizophrenia-spectrum disorders, a higher incidence of mild mental retardation, of neuroses and of asocial behaviour in those relatives of

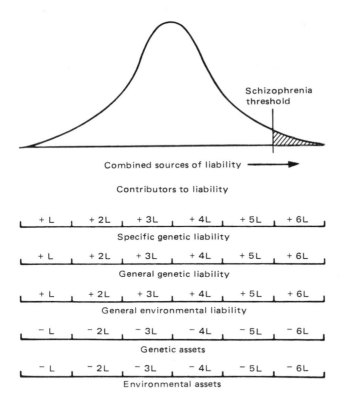

Contributors to liability

| + L | + 2L | + 3L | + 4L | + 5L | + 6L |

Specific genetic liability

| + L | + 2L | + 3L | + 4L | + 5L | + 6L |

General genetic liability

| + L | + 2L | + 3L | + 4L | + 5L | + 6L |

General environmental liability

| − L | − 2L | − 3L | − 4L | − 5L | − 6L |

Genetic assets

| − L | − 2L | − 3L | − 4L | − 5L | − 6L |

Environmental assets

FIG. 7.14   A schematic representation of possible genetic and environmental determinants of the liability for developing schizophrenia. (From I. I. Gottesman and J. Shields (1982) *Schizophrenia: The epigenetic puzzle.* Cambridge University Press, New York.)

schizophrenics who are not actually schizophrenic themselves. In Kety and Rothenthal's study 31.6% of the children of the schizophrenics were in the schizophrenia spectrum, compared with 17.8% in control adopted children born to non-schizophrenic mothers. While the latter figure is much lower, it is still high enough to suggest some pathology in their biological or adoptive families.

Rosenthal (in Mednick *et al* 1974) examines the different adoption designs possible with psychiatric disorders (Fig. 7.15).

The first is the *adoptees's family study* where the interest is in the proportions affected among biological and adoptive relatives of schizophrenics and controls. Rosenthal found a higher incidence (8.6%) of schizophrenia spectrum disorders in biological relatives of the schizophrenic probands, but uniform frequencies (3% approx.) in the other three cells of Fig. 7.15a. The second design is the *adoptee's study design* (Fig. 7.15b) where one looks at the children given for adoption by schizophrenic and normal biological parents. As already mentioned, 31.6% of the children of the schizophrenics had schizophrenia-spectrum disorders even though at the time only 3 out of 76 had actually been diagnosed as schizophrenic. Rosenthal

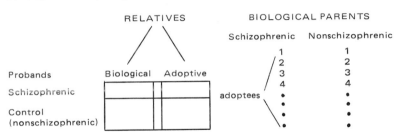

(a) Adoptee's family design

(b) Adoptee's study design

RELATIVES

Probands

Schizophrenic

Control
(nonschizophrenic)

Biological   Adoptive

BIOLOGICAL PARENTS

Schizophrenic   Nonschizophrenic

adoptees

1   1
2   2
3   3
4   4
•   •
•   •
•   •

CROSS-FOSTERING DESIGNS

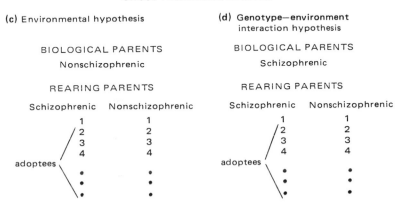

(c) Environmental hypothesis

BIOLOGICAL PARENTS
Nonschizophrenic

REARING PARENTS

Schizophrenic   Nonschizophrenic

adoptees

1   1
2   2
3   3
4   4
•   •
•   •
•   •

(d) Genotype—environment
interaction hypothesis

BIOLOGICAL PARENTS
Schizophrenic

REARING PARENTS

Schizophrenic   Nonschizophrenic

adoptees

1   1
2   2
3   3
4   4
•   •
•   •
•   •

FIG. 7.15   Different adoption designs in the study of psychiatric disorders.
(Derived from D. Rosenthal (1974) In *Genetics, Environment and
Psychopathology* Mednick S. A. *et al* (eds), p. 19. North Holland/Elsevier,
Amsterdam.)

considers that the actual process of adoption may limit the expression
of schizophrenia.

A more elaborate design (Fig. 7.15c) akin to the cross-fostering
studies of rodents discussed in Chapter 5 is one which provides two
types of rearing (schizophrenic or non-schizophrenic adoptive
parent) and two corresponding genetic variables (schizophrenic and
non-schizophrenic biological parent). Of course the chances of a child
having a schizophrenic biological parent and then being adopted by
another schizophrenic is small. Excluding this group, the highest
incidence of pathological behaviour (18.8%) was in children of
schizophrenics adopted by normals. Normal children adopted by
normal or by schizophrenic parents both had about a 10% incidence of
pathology, indicating that the rearing environment provided by
schizophrenics is of little effect on normal children.

Carrying out a biometrical genetic analysis on the various twin and
adoption studies of schizophrenia, Fulker (in Royce and Mos 1979)
estimates the additive genetic variance at 73%, leaving 27% which is
environmental. Essentially none of this is between-family environmental variance, suggesting that the search for environmental

factors in schizophrenia must be within the family not in terms of rearing practices commmon to all family members. Even this estimate of 73% may be too low, since the analysis included Tienari's study which was noticeable in Fig. 1.2 for the very low concordance rate. Although striking mainly early in adulthood, schizophrenia does have a variable age of onset and as Tienari's sample have grown older, the MZ concordance rate has gone up from 0% to 35%. A more elaborate path analysis model summarized by Henderson (1982) gave a similar estimate of genetic variance (71%) but divided the rest into a cultural effect of 20% and a specific twin environment effect of 9%. This analysis used only twin and family data rather than adoption data, so a different pattern of family influences could have been anticipated.

Based on higher concordance rates for twins than for siblings especially where the twins are living together at the time schizophrenia is first diagnosed, it has recently been proposed that a viral infection is involved. While this is compatible with the twin and family data in Henderson (1982), the adoption data indicate that such infection can at best be only a minor contributor to the prevalence of this disease.

Gottesman and Shields (1982) provide much fuller details of schizophrenia and its genetic analysis, including the elegant half–sib study by Kety. Two other facets of schizophrenia research merit special mention:

1 *Genetic heterogeneity* Not all schizophrenics have the same symptoms — does this reflect different genes, different environments or different points on one distribution of liability? Or as Gottesman and Shields put it 'how far are we dealing with clinical diversity in genetic unity, and how far are we dealing with genetic diversity in clinical unity' (p. 218). Tsuang and Vandermey (1980) present a basic classification of schizophrenia in terms of 'atypical', also called 'acute' (good prognosis, schizoaffective features, and obvious environmental precipitating stress) or 'typical' (chronic, deteriorating). Within 'typical' schizophrenia, patients may be paranoid or non-paranoid, the latter including catatonic (stuporous and often immobile), hebephrenic (disturbed thinking and childlike mannerisms) and simple (indifferent and apathetic).

Within 'typical' schizophrenic there is some specificity in that 80–90% of concordant MZ twins have the same subtype. The family studies (Table 7.8a) also suggest genetic distinctions between subtypes, although these results could be interpreted as differences in severity along a continuum rather than specificity. That is, paranoid schizophrenia with a late onset and generally better prognosis is a less severe form than hebephrenia. In contrast Table 7.8b shows that chronic schizophrenics have more relatives with 'schizotypal personality disorders' (SPD) than do other schizophrenics even though the incidence of borderline schizophrenia is similar in both cases. SPD is a specific subset of symptoms in the schizophrenia spectrum disorders, suggesting that relatives of chronic schizophrenics differ qualitatively as well as quantitatively from those of other schizophrenics. The situation is still unresolved and

TABLE 7.8   (a) Risk of schizophrenia (%) in studies of relatives of probands with two sub-types of schizophrenia. Derived from M. T. Tsuang (1975) *Biological Psychiatry* **10**, 465.

(a)

Type of schizophrenia in proband	Relationship			
	Sibs		Children	First-degree relatives
	Schulz (1932)	Hallgren and Sjoren (1959)	Kallman (1938)	Winokur *et al* (1974)
Paranoia	5.4	4.7	8.5	0.8
Hebephrenia	18.0	7.0	17.3	2.8

(b) Percentage incidence of schizotypal personality disorder and borderline schizophrenia in the biological relatives of schizophrenic adoptees. Derived from K. S. Kendler, A. M. Gruenberg and J. S. Strauss (1981) *Archives of General Psychiatry* **38**, 982.

Disorder in relatives	Type of schizophrenia in proband		
	Chronic	Borderline	Acute
Schizotypal personality disorder	15.2	0	5.0
Borderline and uncertain schizophrenia	15.5	19.0	14.3

must await genetic analyses using more detailed diagnostic and biochemical measures.

2 *In half the MZ pairs where one has schizophrenia the other does not* Apart from demonstrating the role of the environment, this provides a good opportunity to see what biochemical, environmental or other differences there may be between the co-twins. In studies where low MAO levels have been associated with schizophrenia (but remember Fig. 7.13), the schizophrenic twin has been found to have even lower levels than the normal co-twin. There is some indication of unusual patterns of brain lateralization in schizophrenics which may explain part of their language impairments and Boklage *et al* (1979) claim that those MZ twin discordant for schizophrenia are also often discordant for hand preference.

Both these examples suggest the role of discordant MZ twins in understanding the biological bases of schizophrenia. As regards environment, comparisons of discordant MZ twins have largely ruled out a common theory that birth stress is a predisposing factor, because birth stress is as common among the non-schizophrenic co-twins. On the other hand in 84% of the cases in one study, the schizophrenic twin was more submissive as a child. While this may suggest the role of parental attitudes in shaping the child's behaviour towards schizophrenia, cause and effect cannot be separated. The child's behaviour may equally well be interpreted as an early indication of schizophrenia outside parental control.

TABLE 7.9   Number of parents of bipolar psychotics showing particular disorders (derived from J. Mendlewicz and J. D. Ranier (1977) *Nature* **268**, 327. Reprinted with permission. © 1977 Macmillan Journals Ltd).

Disorder parents	Adopted probands (N=29)		Non-adopted probands (N=31)	Normal adoptees (N=22)	
	inAdoptive parents	Biological parents	Biological parents	Adoptive parents	Biological parents
Bipolar	1	4	2	0	0
Unipolar	6	12	11	3	1
Schizo-affective	0	2	1	0	0
Alcoholism	2	3	2	0	3

Manic-depressive psychoses

In contrast to schizophrenia, heterogeneity within manic-depressive psychoses on the basis of the bipolar–unipolar distinction has long been recognized. More recent studies cloud this distinction. Table 7.9 gives data on an adoptee's family design where the parents of diagnosed bipolar manic-depressives are examined. The biological parents of the adopted bipolar psychotic are more likely to have an affective disorder but it need not be bipolar. On the other hand there is some apparent evidence for specificity in that MZ twins (Table 7.10) are most likely concordant for the same subtype of psychosis. Bipolar patients respond better than unipolar ones to the use of lithium to treat acute depression, again suggesting some specificity.

TABLE 7.10   Differential diagnosis of bipolar or unipolar psychoses in MZ twins (from C. Perris (1979) In *Genetic Aspects of Affective Illness* Mendlewicz J. and Shopsin B. (eds), p. 7. Spectrum, New York)

Disorders in the twin pair	%
Both unipolar depressive	23.9
Both bipolar or manic	25.4
1 unipolar — 1 bipolar or manic	10.1
Partially concordant	16.7
Not concordant	23.9

In discussing the data in Table 7.9 Mendlewicz and Rainer propose that the bipolar type reflects a more severe genetic and/or environmental load, whereas the unipolar disorder appears when the load is more moderate. In their adoption study, bipolar psychosis had an early onset in all parents. Unipolar psychosis occurred in the biological parents usually *before* the onset in their children but in the adoptive parents usually *after* the onset in the children, suggesting it was a less severe reaction perhaps to the children's illness. On this basis, the concordance for subtype in the MZ twins may reflect their similar genetic and environmental loads along a continuum rather than their having genes for a specific subtype.

This example leads us to the most prolific area of genetic research on manic-depressive psychoses, the search for associations or linkage. Only rarely have both fathers and sons both had bipolar psychoses, suggesting that a locus on the X chromosome is involved. The high incidence in females particularly among those related to manic-depressive probands suggests the allele for psychosis would have to be X-linked dominant. More detailed studies have confirmed a linkage to red-green colour blindness and to the enzyme Xg, whose loci are on the long and short arms of the X chromosomes respectively. However Gershon has pointed out that such a linkage to loci on opposite sides of the centromere is biologically unlikely. While the situation is unresolved it is a potentially important one for genetic counsellors when a father has bipolar psychosis and the son is at risk. Recent efforts to make more specific distinctions among bipolar psychotics suggest that some but not all bipolar psychoses may be linked to Xg.

Just as much searching has gone on for linkage to the HLA loci on chromosome 6. The most controversial evidence is based on differences in the distributions of haplotypes between families with only one or two children affected compared with those where most or all are affected. The argument is that the latter group probably involves parents homozygous for the 'depression susceptibility genes', while at least one parent is heterozygous and carries the normal allele in families when few children are affected. The one conclusion from this study and from attempts to demonstrate linkage to the ABO locus on chromosome 9 is the variability between surveys. Different populations, different sampling procedures and different diagnostic criteria combine to present a very confusing and often contradictory picture (Gershon in Kety *et al* 1983).

An alternative approach has been to screen large numbers of electrophoretic variants (like those in Fig. 5.11) to see if there are

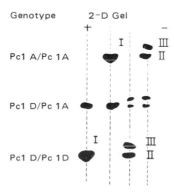

Fig. 7.16   A diagrammatic representation of the three genotypes involving the Pc 1A and Pc 1D alleles. Each homozygote has three polypeptides and the heterozygote has all six. (From D. E. Comings (1981) In *Genetic Research Strategies for Psychobiology and Psychiatry* Gershon E. S. *et al* (eds), p. 59. Boxwood, Pacific Grove, Ca.)

particular polymorphism–disease associations. Comings found an unusual variant Pc 1 Duarte (Pc 1D) of a brain specific protein (perchloric acid extract or Pc) where Pc 1A is the usual allele (Fig. 7.16). Pc 1D was present in 32% of normal brain tissues, 2.6% being homozygous. In brains from schizophrenics the frequency was the same, but in individuals committing suicide with manic-depression, severe depression or alcoholic depression, the frequency was 64% with 18% homozygous. While more work needs to be done with people with affective disorders who have died for reasons other than suicide to determine if Pc 1D is associated with manic-depressive psychosis in general or specifically with suicide in manic-depressives, it adds further support to the connection with single genes. How the Pc 1D allele may operate is unclear. It is also more common in multiple sclerosis and other diseases where viruses may play a part, suggesting that affective disorders are associated with increased susceptibility to viral infections of the central nervous system (Gershon in Kety *et al* 1983).

Alcoholism

With alcoholism being more prevalent than the psychoses, even more elaborate adoption studies are possible. Shuckit *et al* classified the half-sibs of alcoholic probands, according to alcoholism in the

TABLE 7.11   Incidence of alcoholism (%) in half-siblings of alcoholic probands according to their biological and rearing parents (from M. A. Shuckit, D. W. Goodwin and A. Winokur (1972) *American Journal of Psychiatry* **128**, 122)

		Half-sib Alcoholic (%)
Rearing parent	*Biological parent alcoholic*	
	Alcoholic	46
	Non-alcoholic	50
Rearing parent	*Biological parent non-alcoholic*	
	Alcoholic	14
	Non-alcoholic	8

biological and/or rearing parent. This constitutes the full cross-fostering design as described in Fig. 7.15c and d. Table 7.11 shows that it is the alcoholic biological parent who counts rather than the one who rears the child.

The most complete adoption studies have been carried out in Denmark and Sweden. The Danish work summarized in Tsuang and Vandermey (1980) confirmed the result of the half-sib study and in fact found slightly fewer problems among the unadopted sons of alcoholics than among those adopted away.

The Swedish studies have approached alcoholism as a more heterogeneous disorder. First they identified four groups among adopted males in terms of the degree of parental alcohol abuse (none, mild, moderate and severe), incorporating also information on the parents' occupational status and criminal records. For example, the biological fathers of the moderate abusers were distinguished by low SES and frequent criminal convictions in addition to the alcohol abuse. Characterizing genetic and postnatal environmental predisposition in this way for the three groups with alcohol abuse gave the results in Fig. 7.17. What stands out is the high level of moderate abuse irrespective of the environment amongst those with the suitable predisposing genotype. In contrast, mild and severe abusers showed both genetic and environmental effects which were generally cumulative.

On these bases two types of alcoholism can be distinguished. First is the 'milieu-limited' with mild or severe abuse where there are environmental effects and where both parents may be affected. Second is the 'male-limited' with moderate abuse, where

Is predisposition to MILD subtype?		Male adoptees observed
Genetic	Environmental	Percentage with mild abuse
No	No	6.1%
No	Yes	7.1%
Yes	No	10.4%
Yes	Yes	26.7%

Is predisposition to MODERATE subtype?		Male adoptees observed
Genetic	Environmental	Percentage with moderate abuse
No	No	1.9%
No	Yes	4.1%
Yes	No	16.9%
Yes	Yes	17.9%

Is predisposition to SEVERE subtype?		Male adoptees observed
Genetic	Environmental	Percentage with severe abuse
No	No	4.5%
No	Yes	6.7%
Yes	No	7.1%
Yes	Yes	10.3%

FIG. 7.17   The effects of cross-fostering on the incidence of alcoholism in three groups of male adoptees (from C. R. Cloninger, M. Bohman and S. Sigvardsson (1981) *Archives of General Psychiatry* **38**, 861.)

environment plays little role and where alcoholism is characteristic of the father, never of the mother. Comparable analyses of adopted women confirm this heterogeneity of alcoholism — moderate alcoholic fathers had no excess of alcoholic daughters whereas there was such an excess when the father was a mild abuser.

These results indicate the importance of distinguishing within alcoholism. The Danish studies found that it was only among the heavy alcohol abusers with many alcohol-related problems such as criminal convictions where the adopted-away children of alcoholics differed from controls. Among moderate or heavy drinkers without any problems, the groups did not differ. Similarly Finnish twin data showed the MZ concordance increased relative to the DZ concordance with severity of alcohol abuse. The heritability of heavy consumption of alcohol and loss of control over drinking was 66%, very close to the figure of 64% found for recurrent alcohol abuse among Swedish twins (Cloninger and Reich in Kety *et al* 1983). Cloninger and Reich are cautious about overemphasising the similarity of these heritabilities, partly because of the differing roles of genotype and environment in different types of alcoholism (Fig. 7.17) and partly because of national differences in drinking styles, e.g. the Finnish ration-like system of alcohol distribution encourages occasional binges rather than steady drinking.

The distinction between types of alcoholism also explains conflicting results over attempts to demonstrate X-linkage (Goodwin and Erickson 1979). There are sex-specific patterns of transmission but these depend on the type of alcoholism. Searching for associations with alcoholism among alcoholics as a whole will not (and has not) produced clear evidence except for some quite specific cases such as HLA and cirrhosis (Cloninger and Reich in Kety *et al* 1983).

Discussion

We have concentrated on the adoption analyses of psychiatric disorders and their potential for specifying the influence of familial environment. There are many questions still unanswered over the heterogeneity within disorders especially for schizophrenia.

Compared with other human behaviours there has been far more of an emphasis in psychiatric disorders on the search for single genes. This is perhaps a consequence of the 'medical model', since marker genes are of growing importance in diagnosing physical illness. The question arises of whether adoption, family or twin studies are of much help in understanding psychiatric disorders compared with such specific biological approaches. Gershon (in Gershon *et al* 1981) cites the case of juvenile diabetes, where two subtypes turn out through association with particular HLA haplotypes to be inherited at a single locus, making redundant the various models of polygenic inheritance which have been proposed.

Kety (in Gershon *et al* 1981, p. 397) emphasizes the importance of finding such markers:

However, I doubt that any strategy would do more for psychiatric genetics than the discovery of one or more genetic markers, biological or psychological in nature and susceptible to objective evaluation. It was such discoveries in the field of mental retardation which made possible the progress that has occurred in the delineation of homogeneous subtypes, the elucidation of their genetic transmission and their metabolic concomitants. It may not be too sanguine to hope that similar findings with comparable results may occur in the case of the major psychoses.

But he does stress that these have to be tangible markers. Much effort has been put into trying to explain whether the distribution of particular disorders in families is statistically more consistent with one or two locus models as distinct from polygenic inheritance, e.g., Heston used the schizophrenia spectrum concept to explain the distribution of relatives of schizophrenics on an autosomal dominant single gene hypothesis. Detecting a single gene at such a theoretical level is a long way from it being a biological entity useful in diagnosis.

While the revolution in molecular biology and the developments from recombinant DNA techniques (see Gershon *et al* 1981) have created many possibilities for analysis of disorders at a molecular level most applications to date have been with specific single gene disorders such as Huntington's Disease (Housman and Gusella, in Kety *et al* 1983). Few applications to complex disorders have so far been realized except for the brain enzyme polymorphism discovered by Comings (Fig. 7.16). Pc 1D is perhaps the clearest illustration of how molecular techniques could be useful. While the enzyme activity and hence the genotype can only be determined from post mortem brain tissue, a DNA probe for the particular genetic sequence (Chapter 2) would allow the genotype to be determined from any somatic cells of a living individual.

In the meantime the quantitative approach is our main method of studying the environmental determinants of disorders and of determining whether heterogeneity of disorders exists at a genetic as well as a phenotypic level. Recently there has been more contact between the two approaches. Shuckit has identified unusual patterns of the production of acetaldehyde (involved in the biochemistry of alcohol metabolism) in the relatives of alcoholics. Dorus *et al* (1979) have demonstrated a higher concentration of lithium in the red blood cells of relatives of bipolar I psychotics, but *only* among those relatives who themselves had had episodes of affective illness (of 66 relatives, 16 had experienced major and 22 minor affective disorders). They propose that genetically determined abnormalities of the cell membrane leading to high intracellular concentrations of lithium may place certain individuals at risk for affective psychosis. These two results suggest that the integration of biochemical and quantitative methods may combine the advantages of both approaches.

Conclusion

One common theme emerges from these three very different examples of human behaviour genetics. There is far more involved in behaviour genetics than just estimating the 'heritability' of behaviour. The value of these analyses lies not so much in obtaining accurate information on genetics but more on (a) clarifying issues in the behaviour, such as the relation between the different cognitive skills which make up intelligence and between the different levels of personality measurement. Genetic analyses provide unique information on the heterogeneity within the psychiatric disorders which could not come from more refined biochemical or diagnostic procedures; (b) indicating where the search for environmental factors should be directed, such as the within-family effects which are specific to particular cognitive abilities and which predominate in personality measures. While family environment has a role in mental illness, it has to be considered in terms of the heterogeneity of the disorder and the sensitivity of particular genotypes.

Behavioural development adds another dimension to genetic analyses which is only starting to be explored (Plomin 1983). The necessary longitudinal data are formidable to collect but pose many questions for developmental psychology concerning, for example, the predictability from child to adult abilities and the relation (or lack of it) between measures of child and adult personality.

Behaviour genetics has a potential which will be wasted if it only involves blindly 'doing the genetics' of personality, reading disability, schizophrenia or any other character of a complex, heterogeneous nature. Eaves and Young (1981) make this point even more forcefully.

It would seem important to get away from the endless reporting of heritability estimates based on twin studies of innumerable measures collected simply because they exist and to begin a more rational investigation which designs measures with a view to exploring a particular area of biological or social importance. The adoption of the 'heritabiity' criterion as a guide to what is worth studying is of little use since as we have suggested, almost everything is partly heritable. (p. 177).

Discussion topics

1 "While it may be of theoretical interest to know what the relative contribution of genetic and environmental variance might be at a given age for a particular point in time (and this is all the twin studies really tell us) it is of much more concern to what extent developmental change can be accounted for as a function of environmental and preprogrammed maturational factors. Policy consequences drawn from behavior genetic studies, however, would have a much sounder base if they would rely upon data relevant to the issue of developmental change." From K. W. Schaie (1975) in *Developmental Human Behavior Genetics: Nature–nurture redefined* Schaie K. W., Anderson V. E., McClearn G. E. and Money J. (eds),

D. C. Heath and Co., Lexington, Mass. p. 216.

What sort of behaviour genetic study would provide the necessary data; would such a study be feasible and would the results be definitive or open to alternative explanations and what policies could follow?

2 "From an evolutionary point of view, optimal personality development appears to be away from extremes and towards intermediates and balance. One implication for the personality theorist is that explanations of personality development should be based on individual life experiences interacting with genetic make-up rather than on general experiences in the home." (D. W. Fulker (1981) *British Medical Bulletin* **37**, 119.)

Do you feel such general conclusions are justified? What implications do they have for (a) sociobiology (e.g. E. O. Wilson (1978) *On Human Nature*. Harvard University Press) (b) those approaches to personality which already emphasize person × situation interactions (e.g. D. Magnusson and N. S. Endler (eds) (1977) *Personality at the Crossroads: Current issues in interactional psychology*. Erlbaum, Hillsdale, N.J.).

A more complete summary of Fulker's conclusions is in H. J. Eysenck (ed.) (1981) *A Model for Personality*. Springer-Verlag, Berlin. See also Eaves and Young (1981) p. 175 for a similar statement.

3 At the present stage of our knowledge of the diagnostic criteria and of genetics, is genetic counselling for psychiatric disorders legitimate and what are its limitations? (See Gottesman and Shields (1982), Chapter 10 and Tsuang and Vandermey (1980), Chapter 3 for discussions based on practical experience.)

References

Annotated bibliography

Cattell R. B. (1982) *The Inheritance of Personality and Ability*. Academic Press, New York. (A description of Cattell's views on ability and personality, how they develop and how they can be analysed. This book provides the most comprehensive account of the MAVA method but see the review by H. J. Eysenck (1982) *New Scientist* **96**, 374, where he argues it is more valuable just now to define our measures more precisely than to collect the vast data needed for the MAVA method.)

Eaves L. J. and Young P. A. (1981) In *Dimensions of Personality: Papers in honour of H. J. Eysenck* Lynn R. (ed.), p. 129. Pergamon, Oxford. (A detailed account of how genetic analyses of personality can contribute to personality theory — some prior knowledge of Eysenck's theory is helpful.)

Eysenck H. J. (1979) *The Structure and Measurement of Intelligence*. Springer-Verlag, Berlin. (As well as Chapters 5, 6 and 7 — jointly authored with Fulker — dealing with ability and SES, Chapters 8 and 9 are a good introduction to how cognitive abilities are more than just an IQ score.)

Gedda L., Parisi P. and Nance W. E. (eds) (1981) *Twin Research 3:* Part B. *Intelligence, Personality, and Development. Progress in Clinical and Biological Research* **69**. (Many papers dealing with the topics of this chapter — not only in twins — except for psychopathology which is in the companion volume Part C.)

Gershon E. S., Matthysse S., Breakefield X. O. and Ciaranello R. O. (eds) (1981) *Genetic Research Strategies in Psychobiology and Psychiatry.* Boxwood, Pacific Grove, Ca. (A very different perspective from this chapter, describing the newer molecular techniques for mammalian genetics and their potential for behaviour genetics.)

Gottesman I. I. and Shields J. (1982) *Schizophrenia: The epigenetic puzzle.* Cambridge University Press, New York. (A complete course in behaviour genetics from the perspective of schizophrenia. The reprise, p. 242, summarizes the major lines of evidence on the inheritance of schizophrenia.)

Henderson N. D. (1982) Human behavior genetics. *Annual Review of Psychology* **33**, 403. (A recent review touching on many aspects not fully covered in this chapter — the discussion of affective disorders is a good example of the conflicting results regarding the mode of inheritance of psychiatric disorders.)

Herschel M. (1978) Dyslexia revisited — a review. *Human Genetics* **40**, 115. (His detailed examination of the classification of dyslexia contrasts with his review of genetic analyses where all dyslexias are lumped together, irrespective of type.)

Mednick S. A., Schulsinger F., Higgins J. and Bell B. (eds) (1974) *Genetics, Environment and Psychopathology.* North-Holland/Elsevier, Amsterdam. (The whole book deals with one adoption study and is a good illustration of the problems and advantages of such an approach to genetic analysis. Although basically concerned with schizophrenia, it touches on many other psychopathologies.)

Plomin R. (ed.) (1983) Special section on developmental behavior genetics. *Child Development* **54**, 253. (A comprehensive coverage of the main ongoing adoption and twin studies with children as well as major reviews of personality by Goldsmith and of learning disabilities by Pennington and Smith.)

Royce J. R. and Mos L. P. (eds) (1979) *Theoretical Advances in Behavior Genetics.* Sijthoff and Noordhof, Alphen aan den Rijn. (The chapters by Eysenck, Fulker, McClearn and Royce all approach the question of learning more about behaviour through genetic analysis.)

Additional references

Boklage C. E., Elston R. C. and Potter R. H. (1979) Cellular origins of functional asymmetries: evidence from schizophrenia, handedness, fetal membranes and teeth in twins. In *Hemisphere Asymmetries of Function and Psychopathology* Gruzelier J. H. and Flor Henry P. (eds), p. 79. North Holland/Elsevier, Amsterdam.

Coltheart M., Hull E. and Slater D. (1975) Sex differences in imagery and reading. *Nature* **253**, 438.

DeFries J. C., Vandenberg S. G. and McClearn G. E. (1976) Genetics of specific cognitive abilities. *Annual Review of Genetics* **10**, 179.

Dibble E. O. and Cohen D. J. (1980) The interplay of biological endowment, early experience and psychosocial influence during the first year of life:

an epidemiological twin study. In *The Child in His Family* Anthony E. J. and Chiland C. (eds), vol. 6, p. 85.

Dorus E., Pandey G. N., Shaughnessy R., Gaviria M., Val E., Ericksen S. and Davis J. M. (1979) Lithium transport across red cell membrane: a cell membrane abnormality in manic-depressive illness. *Science* **205**, 932.

Eysenck H. J. (ed.) (1982) *A Model for Intelligence.* Springer-Verlag, Berlin.

Finucci J. M., Guthrie J. T., Childs A. L., Abbey H. and Childs B. (1976) The genetics of specific reading disability. *Annual Review of Human Genetics* **40**, 1.

Goodwin D. and Erickson C. (eds) (1979) *Alcoholism and the Affective Disorders.* Spectrum, New York.

Kaplan A. R. (ed.) (1976) *Human Behavior Genetics.* Thomas, Springfield, Ill.

Kety S. S., Rowland L. P., Sidman R. L. and Matthysse S.W. (eds) (1983) *Genetics of Neurological and Psychiatric Disorders.* Raven, New York.

Koch H. L. (1966) *Twins and Twin Relations.* University of Chicago Press.

Lewis M. (ed.) (1976) *Origins of Intelligence: Infancy and early childhood.* Plenum, New York.

Lytton H. (1980) *Parent–child Interaction: The socialization process observed in twin and singleton families.* Plenum, New York.

McGee M. G. (1979) *Human Spatial Abilities: Sources of sex differences.* Praeger, New York.

Mendlewicz J. and Shopsin B. (ed.) (1979) *Genetic Aspects of Affective Illness.* Spectrum, New York.

Mittler P. (1971) *The Study of Twins.* Penguin, Harmondsworth.

Porter R. and Collins G. M.(eds) (1982) *Temperamental Differences in Infants and Young Children, Ciba Foundation Symposium* **89**, Pitman, London.

Sternberg R. J. (ed.) (1982) *Handbook of Human Intelligence.* Cambridge University Press.

Tsuang M. T. and Vandermey R. (1980) *Genes and the Mind: Inheritance of mental illness.* Oxford University Press.

Vandenberg S. G. and Kuse A. R. (1979) Spatial ability: a critical review of the sex-linked major gene hypothesis. In *Sex-Related Differences in Cognitive Functioning.* Wittig M. A. and Petersen A. C. (eds), p. 67. Academic Press, New York.

8 Behaviour genetics in science and society

Topics of this chapter

1 The role of behaviour genetics in the stratification of society into socioeconomic groups where genetic variation in ability contributes to social mobility just as much as do environmental influences.
2 Arguments over racial differences in ability and their consequences for education and employment. While simple explanations of such differences in terms of test bias are unlikely, other genetic and environmental explanations are generally unconvincing. That the greatest diversity is within rather than between ethnic groups has considerable implications for environmental intervention programs.
3 The two futures of behaviour genetics, firstly as a tool in behavioural research in adding extra precision to animal experimentation, in examining behavioural relationships across time and traits and in evaluating the significance of environmental effects. Secondly, its future in genetics is likely to be at the ecological and evolutionary level, understanding the structure of populations and their utilization of resources. Behaviour genetics may come to have practical applications mainly in paediatrics, genetic counselling and the therapy of those with genetic disorders.

In this chapter we examine the implications for society which arise from genetic and environmental variation in behaviour. Once again we have to concentrate on intelligence test performance, the one aspect of behaviour genetics which arouses so much public controversy. At the same time we consider the future of behaviour genetics in science. These two topics are interrelated since the social impact and motivation underlying behaviour genetics research have been called into question in such contentious areas as that of racial differences in ability. The issue is not merely the scientific one of whether or not genetic differences are involved but also the ethical questions of whether people should be permitted to do such research, how the results could be used or abused and whether government agencies should support this work. Similar criticisms have been made of human biology in general (S. Rose, in Rose and Rose 1976) and must also implicate those areas of behaviour genetics involving experimental animals. To the extent that we use animals as a convenient model of behaviour then we hope that, for example, studies of learning in insects or drug effects in mice will generalize to a variety of species including humans, raising the same ethical issues. If they do not generalize, we must question society's need to support such research.

We concentrate on these areas of human differences concerning the structure of society and of socioeconomic groupings and the origins of racial differences and their implications for education and employment. More detailed coverage is provided in such texts as Brody and Brody (1976), Loehlin et al (1975), Vernon (1979) and Willerman (1979).

Socioeconomic status and the structure of society

Socioeconomic status (SES) is defined in terms of the three interrelated variables of social class, occupational status and level of educational achievement. While parental SES and their children's IQ scores correlate between 0.35 and 0.40 in most studies, e.g. the Louisville Twin Study (Chapter 7), the mechanism of this relationship is open to question. Is it as simple as the deprivation model in Figure 8.1a which postulates environmental factors enhancing or inhibiting

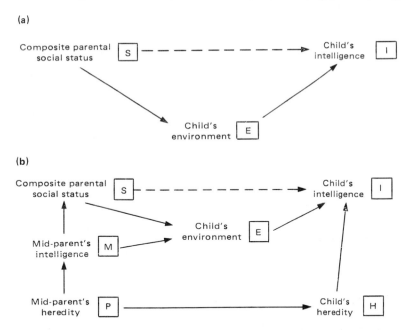

(a)

(b)

FIG. 8.1   (a) The deprivation model and (b) the polygenic model, of possible causal connections between parental SES and child IQ. (Modified from B. K. Eckland (1971)). *Intelligence: genetic and environmental influences*, Cancro R. (ed.), p. 65. Grune and Stratton, New York.)

cognitive development? Or is there a more complex polygenic model (Fig. 8.1b) taking into account parental IQ and both genetic and environmental influences on the child's abilities?

Children do differ in IQ from their parents, a view which we saw in Fig. 7.1 was less compatible with the deprivation than with the polygenic model. Shifts in SES are related to such differences in IQ. In the USA Waller found that sons with higher IQ scores than their father ended up to be of higher SES, with the reverse if sons were of lower IQ. Figure 8.2 comes from a large study of educational and vocational status at the time of military induction of 2000 young Frenchmen.

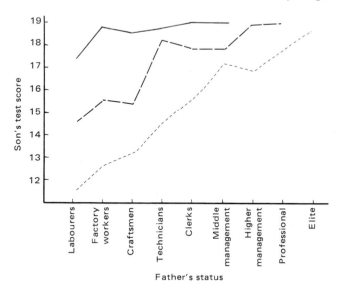

FIG. 8.2   The test scores of young men as a function of their father's occupation and the son's educational/vocational mobility. ——, mobile 1. — —, mobile 2 and 3. ----- remainder. (From C. Lévy-Leboyer (1972) In *Mental Tests and Cultural Adaptation*. Cronbach L. J. and Drenth P. J. D. (eds), p. 311. Mouton, The Hague.)

In Fig. 8.2 they are classified as mobile 1 (strong, upward mobility compared with their father), mobile 2 and 3 (moderate or slight upward mobility) and remainder (same status as father or downward mobility). Lévy-Leboyer summarizes the results thus:

the mobile Ss are not outclassed on tests by the group they join, but they surpass markedly the tests of others in the group of their origin. These differences take on more importance because they reverse, to some extent, inequalities of cultural advantage. Many of the less advantaged overtake and even pass the more advantaged. It appears that, to move upward from his original level, the boy needs to possess greater intellectual powers than it takes merely to hold at the level of one's father. It is as if there were a psychological barrier to surmount, and possession of above-average ability markedly increases one's chances of surmounting it. (p. 315)

In his sample, 27% were upwardly mobile, 58% unchanged and 15% downwardly mobile. Lévy-Leboyer's results show that mobility can occur in spite of and not because of social advantage in that it is social inertia which prevents people of ability advancing:

It appears that success on tests is a virtually obligatory requirement for marked upward mobility. But it is not a sufficient condition. An appreciable number of highly able Ss starting from social levels well below the highest did not make use of their ability to rise in status. (p. 313).

Such movement can also be downwards and indicates that intelligence testing is not a good strategy for preserving social inequality. In his book dealing with inequality in educational opportunity Jencks (1972) states:

Finally, we can say that if an economic elite wanted to pass along its privileges to its children, establishing a system in which privilege depended on test scores would not be a wise strategy. Suppose, for example, that we define the 'upper-middle class' as those families that rank in the top one-fifth in terms of income and occupational status. If access to this elite were strictly random, one upper-middle class child in five would end up in the upper-middle class. If America were suddenly to create a system in which new recruits to the upper-middle class were selected entirely on the basis of test scores, one upper-middle class child in three would be able to maintain his or her parents' privileges. This suggests that a good deal of liberal and radical rhetoric about testing ought to be reexamined. The idea that tests serve mainly to maintain the privileges of the economic elite is exaggerated. (p. 81)

This view of intelligence tests contributing to social change contrasts with the argument of Herrnstein (1973) that society is moving to a future where social class differences become solidly built upon inborn differences in ability, a situation he refers to as a 'meritocracy'. His argument centres around the following syllogism (a) if differences in mental abilities are inherited and (b) if success requires those abilities and (c) if earnings and prestige depend upon success, then combining all of these implies that social standing (which reflects earnings and prestige) will be based to some extent on inherited differences among people.

Herrnstein (p. 129) introduces five corollaries to this syllogism:

1   As the environment becomes more favourable for the development of intelligence, heritability of IQ will increase.

2   All modern political doctrines preach social mobility and should allow people to rise or fall according to their own merits.

3   For many bright but poor people, the social ladder is tapered slightly with far more room at the bottom than at the top (the same point made by Lévy-Leboyer).

4   Technological innovations may change the demands on IQ.

5   The syllogism deals specifically with intelligence.

Apart from perhaps overemphasizing how IQ 'runs in families' and playing down the genetic diversity existing within the different social classes, his argument can be attacked for other reasons (Chomsky, reprinted in Block and Dworkin 1976). Nevertheless, the corrollaries raise some interesting issues. Does society's emphasis on adequate access for everyone to environmental and educational variables mean that genetics is becoming the major source of individual differences? Is the replacement of clerical personnel by business computers changing the relation between SES and IQ?

Fulker (in Royce and Mos 1979) explored the connection between schooling, occupation and income further in a genetic analysis similar in style to Table 7.2. Using data collected by Taubman from twins who were US Army veterans, Fulker summarizes the results thus

... insofar as schooling influences adult status, home environment is almost as important as genetic endowment, but ... large independent genetic and environmental influences unrelated to home environment play the major role. One could hazard a guess that these later genetic influences are related more to temperament and special skills rather than to IQ, which we know has a powerful influence on schooling. The environmental factors probably relate to market imperfections and luck. (p.375.)

That is, apart from E_2 and G components common to the three variables, there are also quite specific E_1 and G effects which influence one variable but not the other. In more practical terms, while E_1 effects on schooling such as illness or bad luck have little later effects, influences of the general family environment (E_2) persist throughout life.

Apart from the consequences for an individual person or family, what are the consequences for society as a whole of familial differences in IQ? Even before formal intelligence tests were developed it was noted that less intelligent people had larger families. With the report of a correlation of -0.3 between IQ and number of children, fears grew particularly in the USA that the 'national intelligence' was declining (Karier, reprinted in Block and Dworkin 1976). It was predicted on this basis that the mean IQ should decline by two to four points per generation but the facts were the opposite. Comparing the scores of Scottish 11 year old children tested on a group IQ test in 1932 with those tested in 1947 revealed a rise of two IQ points. Other studies in stable developed countries (reviewed in Loehlin et al 1975) produce similar results.

TABLE 8.1   Reproduction of people of different IQ levels (derived from J. V. Higgins, E. W. Reed and S. C. Reed (1962) *Eugenics Quarterly* **9**, 84)

IQ range	Reproductive rate excluding unmarried individuals	× Portion of individuals who married	Reproductive rate including married and = unmarried individuals
0–55	3.64	0.38	1.38
56–70	2.84	0.86	2.46
71–85	2.47	0.97	2.39
86–100	2.20	0.98	2.16
101–115	2.30	0.98	2.26
116–130	2.50	0.98	2.45
131 and above	2.96	1.00	2.96

While some critics have made much of the discrepancy between the observed increase and the decline predicted on the genetic hypothesis, the explanation for this paradox is shown in Table 8.1. Working from the pedigrees of mentally-retarded individuals collected by Reed and Reed, it emerges that people of low IQ often had large families if they had any children at all, but fewer of them had any family. Taking this factor into account the most able people were the most fertile. Vandenberg (1973) indicates that this differential fertility is declining as everyone irrespective of ability tends to have smaller families.

Eckland has theorized about how the relationship between IQ and fertility altered in history. Originally (Fig. 8.3a) high intelligence was needed to look after a large family and in many societies intelligent males often had several wives. When knowledge of birth control first became available it was largely confined to the most able (Fig. 8.3b). Nowadays we have the situation in Fig. 8.3c where as Eckland himself quotes Davis 'children have become items of conspicuous con-

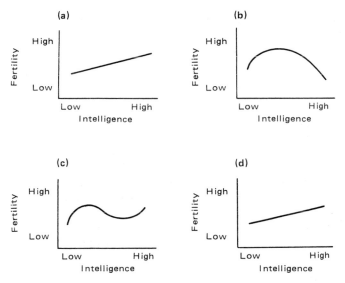

FIG. 8.3   Hypothetical relationships between differential fertility and intelligence in four historical periods. (a) Preindustrial man, (b) early industrialization, (c) late industrialization and (d) post industrialization man. (From B. K. Eckland (1972) *Evolutionary Biology* **5**, 293.

sumption which only the better-off can afford' (p. 300). Least fertile are the middle class 'who are too educated to have children carelessly and too poor to have them abundantly' (p. 300). As birth control becomes common throughout the population we may return again to the original situation (Figure 8.3d).

Like those of Herrnstein, Eckland's speculative views suggest that the role and determinants of IQ are changing in our society. So too are the factors which shape the evolution of intelligence. Until recently natural selection was the major factor. Two out of every five white males born in 1840 did not survive until age 20 so that there was considerable scope for selective advantage. Gradually differential fertility supplanted differential mortality in importance with the mean live births per female dropping from 5.5 for women born in 1839 to 2.2 for those born in 1922. Now things are changing again in that the advent of mass education has made assortative mating of greater importance. People at both extremes of the IQ range (university graduates and those with less than 8 years of education) demonstrate a much higher degree of assortative mating than do those in the middle. Since these are the two groups who tend to be most fertile (Table 8.1), the human gene pool for abilities must be undergoing considerable change.

Discussion

The aim of this section has been to introduce a broader perspective of the genetics of intelligence than just the analysis of individual differences. Darlington (1969) develops this argument much further, suggesting that major historical events such as the Norman Conquest should be viewed in terms of their consequences for the gene pool and not just for history. Even on a microgeographical level it is important to appreciate the historical reasons that can influence the distribution of IQ. For example a study of seven villages in Oxfordshire, England showed that the average IQ varied among the different villages from 100 to 112 in males and 98 to 113 in females. These differences between villages were closely correlated with SES such that in the top two villages 20% of the inhabitants belonged to the highest SES class.

TABLE 8.2   Mean IQs and A_1 and O phenotypes among locally and non-locally born (derived from J. B. Gibson, G. A. Harrison, V. A. Clarke and R. W. Hiorns (1973) *Nature* **264**, 498)

| | Female | | Male | |
	A_1	O	A_1	O
Local	99.3	100.0	100.5	103.4
Non-local	107.5	110.2	113.3	113.0

Many of these high SES people were not locals but newcomers, generally of higher IQ (Table 8.2) and with such genetic differences from the locals as a higher frequency of O blood group phenotypes. Although Gibson et al. suggest that the O allele and also the A_2 allele (the rarer of the A alleles) are associated with higher IQ than other alleles at the ABO locus most notably A_1, Table 8.2 shows that the evidence is ambiguous. A much larger study (Beardmore and Karimi-Booshehri 1983) of blood donors in Yorkshire and in south-west England indicates the opposite, in that people of blood group A are 15% more likely to be found in the top two SES categories (on a scale of five) with O being correspondingly less common. In the bottom two groups the converse applies. Eliminating explanations in terms of historical and migration effects, they conclude that the ABO alleles (or closely linked ones) have pleiotropic effects on occupational type, social mobility and social class. (A recent commentary has queried these results and their generality (*Nature* **309**, 395 (1984)).)

The data in Table 8.2 illustrate the role of social mobility in behavioural and genetic diversity within even a small locality, a point worth considering when we examine racial differences across a whole continent. To illustrate the influence of social mobility further, Thoday and Gibson (1970) modelled the situation using the number of bristles in *D. melanogaster* which also is a polygenic trait. Flies develop more bristles when reared at 20°C than at 25°C (an environmental effect) but there is also some genetic variation ($h^2 = 0.25$). In each generation they took all the flies with more bristles for either genetic or environmental reasons and bred them together at 20°C, that is grouped them in an environment conducive to high bristle number. The other group of low bristle number were bred at 25°C. After nine generations, the overall heritability was much the same ($h^2 = 0.30$) but the *within*-group heritability was only 0.13, while the *between*-group heritability was 0.42. That is, genetic factors contributed far more to differences between rather than within groups. As they conclude

it strengthens the expectation that social mobility related to a heritable variable will give rise to some genetic difference between class means despite strong parent–offspring environmental correlation. We therefore believe that our experimental results support those who hold the view that neither cultural nor genetic approaches alone are likely to lead to adequate explanations of social class phenomena. (p. 992.)

Racial differences in ability

The history of science has many examples of 'scientific racism', defined as 'the attempt to use the language and some of the techniques of science in support of theories or contentions that particular human groups or populations are innately superior to others in terms of intelligence, civilization or other society defined attributes.' (S. Rose in Rose and Rose (1976), p. 113.)

As Rose points out, many of the early efforts in this direction owed as much to personal beliefs as to science. The result was that scientific interest in this topic waned until the late 1960s when A. R. Jensen in an article entitled 'How much can we boost IQ and scholastic achievement?' made the following statement:

There is an increasing realization among students of the psychology of the disadvantaged that the discrepancy in their average performance cannot be completely or directly attributed to discrimination or inequalities in education ... So all we are left with are various lines of evidence, no one of which is definitive alone, but which, viewed all together, make it a not unreasonable hypothesis that genetic factors are strongly implicated in the average Negro–white intelligence difference. (*Harvard Educational Review* (1969) **39**, 82.)

Jensen wrote this in the context of the limited success of the compensatory education projects designed in the 1960s to help disadvantaged children, many of whom were Black. The main project, Head Start, was a massive endeavour involving over 12 000 centres throughout the USA, over 200 000 children each year and an annual budget in the late 1960s of $350 million (White in Hellmuth 1970). Yet the academic gains achieved with all this effort were by no means large, permanent or consistent across the many different behaviour programs tried within the compensatory education schemes (Bronfenbrenner, reprinted in Montagu 1975), leading Jensen to propose this different way of viewing the situation.

Although the suggestion of genetic differences is often interpreted as justifying inadequate educational facilities for Black children, Jensen in fact said the opposite. Starting from the observation that disadvantaged children in school often appear much 'brighter' in nonscholastic ways than middle-class children of comparable IQ, Jensen developed the concept of Level I and Level II abilities. Level I abilities are basic rote learning skills where Black children do just about as well as White irrespective of SES. An example of Level I skills is forward Digit Span, the ability to repeat a sequence of numbers (Fig. 8.4a). Where Black children differ is in Level II abilities which involve more complex problem solving and concept learning tasks and which relate more to g, general ability. Digit Span becomes a Level II task by adding complexity and making the child repeat the numbers backwards. Except for the lowest SES groups, Black children do worse on this task (Fig. 8.4b). While some aspects of the Level I/Level II distinction are open to question (Brody and Brody 1976), Jensen's concluding remarks in his 1969 paper indicate that he is not advocating a denial of resources to disadvantaged children

There can be little doubt that certain educational and occupational attainments depend more upon g than upon any other single ability. But schools must also be able to find ways of utilizing other strengths in children whose major strength is not of the cognitive variety. One of the great and relatively untapped reservoirs of mental ability in the disadvantaged, it appears from our research, is the basic ability to learn. We can do more to marshall this strength for educational purposes.

. . . Accordingly, the ideal of equality of educational opportunity should not be interpreted as uniformity of facilities, instructional techniques, and educational aims for all children. Diversity rather than uniformity of approaches and aims would seem to be the key to making education rewarding for children of different patterns of ability. The reality of individual differences thus need not mean educational rewards for some children and frustration and defeat for others. (*Harvard Educational Review* (1969) **39**, 117.)

The arguments over compensatory education have subsided, as newer more intensive and invasive programs meet with some success (Bronfenbrenner reprinted in Montagu 1975) even though they require massive parental involvement and their long-term success is still uncertain (Darlington 1980). However the issue of racial differences has not disappeared. It is being maintained in the law

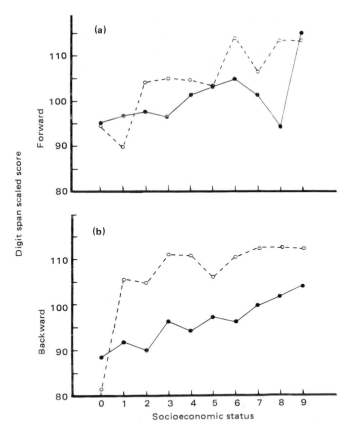

FIG. 8.4   Scores of Black (●—●) and White (o--o) children on (a) the forward and (b) the backward components of the Digit Span test of the WISC-R. (From A. R. Jensen and R. A. Figueroa (1975) *Journal of Educational Psychology* **67**, 882.)

courts by disputes centering around the following two areas.

1   *The use of psychological tests* In *Larry P.* et al. versus *Wilson Riles*, the California Department of Education was banned from using tests to allocate students to classes for the educable mentally retarded (EMR). While Black children comprise 9.1% of all Californian school children, they represent 27.5% of the EMR students, a three-fold excess which Judge Peckham ruled was evidence of intentional discrimination. (In San Francisco where 28.5% of the children in the school district are Black, we saw in Fig. 3.2 that the 66% of Blacks making up the EMR classes is actually slightly less than would be predicted from the mean IQ of the group.) As noted in Chapter 2, the judge emphasized the 'limited, dead-end education' which went with the label EMR and not the possibility that intensive, remedial help resulting from such a diagnosis may enable the child to return to normal classrooms more able to cope than before.

As a result of earlier court action, California had stopped using IQ tests for EMR assignment in 1971 but the new ruling went further and gave them 3 years to eliminate the disproportionate number of Blacks. Jensen (1980, Chapter 2) describes other court cases along similar lines. Their number is likely to increase with such plans as the intended National Teacher Examinations setting minimal levels of teacher competence, which a Virginia study has indicated may pose more problems for predominantly Black teacher training colleges.

2 *Affirmative action programs* Such programs are an attempt to redress the low proportion of Blacks and other minorities in higher education and professions by giving them preference and by relaxing admission requirements. The case *Regents of the University of California vs Bakke* brought to a head the question of whether Blacks and other minorities should be allowed into universities (in this case, Davis medical school) with a lower entrance requirement than Whites such as Bakke. Although Bakke won his case and the Supreme Court ruled that quotas based solely on race were illegal, they still ruled that universities could take race into consideration in admissions, a result which has flowed on to litigation over affirmative action in employment.

How much do Blacks and Whites differ?

Figure 3.1 illustrated the 15 IQ point difference generally found between Blacks and Whites. While this may appear to be a large difference, Table 8.3 shows that SES and race account for only a small proportion of the total variation in IQ, compared with differences between and especially within families. But the effect of race cannot be dismissed. It may have major consequences for those at the higher and lower extremes of the distribution, as was discussed in relation to Fig. 3.2 and the proportions of different racial groups in classes for EMR and the gifted.

A similar contradictory pattern emerges at the biological level when one considers those blood group markers which distinguish the races. About 85% of the variation at each locus is within rather than between the races, so that individuals cannot be assigned to a particular race on the basis of just one locus. Neel (1981) challenged this view arguing that if one considers variation at many loci simultaneously, the races can be clearly distinguished biologically with little overlap. That is, while so few loci show large racial differences, many loci show small differences whose effects can be cumulative. Only eight loci are sufficient to assign individuals to a particular racial group with 87% accuracy. While the biological definition of 'race' is open to debate, this question is not relevant to the behavioural issues. In arguments over bias in testing and admissions policies, race is not defined *genetically* but *socially* in terms of who see themselves and are seen by others as belonging to a particular race.

The next question is whether the IQ difference is a real one or simply an artefact of the test procedure. One can quickly eliminate such explanations as poorer motivation among the Blacks to complete the tests. If this were the case why would the racial groups do equally

TABLE 8.3   The contributions of SES, race and family to variance on the
WISC-R (from A. R. Jensen (1976) *Phi Delta Kappa* Dec. 340)

Source of variance	Percentage variance	Average IQ differences
Between social classes (within races)	8	6
Between races (within social classes)	14	12
Between families (within races and SES)	29	9
Within families (between sibs)	44	12
Measurement error	5	4
Total sample	100	17

well on forward but not on backward Digit Span (Fig. 8.4)? The
evidence for other obvious explanations such as Blacks doing badly
because they do not relate to a White tester or because they are
expected to fail is inconclusive and often conflicting (Jensen 1980).

While Gould (1981) hypothesizes many apparent test biases which
disadvantage minority groups, Jensen (1980) provides experimental
evidence discounting the majority of these biases. For example, such
WISC verbal comprehension items as 'What is the thing to do if a
fellow (girl) much smaller than yourself starts a fight with you?' the
expected answer of not fighting back may seem to reflect white
middle-class values. Jensen points out that if you examine the results
on this particular question Black children are relatively more likely to
get it correct than are Whites. In general there is no connection
between items which appear biased against Blacks and those which
Blacks actually get wrong. It is necessary then to distinguish test items
which are *culture-loaded* and require that degree of exposure to the
mainstream culture which all children are likely to have experienced
from those items which are *culture-biased* and require information to
which not all cultures have equal access e.g. 'Name three parks in
Manhattan?'.

Jensen (1980) examines two potential sources of bias, the first
being the adequacy of the test at predicting educational and
employment success of individuals from different racial groups. Test
scores are as good if not better a predictor of success for Blacks as for
Whites. This still means that the tests reflect the values of the
dominant culture as many critics e.g. Karier (in Block and Dworkin
1976) mention, but it is precisely success in this situation which the
tests have been designed to predict. The most convincing support for
tests is the difficulty of replacing them — given that society must
make some selection, is there any alternative which is more objective
and less sensitive to individual prejudices?

The second is the internal consistency of the test, considering such
issues as whether Black and White children find the same items
difficult. If they made different errors one could argue that the tests
were biased and reflected their different experiences. In contrast
Jensen found on both verbal and performance tests that Black
children made precisely the same mistakes as White children 2 years
younger. This racial difference applies also to Piagetian tests such as
judging the water-line in a tilted bottle (Fig. 8.5) where it is likely that

US children of different races have had similar experience (Ginsburg 1972). Among 6–8 year olds in California, 43% of Orientals had grasped the concept of horizontality necessary for this task compared with 35% of Whites and 13% of Blacks.

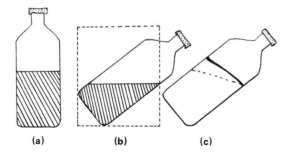

(a) (b) (c)

FIG. 8.5   Piaget's test of the concept of horizontality. The child is shown the bottle (a) and, after it has been hidden with a card and tilted (b), is asked to draw the water level (c) (From A. R. Jensen (1975) in *Racial Variation in Man* Ebling F. J. (ed.) Symposia of the Institute of Biology **22**, 71.)

What causes the racial difference?

If the racial differences cannot be attributed to the test, then are they a product of genes or environment or both? Vernon (1979, p. 320) presents 30 lines of evidence which can be used for one side or the other and we examine only a few.

1   *The difference is a product of Blacks' low SES*   Sanday (reprinted in Montagu 1975) reviews several studies which show that by statistically taking into account differences in education, SES and related variables, Blacks have the same or a higher IQ than Whites. Jensen calls this the 'sociologists' fallacy', since if education and SES depend partly upon polygenic characters (Fig. 8.1b) one is controlling for genetic as well as environmental differences and cannot claim that this technique demonstrates that racial differences are purely environmental.

2   *The moderate within-group heritability for IQ suggests that there must be some between-group heritability*   Jensen (1980) states this is not the case — within-group heritability can never *demonstrate* between-group heritability but only limits the possible environmental factors contributing to the differences between groups. For example since E_2 effects on IQ are not of major consequence within groups (Chapter 6), then any search for environmental explanations of between-group differences must involve some other environmental factors which do not enter into within-group differences.

3   *If intelligence is an important attribute it must have been subject to similar selection pressures in different civilisations making genetic differences unlikely*   However, Loehlin *et al* (1975) explain how, given Negroid and Caucasoid races separated about 33 000 years ago, it would require only a very slight differential selection for there now to be a considerable racial difference. Such mathematical explanations may be misleading, ignoring the possibility that the slaves taken to America were a selected sample of the African population. While it is simplistic to accept such explanations as slaves being the only Blacks stupid enough to be caught by slave traders, there is no doubt that some West African tribes regularly procured and in some cases bred less fortunate inland groups for sale to the slave traders.

We can only speculate on who was taken as a slave, but the situation of the US Black must be distinguished from that of Blacks in the UK who were generally sufficiently well-educated to move from Africa or the West Indies by choice, paying their own fare. Arguments such as those of Tizard (in Ebling 1975) which extrapolate to the American situation from Blacks in Britain (who score little different from Whites) must be viewed with caution since the gene pools may be very different.

4 *Twin and adoption studies* While conventional twin studies tell us only about the heritability within groups, the work of Scarr and Barker (reprinted in Scarr, 1981) sought to compare the genetic architecture in Black and White twins. We saw in Fig. 3.1 that the variance in Black IQ was some 25% less than in Whites and comparative genetic analyses in Black and White twins can determine if this difference is due to less genetic or less environmental variance in Blacks. Scarr and Barker found that the MZ correlations were similar in both races but the DZ correlations for Blacks were larger. To put it in another way, differences within DZ twin pairs were smaller for Blacks than for Whites.

This result could be interpreted as there being less genetic variation among Blacks — Scarr cites mouse experiments which show reduced behavioural genetic variation among animals reared in deprived conditions — or that family membership and E_2 is a more important variable for Blacks. These results should be interpreted with caution given the low participation rate (457 out of 702 twin pairs in public (i.e. state) schools refused) and the relatively small sample size. Eaves's computer simulation data discussed in Chapter 6 suggest there are insufficient numbers for the breakdown of MZ-DZ correlations by race and SES on which Scarr and Barker based their conclusions.

Scarr and Weinberg (reprinted in Scarr 1981) have also carried out a study of Black children adopted into advantaged White homes. Using the reaction range concept introduced in Fig. 7.2, they summarize their results thus:

In other words, the range of reaction of socially classified black children's IQ scores from average (black) to advantaged (white) environments is at least 1 standard deviation (15 IQ points). Conservatively, if we consider only the adopted children with two black parents (and late and less favorable adoptive experiences), the IQ reaction range is at least 10 points between these environments. If we consider the early-adopted group, the IQ range may be as large as 20 points. The level of school achievements among the black and interracial adoptees is further evidence of their above-average performance on standard intellectual measures.' (Reprinted in Scarr 1981, p. 130.)

These results indicate that the racial difference in IQ is malleable. But we saw in Chapter 6 that the study of means is only one way to analyse adoption, the other way being to examine the causes of variation among the adopted children. While 33% of this variation is accounted for by mother's race and mother's and father's education, these factors are confounded with social influences rendering useless any estimate of the extent of genetic variation. For example, the children from inter-racial unions where the mother was white and usually better educated were adopted earlier, had fewer and better quality placements before adoption and had spent more time with the adoptive families. Other criticisms of this project are reprinted in Scarr (1981).

5 *Studies of racial admixture* Apart from the issues raised above in connection with inter-racial adoptions, studies of this sort have other difficulties. Flynn (1980) examines the data that illegitimate children born to White German mothers and Black US servicemen had the same IQ as children born in similar circumstances to White fathers. This lack of any racial difference need not exclude genetic differences for the population as a whole. Loehlin *et al* (1975) point out that Black servicemen in general were subject to more positive selection for IQ than were Whites in that 30% of Blacks and 3% of Whites were rejected on the pre-induction mental tests. An additional problem is that nothing is known about the servicemen who were the fathers in these particular cases. Dyer (1974) provides a comprehensive account of racial admixture and its many biological as well as behavioural consequences.

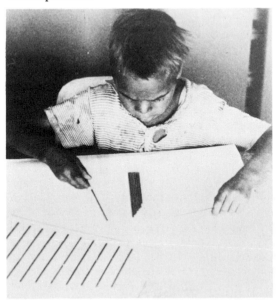

FIG. 8.6   Testing Aboriginal children on the concept of seriation (the ability to handle the relationships of 'greater-and less than' in ordering lengths). (From G. N. Seagrim and R. J. Lendon (1980) *Furnishing the Mind: A comparative study of cognitive development in Central Australian Aborigines.* Academic Press, Sydney.)

Most of the argument over racial admixture concerns a very different population, the Australian Aborigine. In work in the early 1970s (summarized in Vandenberg 1973), De Lemos and Dasen obtained conflicting evidence over whether Aboriginal children with some White admixture acquired Piagetian concepts similar to that shown in Fig. 8.5 earlier than 'full-bloods' with no history of admixture. ('White' here also includes Afghan from some of the camel-drivers who used to operate in the area). Their supervisor, Seagrim, has repeated their studies with the result shown in Table 8.4.

TABLE 8.4   Performance on two conservation tests of Aboriginal children with differing degrees of White admixture (16/16 = full Aboriginal descent). Data are number of children in category (from G. N. Seagrim and R. J. Lendon (1980) *Furnishing the Mind: A comparative study of cognitive development in Central Australian Aborigines.* Academic Press, Sydney. Data were obtained in 1972 and 1973 from two overlapping samples — these are the 1973 data which they consider to be more reliable since the children were familiar with the testing situation.)

Ancestry	Conservation of quantity		Conservation of weight	
	Conservers	Non-conservers	Conservers	Non-conservers
10/16	2	0	1	1
11/16	—	1	—	1
12/16	1	4	1	4
13/16	3	11	5	9
14/16	7	23	10	20
15/16	2	17	3	16
16/16	2	20	5	17

Thus full Aboriginal children are somewhat less likely to develop conservation of quantity of liquid (e.g. does a tall, slim beaker hold the same as a short, squat one?) and of weight (e.g. does the child realise that altering the shape of a ball of plasticine does not change its weight?). But the relationship with ancestry varies across tests. While on horizontality (Fig. 8.5) and seriation (Fig. 8.6) children of mixed ancestry do better than full Aboriginals, this is less consistently the case with other Piagetian tasks designed to measure the child's ability to classify objects on the basis of certain attributes.

Seagrim explains the discrepancy between the three investigations as an effect of sampling. In the particular community involved there are several extremely able full Aboriginal families who, while numerically small, provided the majority of the most successful children in some of the studies. At the time other studies were being conducted, none of these children were the right age to be included. There are other lines of evidence indicating the success of these particular families beyond their children's performance e.g. relatives of Albert Namatjira, the famous landscape painter accounted for the majority of the children in Seagrim's study who showed unambiguous conservation.

6   *The deficits of Black children are cumulative*   so that minor early deficits create a cycle of increasing disadvantage e.g. if one is slow to develop speech or reading, then whole areas of education are inaccessible. The term 'cumulative deficit' was initially developed for English canal boat children and for children in the remote areas of the Blue Ridge Mountains of Kentucky where the relative inaccessibility of education through constant travel and remoteness respectively led to a decline in IQ.

As mentioned in Chapter 5 there is some, albeit conflicting, evidence that Black infants develop as fast if not faster than White infants. This fast development is not maintained. In one study the average IQ of southern Negroes dropped from 86 to 51 between Grades 1 and 6. Jensen (1974) suggests that this may not indicate cumulative deficit so much as a sampling problem where the average IQ at Grade 1 is inflated by some underage children who are so able that they have been admitted early, while the Grade 6 average is similarly deflated by some below average children whose abilities are so poor that they cannot advance to further classes.

Jensen's solution to this sampling problem has been to use only within family differences, comparing older children with their younger siblings to see if the older siblings do worse the greater the age difference. In this way he found some slight evidence of cumulative deficit in Californian Black children on verbal but not on nonverbal IQ. In contrast very low SES Black children from rural Georgia showed substantial cumulative deficit for nonverbal as well as verbal IQ. For example, with siblings 4 years apart, the older was lower by 6.3 points on verbal IQ and 6.4 points on nonverbal IQ, a result opposite to the usual birth order effects (Fig. 6.6). Thus cumulative deficit is an environmental contributor to racial differences in ability for some but *not* for all Blacks.

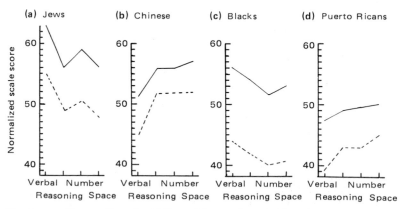

FIG. 8.7   The profiles of middle and lower class children from four ethnic groups on the Hunter College Aptitude Scales for Gifted Children. These tests are standardized to a mean of 50. Solid line, middle class. Broken line, lower class. (From G. S. Lesser, G. Fifer and D. H. Clark (1965) *Monographs of the Society of Research in Child Development* **30**, No. 4.)

7 *Racial differences in the profile of performance across different tests or sub-tests* (Fig. 8.7) are discussed by Vandenberg (1973) and

Loehlin *et al* (1975) among others. It is less obvious how a purely environmental hypothesis would explain why Blacks do better on the culture-loaded verbal IQ tests than on the seemingly more culture-fair performance IQ measures. Different SES groups from the same ethnic group have similar profiles but different overall levels, indicating that SES and ethnic group differences cannot be attributed to the same causative factors.

While such evidence may seem to support a genetic hypothesis, Jensen (1980) launches a scathing attack on the analysis of profile differences, arguing they are more 'arbitrary artifacts of the particular groups, tests and methods used than they are "facts of nature"' (p. 731). Apart from their total reliance on the samples on which the tests were developed (usually White middle-class) and the scaling of the different subtests developed therefrom, his main point is that the profiles reflect not ethnic group differences so much as the extent to which the subtests correlate with the general factor g. So Blacks do better on Verbal than on Reasoning tests (Fig. 8.7), where the respective correlations with g are 0.55 and 0.86. In turn Jensen's own emphasis on g is fiercely attacked by Gould (1981) with such statements as:

Jensen would not only rank people; he believes God's creatures can be ordered on a g scale from amoebae at the bottom to extraterrestrial intelligences at the top. I have not encountered such an explicit chain of being since last I read Kant's speculations about higher beings on Jupiter that bridge the gap between man and God. (p. 317.)

These seven approaches to the determinants of racial differences are not being offered as scientific evidence so much as to show that every argument for one stance or the other can be found wanting. With exceptions such as Loehlin *et al* (1975) and Vernon (1979), it often happens that one particular line of evidence is presented as *the* explanation of racial differences. The one thing this section should do is to encourage critical examination of such simplistic accounts.

Will there ever be definite evidence as to the relative role of genetic and environmental contributions to the Black–White or any other racial difference? We can best quote from someone who features so prominently in this debate:

Without a true genetic experiment involving cross-breeding of racial samples and the cross-fostering of progeny of such random samples of every race × sex combination of the two populations in question, all other types of behavioral evidence can do no more than enhance the plausibility (or implausibility) of a genetic hypothesis. Whatever social importance one may accord to the race-genetics question regarding IQ, the problem is scientifically trivial, in the sense that the means for answering it are already fully available. The required methodology is routine in plant and animal experimental genetics. It is only because this appropriate, well-developed methodology must be ruled out of bounds for social and ethical reasons that the problem necessarily taxes scientific ingenuity, and is hence probably insoluble. (A. R. Jensen (1980) *The Behavioral and Brain Sciences*, **3**, p. 361.)

Other racial differences — the myth of White supremacy

Since intelligence testing is often criticized as a means of confirming White supremacy (Gould 1981) it is important to note that there are Jewish and Oriental groups with average IQs above 100. Willerman (1979) discusses some of the research on Jews. Apart from the above average performance of Jews particularly in verbal tests which was seen in Fig. 8.7, there are also differences between Jewish groups. The Ashkenazi Jews who originate from eastern Europe differ genetically from the Sephardic or Oriental Jews in several respects — Tay-Sachs disease (Chapter 2) is prevalent only among the Ashkenazi Jews. In a study of a Brooklyn neighbourhood, the Ashkenazi adults were found generally to have had more education while their children consistently outperformed the Sephardic children of the same neighbourhood on the Stanford-Binet Intelligence Test (means of 116 compared with 110) and on the Peabody Picture Vocabulary Test (113 compared with 96).

It is uncertain whether these differences are genetic. The constant Ashkenazi emphasis on learning and academic achievement can be interpreted as an environmental factor leading to higher ability or as part of a genotype-environment correlation. The same argument has been applied to Jewish-Caucasian differences where the particularly high Jewish scores on verbal abilities are perhaps more consistent with an environmental explanation. The effects of family environment are most important for verbal skills (Table 7.2) which are also more open to scrutiny by parents anxious to encourage their children's success.

A racial difference which is attracting much attention because of its noticeable effect on education and professional employment is that between Whites and Orientals. Vernon (1982) gives a very comprehensive account of Oriental success and to take just three examples, relative to their numbers in the US population: (a) Orientals produce three times as many PhD students as Whites, (b) ten times as many Orientals have been elected to the US National Academy of sciences, (c) two and a half times as many Orientals are admitted to the University of California as the other ethnic groups combined.

Vernon provides an overwhelming amount of information on the high performance of Japanese in the USA, although their achievement has diminished with acculturation. After the first generation immigrants (the Issei) were disrupted by World War II and internment, the second generation (the Nisei) were outstanding and their children (the Sansei) did not do so well but fitted far better into school activities.

Why do the Orientals do better? The Hawaii Family Study (Chapters 6 and 7) suggests that the same abilities are tapped in tests of both groups, in that their 15 tests contribute exactly the same to the four factors (verbal, spatial, perceptual speed and visual memory) in people of European or Japanese ancestry. This would have been unlikely if the tests had been measuring different skills in the two groups. Having excluded test bias, is there any evidence for genetic differences? Vernon proposes two lines of evidence, the first being differences in temperament between new-born Orientals and Whites which he considers occur so early that they must be genetic. However we saw in Chapter 7 that extrapolation from infant temperament to later behaviour must be done with caution. The second point is that while the emphasis on education by the Orientals has often been stressed as a factor in their success, their better nonverbal than verbal performance (Fig. 8.7) is at variance with our earlier discussion of how comparable emphasis by the Jews may result in a verbal IQ advantage. As Vernon says: 'It would be more explicable if literacy and numeracy are partly genetic, since we could hypothesize that Caucasians (including Jews) have stronger genes on the one, Orientals on the other.' (p. 273.)

Vernon also discusses personality and other characteristics which are potentially important. But what should be stressed is that here is a group in the USA with a history of environmental hardship comparable to the Blacks yet who are of high ability. To quote from his preface

The experiences of oriental immigrants in the United States and Canada — Chinese and Japanese — provide a remarkable example of adverse environment *not* affecting the development of intelligence. There is no doubt that, in the past, they were subjected to great hardships, hostility, and discrimination. They were regarded as a kind of inferior species, who could be used for unskilled labor and menial jobs, but could never be accepted as equals into the white community. And yet the Orientals survived and eventually flourished until they came to be regarded as even higher achievers, educationally and vocationally, than the white majority. (p. ix.)

What's wrong with an environmental explanation?

If the role of genetic determinants of racial differences can never be adequately established, is there anything wrong with concentrating solely on environmental effects, given that compensatory education is likely to be the only practical remedy for inequality? There are at least five reasons against an uncritical acceptance of large-scale educational intervention.

1 If intervention at the preschool or primary school level is not sufficiently effective, then more intensive involvement earlier in life is attempted. It may become totally invasive into family life such as Bettelheim's suggestion that since Black parents cannot bring up their children properly, the children should be reared away from their family in kibbutz-like conditions. Seagrim and Lendon (1980) showed that Aboriginal children brought up away from their families in what they called 'a total immersion in White culture' did develop adequate conservation skills, but such cognitive advantages must be weighed against the disruption of their social and cultural upbringing within the tribal group. de Lacey (1974) raises a related issue with Aboriginal children — does compensatory education help to destroy their culture and move them instead into a White culture where they are bound to be the most inferior group?

2 More contact with White children may not be the answer. Sanday (reprinted in Montagu 1975) examined the simple *quantity* of contact achieved through school integration and 'busing' programs (transporting children to schools in other areas to reduce the racial imbalance). She suggested that this approach may not be effective in encouraging Blacks to 'adopt the cognitive styles associated with occupational success' (p. 248) if it is *quality* of contact and the absence of racial hostility which counts.

While forced integration has been successful in some US cities (Rist, 1979), in others it has led to the phenomenon of 'White flight' away from State schools. For example the Boston school enrolment in 1973 (the year before cross-city busing began) was 93 221 comprising 53 328 Whites, 31 781 Blacks and 8111 other minorities. The corresponding 1982 total of only 58 540 includes 18 720 Whites, 27,807 Blacks and 12 013 other minorities. That is, a policy designed to foster racial understanding may accentuate rather than diminish differences.

3 Is an environmental explanation necessarily less denigrating than one in terms of 'genetic differences'? For example Montagu (1972) explains the Black–White difference in terms of sociogenic (socially-induced) brain damage, a view which Valentine and Valentine (1975) criticize in that it makes inequality just as inherent and permanent among the disadvantaged as any genetic explanation. Even the affirmative action programs can be considered demeaning in that they reinforce the stereotype of minorities as incapable of achieving anything unless someone else helps them.

4 Environmental intervention may accentuate racial differences. In his review, Bronfenbrenner (reprinted in Montagu 1975) includes two general observations about the effectiveness of such programs. Firstly 'the less they have the less they learn', in that the most disadvantaged children improve the least. Secondly unless the programs continue, effects 'wash-out'.

Figure 8.8 demonstrates these two results in White and Aboriginal children in the Bourke Preschool Project. Bourke is a remote rural Australian community where most children do poorly — the average IQ in this project was 68 for Aboriginal children and 77 for Whites. Two preschool programs were tried, a 'traditional' one based on standard kindergarten practice and a structured language-stimulation program based on the Bereiter–Engelmann techniques (with some similarities to the TV program 'Sesame Street') which proved more effective. Both Aboriginal and White children improved in language skills but the latter improved more and retained more of the gains. What has since been introduced is a home liaison program, seeking the support of the Aboriginal parents to complement and consolidate the new concepts introduced at school. Seagrim and Lendon (1980) suggest that otherwise Aboriginal parents may not only fail to provide a suitable model but may actively discourage 'different', i.e. White, ways of thinking in the children.

5 'Compensation in the educational process was to be an affluent society's payoff for two and a half centuries of exploitation and humiliation, the reward for a history of servitude and neglect' (J. F.

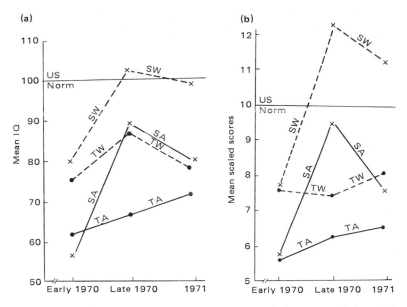

Fig. 8.8   (a) IQ scores on the Peabody Picture Vocabulary Test and (b) scaled scores (average is 10) on the WPPSI Vocabulary subtest for Aboriginal (A) and White (W) children, given structured (S) or traditional (T) preschool programmes. (From B. Nurcombe (1976) *Children of the Dispossessed.* University of Hawaii Press.)

Winschel in Hellmuth 1970, p. 6). Is compensatory education just an easy way for society to salve its conscience about the disadvantaged? Giving money for education is a lot less painful than trying to rectify the inequalities in employment, health and housing faced by minority groups. Jencks (1972) and Valentine and Valentine (1975) see the only possible solution in a major restructuring of society if real 'equality' is to be achieved

In America, as elsewhere, the general trend over the past 200 years has been toward equality. In the economic realm, however, the contribution of public policy to this drift has been slight. As long as egalitarians assume that public policy cannot contribute to economic equality directly but must proceed by ingenious manipulations of marginal institutions like the schools, progress will remain glacial. If we want to move beyond this tradition, we will have to establish political control over the economic institutions that shape our society. This is what other countries usually call socialism. Anything less will end in the same disappointment as the reforms of the 1960s. (Jencks 1972, p. 265.)

Discussion

The methodologies described in this text can never provide conclusive information on the role of genetics in racial differences. On the other hand, the solely environmental viewpoint which stresses that each child is potentially equal in ability given adequate intervention is not without problems. Unfortunately there is so much conflicting material published on this topic that as Winschel (in Hellmuth 1970, p. 22) says 'It becomes increasingly difficult to tell truth from half-truth, half-truth from fiction, and fiction from falsehood'. Eysenck versus Kamin (1981) provides a classic example of diametrically opposed interpretations of the same information.

One of the disturbing features of this controversy is that it puts so much emphasis back on IQ at a time when the rest of behaviour genetics is moving to a broader investigation of cognitive skills and their development (Chapter 7). Contrary to what Gould (1981) suggests, the emphasis on IQ is not confined to those with a genetic bias. The system of performance-contracting employed in some of the 1960s compensatory education projects paid companies for every point the mean IQ was raised, reflecting an absolute faith in IQ which few would or should share.

Zigler and Trickett (1978) suggest that the long-term effectiveness of intervention programs should be considered in terms of such aspects of social competence as the incidence of delinquency, teenage pregnancies, child abuse and being self-supporting rather than on welfare. As one step in the right direction, Darlington (1980), rather than concentrating on their IQ scores, studied the locations in schools of children who several years previously had participated in Head Start programs and found fewer had been kept back a grade because of poor performance or assigned to remedial classes. But this effect of Head Start may not be on the children. Glass and Ellwein (1984) argued that the attitude of parents and teachers may have been changed, so that they are less likely to accept transfer of Head Start children to remedial programs.

The distinction between IQ and social or educational competence has long been a concern in mental retardation (Chapter 2) and Mercer's System of Multicultural Pluralistic Assessment (SOMPA) is an attempt to extend such a method of assessment to the culturally disadvantaged. The question being asked is no longer 'how far is the child behind the norms?' but 'how well has the child managed to cope with life, given her/his disadvantages?', where disadvantages include social milieu, medical history and sensory impairment.

Appreciating the child's environment in this way means that the assessment of different racial groups can be instructive and not just inflammatory. Thus Vernon (1969) carried out extensive cognitive assessment of 11 year old English, Jamaican, Scottish, Ugandan as well as Canadian Indian and Eskimo boys. By combining these results with detailed information on the home and school environments, he demonstrated for example that both Gaelic speaking boys from the Scottish Hebrides and Ugandan boys were handicapped in oral vocabulary and Piagetian skills as a result of rigid, formal education systems which stressed rote memory rather than flexibility. Ignoring mean IQ, racial comparisons can thus provide important natural experiments in the analysis of behavioural development.

A similar point emerges from the poor Piagetian skills of Aboriginal children. Seagrim and Lendon (1980) do not consider these a deficiency compared with White children, but rather a difference which reflects the structure of their language and society. Neither adult nor child Aborigines make much use of comparatives. They may say 'big' or 'little' but rarely 'more' or 'less'. The concept of number is very limited as is any interest in causation, e.g. the children often asked the researchers for personal information but not about why they were in the community. All these factors together make acquisition of Piagetian concepts unlikely and in many ways unnecessary for traditional Aboriginal society.

The question remains as to the motive behind research on Black–White differences. It is interesting that this difference has attracted all the attention rather than the similar-sized difference in favour of certain Oriental and Jewish communities compared with Whites. As long as there is affirmative action for particular groups then there must be value in researching the causes of the disadvantages which make such affirmative action necessary in the first place. It is better that views and policy be based on research rather than personal and political opinion which are the only alternatives.

Affirmative action is on the basis of sex as well as race. But as we saw in Chapter 7 in connection with spatial abilities, sex differences are as difficult to study genetically as racial differences. This augments the problem because of what Jensen (in Cancro 1971) has called the race $\times$ sex $\times$ ability interaction. While in Whites there is only a small sex difference in cognition in favour of females, superiority of females is twice as large among Blacks. The result is that any academic or professional selection based on high ability will show a much larger sex difference among Blacks where the mean is lower (Fig. 3.2). Jensen reports that 67% of the finalists in the National Merit Scholarship Program were males whereas in the comparable National Achievement Scholarship Program restricted to Blacks only 43% were male. Jensen does not explain this sex difference beyond suggesting that it could either have a genetic (X chromosome) basis or reflect the greater susceptibility of Black males to the physical and psychological stress often experienced by their race.

The research poses one question which remains unanswered. If racial differences only contribute a small part of the total variation (Table 8.3) and interact in a complex way with sex differences, why have policies been based solely upon racial and sex quotas? In the Bakke case, Justice Marshall of the Supreme Court justified it thus 'The racism of our society has been so pervasive that none, regardless of wealth or position, has managed to escape its impact'. While science finds it difficult to identify how 200 years of discrimination makes a professional Black family more disadvantaged than an impoverished White family not eligible for help by affirmative action, writers of fiction have captured the distinction

'Yes', Randolph resumed, 'a fear so deeply ingrained that there is in this country no such thing as a secure black man — no matter how educated, wealthy or accepted by the public he may be. Thus, every lower-class black is a constant reminder and a threat that the middle-class black could easily be forced back into the ghetto by a repressive society which avenges itself on the lower-class black man.' (H. Denker (1973) *A Place for the Mighty* W. H. Allen, London, p.157.)

What is needed are research programs to determine individuals' strengths and weaknesses and to establish appropriate remedial help, taking race into account but not making it the exclusive variable. For example compensatory education programs, like those for some of the mental retardation syndromes (Chapter 2) may have to be tailored more to the needs of specific children rather than being thought of as a blanket approach, equally appropriate for everyone. Nurcombe (1976) explains how a less structured program more meaningful to the Aboriginal children, gave less improvement but better retention than the formal programs of Fig. 8.8. The role of genetics in such research must not simply be to examine why children or races differ in abilities but more to understand why some people benefit more than others from particular programs. As we saw in Chapter 7 there is no reason why the genetic or the environmental explanations of ability in young children need say anything useful about how these abilities change with intervention.

Loehlin *et al* (1975) give a much more comprehensive discussion of potential research but add the important caution that it should be accompanied by extensive public education as to what is being attempted and why. The issues are complex scientifically as well as in their political implications and the message must be that people cannot expect simple explanations or solutions.

The future of behaviour genetics

This book presents a brief but wide-ranging view of behaviour genetics which may leave the reader uncertain as to what behaviour genetics is and where its future lies. Is it the analysis of membrane mutants in *Paramecia*, of the response of mouse strains to ethanol or of the factors determining the course of cognitive development in children? Behaviour genetics is all of these in that it involves the *understanding* and *utilization* of individual differences in behaviour through whatever means and species available.

As stressed in Chapter 1 it is important that behaviour genetics be more than just the documentation of genetic effects whether they be at the level of mutants affecting *Drosophila* courtship or the interplay of genotype and early environment in alcoholism. It needs to be a tool which all behavioural scientists can use. Festing (Chapter 5) stressed the need for all researchers on rodent behaviour to consider genetic differences in efficient experimental design, while Eaves and Young (Chapter 7) emphasized how genetic and behavioural theory should interact in the development of an adequate model of human personality.

Viewed in such ways behaviour genetics has no glamorous role in new scientific innovations but only a pedestrian function in improving both experiment and theory. However it would be wrong for psychologists to consider genetics in their experiments merely as a means to reduce error variation between subjects (Chapter 3) or for geneticists simply to think of behavioural traits as a convenient and quick means of screening for mutations (Chapter 4). We can

summarize in turn some wider implications for both disciplines
introduced in the preceding chapters.

In the case of psychology Table 8.5 shows that current research
interests in behaviour genetics are non-specist as well as non-racist.
That is, there is a continued commitment to studying the principles of
behaviour in a wide variety of organisms and not just the white rat
and college student traditional in psychology (Chapter 1). Three major
trends, apparent in research on all species, have considerable
implications for psychology.

TABLE 8.5   Research subjects listed in the abstracts of papers presented at
the 1981–84 annual meetings of the Behavior Genetics Association

Research subjects	Number of papers*			
	1981	1982	1983	1984
Invertebrate	14	10	12	9
Vertebrates other than humans	29	29	31	26
Human (within ethnic groups)	18	36	48	26
Human (between ethnic groups)	2	2	2	2
Theoretical	3	3	1	2

*Several papers are listed twice, e.g. if they include both insects and
mammals or studies both within and between ethnic groups.

1 *The multivariate approach, integrating data on different aspects of
behaviour and physiology* This is readily achieved in rodent
research where data are compiled on the various strains. In Chapter 5,
Henderson, Oliverio and Wahlsten all provide differing but equally
valuable approaches to the relationship between learning in mice and
other behaviours. As a corollary this approach is also useful in
allowing research to concentrate on specific genotypes with a
particular set of characteristics relevant to that experiment.
Comparisons among selection lines differing in learning or alcohol
tolerance are a much more efficient approach to the mechanisms
underlying these behaviours than any methods using non-selected
animals and are probably the major way in which behaviour genetics
is of relevance to those psychologists otherwise not interested in
genetics.

The same emphasis can be found in human behaviour genetics
with increasing attention to the nature and development of cognition
and of personality (Chapter 7). Psychology has made much of the
correlations typically around 0.7 between child and adult IQ or
between IQ and performance both on other cognitive tasks, e.g.
Piagetian ones, and on measures of academic achievement. But with
such correlations between variables, a value of 0.7 means that only
49% (0.7×0.7) of the variance is in common. As a first step in
explaining the 51% on which they differ, genetic analyses can
determine whether different genes or different environmental factors
are involved, since the two have different consequences for the
predictability and modification of behaviour across time and traits.
2 *A greater appreciation of the relevance of particular environmental
factors* Behaviour genetics can show not only how much of the

variance is contributed by a certain environmental variable, but also how consistent any effect is across individuals in terms of genotype-environment interaction. The examples of Zajonc's model of birth order effects (Chapter 6) and of compensatory education (this chapter) demonstrated the importance of these considerations in human research, while Chapter 3 provided similar examples from rodent behaviour.

With genotype-environment interaction such a common feature of drug effects (Fig. 3.14), the recent growth of both human and animal psychopharmacology and behavioural teratology (the study of behavioural effects on individuals whose mothers received drugs while pregnant) must be accompanied by an increased appreciation of genetics. The fact that individuals given the same drug treatment may respond in completely opposite ways means that genetic differences cannot be ignored but must be an integral part both of experimental design and of any theory to explain the mechanisms by which the drug action mediates behaviour.

There will also be more attention to the effects of specific environmental variables rather than the often unhelpful division simply into within- and between-family environmental effects. In human behaviour genetics such efforts to be more precise have so far met with little success but potentially provide a very powerful technique in terms of understanding genotype-environment correlation and interaction (Chapter 6). The common practice of merely administering a questionnaire on the home environment that parents provide may not only be insensitive to the crucial environmental variables but also ignores the important sources of environmental variation between children within the family.

3 *Behavioural genetics brings an increased scrutiny of behavioural measures* Understanding of all three behavioural dimensions considered in Chapter 7 (cognition, personality and psychiatric disorders) has been increased through genetic analysis. This results from much more than merely the need to examine every behavioural measure carefully before committing oneself to its inclusion in an expensive and protracted genetic analysis. Genetic analysis provides a unique perspective on the structure of behaviour and on the interrelationship or otherwise between different traits, e.g. in animals between various measures of learning and non-cognitive factors involved in their performance and in humans between verbal and spatial abilities, between the various measures of temperament or between different psychiatric disorders. Given the demonstrated success of genetic analyses with such behaviour, then the major expansion of behaviour genetics is likely to be the application of these techniques to a wider range of behaviours. Infant temperament, reading disabilities and pre-senile dementia are examples of the sorts of behavioural traits now under genetic investigation and many more are possible. Fuller (in Royce and Mos 1979) provides a comprehensive list of psychological and genetical criteria for the choice of good 'psychophenes', his term for behavioural phenotypes with potential for genetic analysis.

The future of behaviour genetics within genetics is less certain for two reasons. Firstly geneticists do not share the same interest as psychologists in particular behavioural traits. For example (unless one is especially interested in the behaviour) why choose to do genetic analysis of something as inconvenient and as awkward to measure as infant temperament or something as potentially heterogeneous as manic-depressive psychosis (Chapter 7)? Secondly any person who studies behaviour is forced to cope with individual differences and potential genetic variation while the converse does not apply — the geneticist who studies individual variation need not be measuring behaviour. In 1975, Lewontin (Discussion topic 1, Chapter 6) used these arguments to attack studies of cognition for their failure to provide information on the mechanisms of genetics.

It might be argued that the influence of genes on any trait in any species is a valid subject of scientific enquiry, since, after all, geneticists are objective scientists who are simply curious about nature in all of its manifestations. But that surely cannot be the answer since, for example, genetic aspects of snout length of pigs is not a subject that receives much attention. Perhaps the genetics of intelligence is intrinsically more interesting than snout length in swine because somehow research in this field has revealed some fundamental information about the way genes mediate the development of behavior. But that is not true, clearly. All research on the genetics of normal human intelligence has been of a statistical nature, using the techniques of biometrical genetics to estimate genetic and environmental sources of variation in specific populations. There has not been, and in the present state of developmental and neural biology cannot be, any attempt to analyze cellular and developmental mechanisms of gene action in influencing cognitive traits. (*Annual Review of Genetics* **9**, 400.)

With the exception of research on single-gene disorders, Lewontin is essentially correct within a very narrow definition of genetics. In a broader context behaviour can modify the gene pool in animals (Chapters 4 and 5) and, through assortative mating, in humans (Chapter 6). Thus behaviour is more likely to be important in population than in molecular genetics except for advances in neurobiology (Chapter 4).

Ehrman and Parsons (1981, Chapter 14) provide many examples of how behaviour genetics could proceed in this direction particularly as regards the process of speciation. Parsons (1983) develops this theme further with an examination of the role of behaviour in colonization. His concept of the ecobehavioural phenotype emphasizes the fact that behaviour is one aspect of the phenotype which is finely-tuned to features of the environment (Chapter 4) and is therefore an excellent choice for studies of evolution in response to particular external pressures. As Ehrman and Parsons (1981) indicate, assortative mating is the rule rather than the exception across the majority of species and much more needs to be known about the genetic consequences of such behavioural phenomena.

The presence of assortative mating in human behaviour may be seen as a disadvantage. To continue the quote from Lewontin (1975):

Nor can it be maintained that work on the biometrical genetics of intelligence has somehow led to progress in biometrical genetics as a general approach. On the contrary, because of the impossibility of control and manipulation of human environments and mating, man is among the worst choices of experimental organisms for testing the methods of quantitative genetics. (p. 401.)

On the other hand, human behaviour genetics is crucial to genetics precisely because of assortative mating and the complications such as cultural transmission where the transfer of information from parent to offspring mimics some aspects of Mendelian inheritance. It may not help to *test* the methods of quantitative genetics as Lewontin requires but rather to *extend* these methods to populations with such complex structures as those suggested by Eckland (Fig. 8.3).

The study of behavioural evolution in this context must be clearly distinguished from sociobiology. Table 8.5 showed that very few behaviour geneticists are concerned with the theoretical issues in sociobiology and instead are interested in the extent of genetic variation from a genetical or psychological viewpoint. Such an interest is the antithesis of sociobiology. To quote Dawkins, one proponent of sociobiology, in a review of M. Ruse (1979) *Sociobiology: Sense or nonsense?* (Reidel, Dordrecht):

Whoever the sociobiologists may be, nobody would deny that they are interested in behaviour as Darwinian adaptation. In order to talk about the Darwinian evolution of anything we have to postulate the sometime existence of genetic differences between individuals — otherwise there would have been nothing for selection to choose between. This does not mean there has to be genetic variance any longer: any characteristic that has been strongly favoured by selection is likely now to have low heritability, the genetic variance having been used up. (*Nature* (1979) **280**, 427.)

So sociobiologists are interested more in those behaviours for which there is no genetic variation while such variation is the essential tool of behaviour genetics. It does not mean that the two disciplines should remain isolated as there are many useful ideas on behavioural evolution to be gained from sociobiological research on the significance of animal social behaviour. Conversely the existence and nature of genetic variation can say something about selective pressures (Chapters 4 and 5) rather than implying as Dawkins does that selection has been unimportant for that particular trait.

Human sociobiology may be different. Rightly or wrongly it has received considerable criticism over some of the more extreme conclusions that can be drawn. Dawkins selects one such example from a review by Rose of E. O. Wilson's *On Human Nature*, '... human males have a genetic tendency towards polygyny, females towards constancy (don't blame your mates for sleeping around, ladies, it's not their fault they are genetically programmed)' (*Nature* (1981) **289**, 528.) As pointed out in Chapter 6, one main strength of current human behaviour genetics is its emphasis on testable hypotheses and little would be gained by being associated with such untestable speculation.

Sociobiologists may also feel the same way about behaviour genetics. To quote Dawkins again:

It is much wiser to say: 'I don't think *X* is scientifically valid, but even if it is so what?' This message is nowhere more important than in our dealings with the other controversy mentioned in Rose's letter, the race/IQ controversy. As Stephen Jay Gould puts it with his customary cogency: 'I do not claim that intelligence, however defined, has no genetic basis — I regard it as trivially true, uninteresting, and unimportant that it does ... It is just as likely that blacks have a genetic advantage over whites. And, either way, it doesn't matter a damn.' (*Nature* (1981) **289**, 528.)

With Lumsden and Wilson (1981), sociobiologists are trying to relate developmental and cognitive psychology to evolutionary biology but a comparison of their text with the material in Chapter 7 demonstrates just how large a gap remains between the disciplines. In his chapters dealing with ethology and sociobiology in relation to behaviour genetics, Fuller (in Fuller and Simmel 1983) has started to bridge the gap but also indicates how much there is still to go.

Can there be an applied behaviour genetics?

Molecular biology has moved rapidly from a laboratory discipline to a commercial venture with the advent of genetic engineering. Will behaviour genetics ever find such practical application? There is some attention paid to behaviour in livestock breeding and population structure is being examined as a factor in the mouse plagues which devastate Australian cereal crops. However any major application is likely to be in the human area, firstly through providing more adequate empirical risk factors for the chances of any specific behavioural disorder in relatives of an affected individual (Fig. 2.23). Genetic analysis is often the best means of specifying who is likely to have a particular disorder and of determining which factors contribute mostly to this disorder. For example, what is the best course of action with children whose parents have a major psychiatric disorder? The evidence in Chapter 7 would suggest that as long as the parents can look after them adequately, there is little to be gained as regards the children's prognosis by adoption or fostering.

Secondly molecular biology and a greater understanding of single gene and chromosome disorders should result in a whole new field for behaviour genetics. As new methods are developed for prenatal diagnosis and/or postnatal drug (and in the future, gene) therapy for the multitude of genetic disorders, then two questions posed in Chapter 2 become vital: (a) what causes variability within a disorder and what future is predicted for that individual sufferer? (b) what specific behavioural disabilities exist for that disorder and which disabilities may remain or even result from a particular course of therapy?

The enormous enthusiasm in contemporary science for molecular genetics overlooks such questions but a major step in the right direction has come from Schmid and Nielsen (1981). Although dealing mainly with sex chromosome disorders, they make one of the first attempts to bridge the gap between medical geneticists and social scientists, psychiatrists and psychologists. Schmid raises three questions for behavioural intervention:

1   What is to be done if one of these conditions is found for some reasons in a mentally so far normal young patient. Can or must we do something of a preventative nature?

2   What is the optimal treatment if the patient already is mentally or socially deviant?

3   What is the balanced and objective information we can give to couples if one of these anomalies is discovered in second trimester prenatal diagnosis? (p. 3)

While treatment for any physical symptoms has advanced considerably, help at the behavioural level lags far behind. Schmid gives one example which aptly summarizes the situation:

A young boy with the XYY condition became very troublesome from the age of about 4 years. His mother had to transfer him from one school to the other, later from one institution to the next; with about twelve (sic) he began to set fire when things did not go according to his will. At the time I saw him, at the age of fourteen, he had to be kept under constant surveillance in the ward of a big mental hospital, together with all the incoming adult cases of acute psychoses etc. Although of practically normal intelligence this boy was so difficult to manage that no home for juveniles was ready to accept him any longer. I tried to draw the attention of numerous psychiatrists to this special patient with his dark social prognosis, but I had no success. I suspect that, among other reasons, nobody felt competent. (p. 3.)

There is yet another level of behaviour genetics which has been even more neglected, namely the behavioural reactions *to* a genetic disorder as distinct from those physical and behavioural effects which result *from* the disorder. An example in Chapter 2 was Turner's Syndrome. While one physical symptom is failure to reach puberty, the relevant question here is how these girls cope with being shown to be so different from their peers. We are interested not in the symptom of the disorder but in how the affected individual reacts to this symptom. Another example from Chapter 2 was how the PKU child reacts to the monotony of the necessary diet.

Such behavioural problems do not only affect the patient. Mikkelsen *et al* (in Schmid and Nielsen 1981) quote Rundquist:

The mother described the situation with the words: 'We do not have a handicapped child, we are a handicapped family. The problems are not solely genetical. The process of accepting one's handicapped child is never brought to an end. When an unwanted or serious event occurs in the life of the child, the process of acceptance goes through all its phases again. The death wish for the child comes again and again and is subconsciously present even if one believes that the situation has been accepted.' (p. 10.)

To date most work in this direction has focussed on cystic fibrosis, the autosomal recessive disorder which is the most common lethal or semi-lethal genetic disease among Caucasians (see Chapter 2). The abnormalities of mucous secretion lead to chronic lung dysfunction requiring daily, time-consuming therapy. This affects not only the child with the disease who becomes aware of the sterility (in males) and early death often associated with the disorder, but also the parents who have to provide the therapy and the siblings who find their needs subordinate to the demands of the affected child and the constant therapy. Earlier subjective reports of extreme disturbance in all family members and a very high divorce rate in such families are not wholly supported by recent, objective studies. At the same time the average age when the affected child dies has risen from 7 to 19 years and it is difficult to separate alterations in behavioural effects from these changing physical consequences. Not only the family faces problems. Andrews and Elkins (1981) discuss the school placement of spina bifida children with inadequate planning for their physical and behavioural needs.

Attention to behavioural problems caused by genetic disorders is thus a major new potential combination of behaviour and genetics. It is encouraging that a recent text on inborn errors of metabolism (Cockburn and Gitzelmann 1982) devotes a significant section to the problems of the individual and the family and that genetic counselling has an increasing awareness of the behavioural issues involved (Hsia *et al* 1979; Omenn in Fuller and Simmel 1983).

All this discussion of discrete disorders may seem very different from the theme of polygenic inheritance which has occupied so much of this text, but it is the overlap rather than the distinction between the two approaches which matters. In the past the practical application of polygenic inheritance has always been thought of in terms of education and employment. Unfortunately, as we have seen in this chapter, there has been as much scope for abuse as for use of knowledge. While polygenic inheritance should come to play a major role in our understanding of the structure and development of abilities in the normal population and practical applications should follow, help for those with specific genetic disorders can provide a more immediate and less controversial application. We cannot label someone as 'abnormal' until we understand more about causation and variation among the 'normal' segment of the population and can use this information to predict the extent of abnormality and the likely consequences of any therapy e.g. the issue of parental education and ability in the response of Downs' Syndrome infants to intervention (Fig. 2.13) and of PKU children to dietary therapy (Fig. 2.18).

Conclusion

The major difficulty with behaviour genetics is that people often approach it already convinced of the answers and of the motivation behind the research (e.g. Lewontin *et al* 1984). In a discussion of new approaches to IQ testing Cohen and Shelley suggest such pre-conceptions are common in psychometrics:

The dispute about IQ seems to continue in an equally zealous manner. It is not clear what sort of evidence would make psychologists change their minds. You either believe that IQ tests measure some innate ability to make sense of the real world, or you do not: you are either for or against IQ. We have found no sign of any leading psychologist in the UK changing his or her position on the issue in the past few years. The perfectly rational psychologist seems to be a mythical beast. (*New Scientist* (1982) **95**, 773.)

The history of psychology and of behaviour genetics has seen phases of both general acceptance and condemnation of the role of genetics in behavioural variation.

From the beginning of the thirties onwards scarcely anyone outside Germany and its allies dared to suggest that any race might be in any respect or in any sense superior to any other, lest it should appear that the author was supporting or excusing the Nazi cause. Those who believed in the equality of all races were free to write what they liked, without fear of contradiction. They made full use of their opportunity in the decades that followed . . . (J. R. Baker (1974) *Race*, p. 61. Oxford University Press.)

While this quote concerned group differences, Cronbach (1975) suggests that since the mid-1960s we have entered a phase where scientists, the media and the public decry any emphasis on individual differences in performance and their possible biological bases. Perhaps this bias is seen most clearly by comparing the response to two scandals over the determinants of intelligence. The controversy over Burt's MZ_A data (Chapter 6) has been extensively documented and used to condemn behaviour geneticists. In contrast, Herrnstein (1983) argues that very little attention has been paid to the jailing in 1981 for the misappropriation of research funds of Heber, whose Milwaukee Project is most often cited as proof for the effectiveness of early intervention — and by inference, the total importance of the environment. Herrnstein points out that the revelations about Heber's behaviour add to existing misgivings about the authenticity of the results of his research. Yet the reaction of the media and of scientists to these disclosures has been minimal.

No one would deny the value of informed scientific debate during this period of rapid expansion of behaviour genetics but constant, ill-informed sniping may have a highly negative effect. In his suitably titled article 'Disciplinary barriers to progress in behavior genetics : defensive reactions to bits and pieces', Elias (1973) argues that the development of behaviour genetics has been greatly hindered by the need to be constantly on the alert for extremist criticisms of methodology, interpretation or motivation.

The reader will be aware that in presenting the methodologies of animal and human behaviour genetics there has been a constant attempt to at least introduce alternative explanations. Although these may seem confusing at the time, they do help the reader to be forewarned if not forearmed against the conflicting information he or she is likely to encounter. At the present time when behaviour genetics is expanding, it is more appropriate to concentrate on the methodology than on the results and their implications, even though Elias (1973) argues:

... the most severe limitations on the progress of behavior genetics are not necessarily methodological. Rather, they may be imposed by the emotional and ideological resistance which is inevitably directed toward research with highly visible implications for human beings (p. 119).

One feature of concern is the scant attention behaviour genetics has received from other disciplines, most notably sociobiology and, in complete contrast, sociology with its emphasis on environmental theories of behaviour. While Eckland (1967) suggests that sociology can justify its position because of the practical applications of its ideas, it is unfortunate that the four specific areas where he proposed genetics to be 'sociologically relevant' have received no more emphases than when he proposed them 15 years ago. One reason as seen in Chapter 6 may be the ease with which specific methodological criticisms may be made of behaviour genetics. The criticisms may not be valid, but if they are said often enough and loudly enough, they will be believed to be important.

If there is one theme underlying this text it is the need to be both open-minded and cautious of the ways in which animal and human behaviour genetics are expanding. It is worrying but common to see any genetic explanation being regarded as somehow disreputable and instead a purely environmental hypothesis being favoured, no matter how flimsy its supporting evidence. Using Cattell's term the 'ignoracists' for the latter group, Jensen makes a plea for more adequate education:

The racists may be popularly perceived as the 'bad guys' and the ignoracists as the 'good guys', but in principle they are much the same: they are both equally wrong and in the long run probably equally harmful. The solution, I repeat, is to think more genetically. The problem on both sides is fundamentally a matter of ignorance, the cure for which is a proper education about genetics. (A. R. Jensen (1973) *Educability and Group Differences*, p. 11. Methuen, London.)

The present text is a step in this direction but balances knowledge with criticism. Behaviour genetics is a growing science and we are unsure which are the best paths to nurture and which should be pruned. If this book has made people give behaviour genetics a more balanced but also a more critical consideration, then it has achieved its goal.

Discussion topics

1 *New Scientist* (1979) **81**, 849 asked Eysenck and Rose to respond to four fundamental questions on the validity of IQ tests, their heritability, racial differences and implications for educational policy. Can you provide a third expert opinion? (You may choose to reply to points made by Eysenck and Rose, something they were not permitted by the editors but which has been criticized in relation to another debate involving Eysenck — S. Sutherland (1981) *Nature* **290**, 636.)

2 "Street crime, failing schools, decaying cities, high unemployment, soaring welfare costs... the fuse is lit." (Cover of P. Ehrlich (1977) *The Race Bomb*. Ballantine, New York.) Ehrlich sees the study of genetic differences in behaviour between socioeconomic and racial groups as making the fuse burn faster. Do you agree?

3 R. Lynn (*Nature* (1982) **297**, 222) found that the mean IQ in Japan has risen seven points in a single generation and that the average score of Japanese children on a translated version of the Wechsler Intelligence Scale for Children was 111. The rise has been criticized on two grounds (T. B. L. Kirkwood *Nature* (1982) **299**, 8, and E. J. Pearson *Nature* (1982) **299**, 574) and alternative explanations offered (A. M. Anderson *Nature* (1982) **297**, 180 and A. Sibatani (1982) *Nature* **299** , 102). J. R. Flynn (*Nature* (1984) **308**, 222) argues the Japanese IQ mean is really between 102 and 104, taking sampling biases into account. While Lynn claimed the rise was unlikely to be the result of 'a change in the genetic structure of the population', H. J. Eysenck (*New Scientist* (1982) **94**, 803) comments on the racial differences thus: "The probability of genetic causes becomes more likely the younger the age at which differences appear. The fact that these differences were already observed at the age of 6 . . . suggests that those who wish to look for environmental conditions ... will have a difficult time."
Do you think Lynn's data are sufficient to say anything about such causes? (A wide-ranging, non-technical discussion of the issues is provided by M. Mohs (1982) *Discover* **3**, 18).

4 G.M. Harrington (*Nature* (1975) **258**, 708) carried out a study of maze-learning in six rat lines which he considered a direct demonstration that the construction of human intelligence tests is biased in such a way that majority groups will inevitably score higher. C. R. Reynolds (*Behavioral and Brain Sciences* (1980) **3**, 352), sees this as a potentially very important result: "If the cultural test bias hypothesis is ultimately shown to be correct, then the 100 years or so of psychological research in human differences . . . must be dismissed as confounded, contaminated, or otherwise artifactual."

However, in the same article A. R. Jensen (p. 366) is very critical of the extrapolation of Harrington's results to humans. Could Harrington's results be dismissed as depending on the many extraneous factors such as activity which may influence performance of rodents on learning tasks (Fig. 5.5) but which are less relevant in the human situation or do you regard his experiment more impressive than the evidence against cultural bias indicated in this chapter?

References

Annotated bibliography

Eckland B. K. (1967) Genetics and sociology : a reconsideration. *American Sociological Review* **32**, 173. (This article remains the most detailed examination of areas of sociology where potential genetic influences should be considered.)

Eysenck H. J. versus Kamin L. (1981) *Intelligence : The battle for the mind*. Pan, London. (An illuminating but also confusing presentation of two conflicting views on intelligence. The sections where they offer different interpretations of the same results are particularly useful.)

Gould S. J. (1981) *The Mismeasure of Man*. Norton, New York. (A history of racial bigotry as applied both to physical and behavioural traits. His interpretation from the viewpoint of taxonomy is unusual. This book is clearly biased towards one viewpoint — J. R. Baker (1974) *Race*. Oxford University Press provides contrast, biased in the opposite direction.)

Hellmuth J. (ed.) (1970) *Disadvantaged Child*, Vol 3. Brunner/Mazel, New York. (The third volume of a series dealing with the compensatory education projects of the 1960s. This one is important because it contains the evaluation of the success or otherwise of different programmes.)

Jencks C. (1972) *Inequality*. Penguin, Harmondsworth, Middlesex. (A controversial examination of the implications for educational and economic success of genetic and environmental differences in intelligence test performance. His genetic analysis has been criticized on several grounds — J. L. Jinks and L. J. Eaves (1974) *Nature* **248**, 287.)

Jensen A. R. (1980) *Bias in Mental Testing*. Methuen, London. (The most complete examination of whether tests are biased. A summary of this text along with many critiques is in *Behavioral and Brain Sciences* (1980) **3**, 325. This book examines bias rather than the causes of racial differences which are covered more in another text — A. R. Jensen (1981) *Straight Talk about Mental Tests*. Free Press, New York.)

Loehlin J. C., Lindzey G. and Spuhler J. N. (1975) *Race Differences in Intelligence*. Freeman, San Francisco. (The widely quoted general introduction to the topic. While remarkably free from bias to either a genetic or environmental viewpoint, its stance has been criticized. I. R. Savage (1975) *Proceedings of the National Academy of Education* **2**, 1.)

Scarr S. (1981) *Race, Social Class and Individual Differences in I.Q.* Erlbaum, Hillsdale, N.J. (A collection of Scarr's research on twin and adoption studies of race and SES. Particular attention should be paid to the critiques of her research and to the final chapter where she justifies her approach.)

Vernon P. E. (1979) *Intelligence : Heredity and environment*. Freeman, San Francisco. (A balanced text similar to Loehlin, Lindzey and Spuhler but with more emphasis on testing and on education. The list of 30 lines of evidence for and against racial differences on p. 320 is useful.)

Vernon P. E. (1982) *The Abilities and Achievements of Orientals in North America*. Academic Press, New York (This book is important in scrutinizing another area of racial differences apart from the Black–White one and in attending to differences in personality as well as in ability and achievement. The book is devoted almost completely to describing the differences not to discussing their possible causes.)

Willerman L. (1979) *The Psychology of Individual and Group Differences*. Freeman, San Francisco. See also L. Willerman and R. G. Turner (1979)

Readings about Individual and Group Differences. Freeman, San Francisco. (General psychology texts which provide a broad perspective on individual differences in many behaviours, including age and sex differences which are not covered here.)

Additional references

Andrews R. J. and Elkins J. (1981) *The Management and Education of Children with Spina Bifida and Hydrocephalus*. Education Research and Development Committee, Canberra. Report No. 32.

Beardmore J. A. and Karimi-Booshehri F. (1983) ABO genes are differentially distributed in socio-economic groups in England. *Nature* **303**, 522. (See also commentary, *Nature* (1984) **309**, 395.)

Block N. and Dworkin G. (eds) (1976) *The IQ Controversy: Critical readings*. Random House, New York.

Brody E. B. and Brody N. (1976) *Intelligence : Nature, determinants and consequences*. Academic Press, New York.

Cancro R. (ed.) (1971) *Intelligence : Genetic and environmental influences*. Grune and Stratton, New York.

Cockburn F. and Gitzelmann R. (eds) (1982) *Inborn Errors of Metabolism in Humans*. A. R. Liss, New York.

Cronbach L. J. (1975) Five decades of public controversy over mental testing. *American Psychologist* **30**, 1.

Darlington C. D. (1969) *The Evolution of Man and Society*. Allen and Unwin, London.

Darlington R. B. (1980). Pre-school programs and later school competence of children from low-income families. *Science* **208**, 204.

de Lacey P. (1974) *So Many Lessons to Learn : Failure in Australian education*. Penguin, Harmondsworth, Middlesex.

Dyer K. F. (1974) *The Biology of Racial Integration*. Scientechnica, Bristol.

Ebling F. J. (ed.) (1975) *Racial Variation in Man*. Institute of Biology Symposium **22**.

Ehrman L. and Parsons P. A. (1981) *Behavior Genetics and Evolution*. McGraw-Hill, New York.

Elias M. F. (1973) Disciplinary barriers to progress in behavior genetics : defensive reactions to bits and pieces. *Human Development* **16**, 119.

Flynn J. R. (1980) *Race, IQ and Jensen*. Routledge and Kegan Paul, London.

Fuller J. L. and Simmel E. C. (eds) (1983) *Behavior Genetics: Principles and applications*. Erlbaum, Potomac, Md.

Ginsburg H. (1972) *The Myth of the Deprived Child : Poor children's intellect and education*. Prentice Hall, Englewood Cliffs, N.J.

Glass G. V. and Ellwein M. C. (1984) Review of 'As the Twig is Bent: Lasting effects of the pre-school programs'. *Science* **223**, 273.

Herrnstein R. J. (1973) *IQ in the Meritocracy*. Allen Lane, London.

Herrnstein R. J. (1983) IQ encounters with the press. *New Scientist* **98**, 230.

Hsia Y. E., Hirschhorn K., Silverberg R. L. and Godmilow L. (eds) (1979) *Counseling in Genetics*. A. R. Liss, New York.

Jensen A. R. (1974) Cumulative deficit : a testable hypothesis? *Developmental Psychology* **10**, 996.

Lewontin R. C., Rose S. and Kamin L. J. (1984) *Not in Our Genes: Biology, Ideology and Human Nature*. Pantheon, New York.

Lumsden C. J. and Wilson E. O. (1981) *Genes, Mind and Culture and the Evolutionary Process*. Harvard University Press, Cambridge, Mass.

Montagu A. (1972) Sociogenic brain damage. *American Anthropologist* **74**, 1045.

Montagu A. (ed) (1975) *Race and IQ*. Oxford University Press, New York.

Neel J. V. (1981) The major ethnic groups : diversity in the midst of similarity. *American Naturalist* **117**, 83.

Nurcombe B. (1976) *Children of the Dispossessed*. University Press of Hawaii.

Parsons P. A. (1983) *The Evolutionary Biology of Colonizing Species*. Cambridge University Press.

Rist R. C. (ed.) (1979) *Desegregated Schools : Appraisals of an American experiment*. Academic Press, New York.

Rose H. and Rose S. (1976) *The Political Economy of Science*. Macmillan, London.

Royce J. R. and Mos L. P. (eds) (1979) *Theoretical Advances in Behavior Genetics*. Sijthoff and Noordhof, Alphen aan den Rijn.

Schmid W. and Nielsen J. (eds) (1981) *Human Behavior and Genetics*. North Holland, Amsterdam.

Seagrim G. N. and Lendon R. J. (1980) *Furnishing the Mind : A comparative study of cognitive development in Central Australian Aborigines*. Academic Press, Sydney.

Thoday J. M. and Gibson J. B. (1970) Environmental and genetical contributions to class differences : a model experiment. *Science* **167**, 990.

Valentine C. A. and Valentine B. (1975) Brain damage and the intellectual defence of inequality. *Current Anthropology* **16**, 117.

Vandenberg S. G. (1973) Current directions in behavior genetics. In *Evolutionary Models and Studies in Human Diversity*, Meier R. J., Otten C. M. and Abdel-Hameed F. (eds), p. 63. Mouton, The Hague.

Vernon P. E. (1969) *Intelligence and Cultural Environment*. Methuen, London.

Zigler E. and Trickett P. K. (1978) IQ, social competence, and evaluation of early childhood intervention programs. *American Psychologist* **33**, 789.

Name index

Names are listed only where they appear in the text not the references. All authors are listed even where only the abbreviated citation *et al.* appears in the text.

Scott, J. P.   165, 166
Seagrim, G. N.   308-9, 313-4, 316
Searle, L. V.   122
Sedano, H.   39
Seiger, M. B.   13
Seligman, M. E. P.   7, 193
Sells, C. J.   49
Shaughnessy, R.   289
Shear, C. S.   60, 62
Shelley, D.   326
Shields, J.   223-4, 275, 279-80, 282, 291
Shopsin, B.   275
Shuckit, M. A.   286
Shuster, L.   179, 180
Sibatani, A.   328
Siegel, R. W.   160
Siegler, R. S.   242
Sigvardsson, S.   287
Silverberg, R. L.   325
Silverman, P.   15, 192
Simmel, E. C.   192, 323, 325
Simon, N. G.   192
Singleton, G. R.   202-3
Sinnott, E. W.   101
Skeels, H. M.   212-5, 217, 238
Skinner, B. F.   20, 26
Skodak, M.   212-5, 217, 238
Slater, D.   262
Smith, S. D.   41, 61, 263
Snow, R. E.   221
Sokal, R. R.   86
Spain, B.   72
Spearman, C. S.   16, 19
Spence, K. W.   193
Spiker, D.   50, 51
Spuhler, J. N.   295, 298, 306, 308, 311, 318
Staats, J.   91
Stein, Z. A.   46
Stene, E.   324
Stene, J.   324
Sternberg, R. J.   209, 229, 242, 264
Sterritt, G. M.   82
Stine, G.   60, 65, 67, 74, 76, 79, 80, 82
Stone, C. P.   156
Strauss, J. S.   283
Susser, M.   46
Sutherland, S.   328
Sziber, P. P.   148-9

Taubman, P.   298
Taylor, H. F.   1

Teasdale, T. W.   243
Terman, L. M.   20, 117
Thiessen, D. D.   26, 55, 158, 168, 172
Thoday, J. M.   301
Thomas, A.   270-3
Thompson, W. R.   1, 3, 8, 180, 183, 190, 198
Thurstone, L. L.   16, 19, 258-9
Tienari, P.   282
Tinbergen, N.   22
Tizard, B.   307
Tolman, E. C.   118, 193
Tompkins, L.   160
Torgersen, A. M.   272, 273
Trickett, P. K.   316
Tryon, C.   119
Tsuang, M. T.   275-6, 282-3, 286, 291
Tully, T.   160

Urbach, P.   209, 230, 244

Val, E.   289
Vale, J. R.   4
Valentine, B.   314-5
Valentine, C. A.   314-5
Van De Reit, V.   87
Van Houten, J.   132-3
Vandenberg, S. G.   209, 222, 226, 230, 251, 257, 260-1, 295, 309, 310
Vandermey, R.   275-6, 282, 286, 291
Vernon, P. E.   1, 16, 19, 265, 295, 306, 311-3, 316

Wachs, T. D.   251
Wahlsten, D.   173-4, 178, 183-5, 188, 193-4, 319
Waller, J. H.   296
Walsh, R. N.   14
Walters, J. K.   192
Walton, A.   277
Ward, S.   134-5, 145, 162
Watson, J. B.   19, 20
Watson, J. S.   24
Webb, G.   38, 44, 45, 52
Webber, P. L.   269
Weinberg, R. A.   74, 269, 307
Weir, M. W.   169
Weiss, J. M.   152
Wellman, N. S.   60, 62
Wenz, E.   62
White, S. H.   302
White, J. C.   87

Whitsett, M. 172
Wilcock, J. 5, 82, 167–8, 170, 178–9, 198, 205
Will, B. E. 190
Willerman, L. 19, 216, 295, 312
Willows, A. O. D. 162
Wilson, E. O. 24, 291, 322–3
Wilson, J. R. 127, 209
Wilson, R. S. 226, 252–3, 257–8, 270
Winokur, A. 286
Winschel, J. F. 315–6
Witkin, H. A. 42
Wittig, M. A. 269

Wright, S. 86, 90
Wright, T. 140

Yalow, E. 229
Young, P. A. 244, 266–7, 269, 274, 290–1, 318
Yunis, J. J. 32, 34, 81

Zajonc, R. B. 233, 320
Zazzo, R. 223, 233
Zellweger, H. 36, 68
Zigler, E. 30, 69, 81, 316
Zubek, J. P. 205
Zuill, E. 141
Zweep, A. 203

Subject index

For all entries on specific rodent strains, see under 'MOUSE STRAINS' and 'RAT STRAINS'; and for specific *Drosophila* and mouse mutants, see under *Drosophila melanogaster* MUTANTS' and 'MOUSE MUTANTS'.

polygenic limitations   321
behaviour modification, in
    Lesch-Nyhan syndrome   63–4
    in mental retardation   82
behavioural diversity and
    ecology   155
behavioural phenotype   192,
    194
behaviourism   19, 26
benchmark tests   243
bent-headed   134–5
Bereiter-Engelman
    technique   314
Bergmann glial fibre   176
β-alanine   142–3
bias, adoption samples   215
    in screening   37, 39, 41–2
    twin samples   223
biochemistry and
    learning   190–1
biochemistry, *Drosophila*
    courtship   143
biological adaptation   7, 26
biological determinism   25
biological preparedness   7, 25,
    137, 160
biometrical genetics see
    polygenic inheritance
biometricians   124
biosocial norm   237, 266
bipolar psychosis see
    manic-depressive psychosis
birds   22, 240
birth order   115, 225, 233, 310
birth stress   283
birthweight   270
Blacks, ability   30, 87ff
blastula   145–6
'blending' theory of
    inheritance   53
block design   260, 263
blood group, ABO   53, 275, 286
blood groups, racial
    differences   303–4
blood-pressure   122–3
blowfly   160–2
Bourke Preschool Project   314
brain lateralisation, in
    Klinefelter's syndrome   40
    in mice   94, 198
    in schizophrenia   283
    in Turner's syndrome   40
    in twins   263
brain structure   56
    in albinos   172–3
    in mouse strains   94

brain weight   188, 190
British ability scales   16

Caenorhabditis elegans see
    nematode
caffeine   149
calling song, cricket   139
Cape Teal   22–3
carrier   46, 58
cat eye syndrome   43–4
catatonia   282
caudate nucleus   56
cell division   34
central excitatory state   162
central nervous system
    malformations   66, 73
centromere   32
cerebellum   175, 176–7
cerebral cortex, in Huntington's
    disease   57
    in mice   187
    in phenylketonuria   57–8
cerebro-spinal fluid   72
chemoreception, bacteria   128
    Drosophila   149
    honeybees   137
chemotaxis, nematode   134–5
    paramecia   131
chi-squared test   99
child rearing   20
childhood schizophrenia   275
chimera   177
chlorpromazine   108
choline acetyltransferase   190–1
chorion   227–8, 263
chromosomal variants,
    mice   167
chromosome   32ff
    aberration   32–4
    arm   32
    banding   34
    deletion   34
    duplication   34
    inversion   34
    non-disjunction   35
    notation   32
    number   32
    staining   32–3
    translocation   34
chromosome disorders and
    psychology   323–4
chromosome disorders, cat eye
    (+der(22))   43–4
    Cri du chat (del(5)(p13))   32,
    43–4

Down's, translocation
(−14,+t(14q21q)) 46
trisomy(+21) 34–5, 44ff
fragile X 51–2
Klinefelter's (XXY) 35–6, 40,
43
Turner's (XO) 35–6, 37ff, 43
XXX 35–36, 40–1
XXXXY 37–8
XYY 25, 41–2, 82, 324
chromosome inversion,
Drosophila 150–1
chromosome substitution
line 124, 153–4
cilia, Paramecia 131
circadian rhythm 127, 187
cirrhosis 288
classical conditioning 160
clomiphene 218
codominance 55
coordination mutants,
nematodes 135
cotwin study 217, 275, 283
coat colour mutants,
rodents 167
cognition see intelligence and
specific cognitive abilities
cognitive development see
development of intelligence
cohort 210
cold-narcosis 150
Collaborative Perinatal
Project 226–8, 242
colonization 321
Colorado Adoption Project 217,
242, 256, 270
Colorado Family Reading
Study 262
colour blindness 53, 285
comparative psychology 26
compensatory education 26, 73,
302, 315, 318
competition, Drosophila 159
in twins 266
computerized tomography 56
concrete operations 256
conditioned inhibition 148
confluence model 233
consanguineous marriages see
inbreeding
conservation, Piagetian 256
of quantity 309
of weight 309
consolidation 150, 168, 189
contraception 219, 299
copper metabolism 173

corpus callosum 94
mice 177–9
correlated response 121–3
correlated traits 237
correlation 113
intraclass 219
correlation vs. causation 122,
167, 193
correlations, twin 226, 231
counselling, genetic 77ff, 285,
291
countercurrent
distribution 144, 148
courtship and learning 150, 158
Drosophila 121, 140–3, 147,
150
ducks 22
cousin marriages 240–1
covariance 112
genetic 114–5
Cri du chat syndrome 32, 43–3
cricket 139
criminality 20, 25, 275, 287–8
in XXY 42
critical periods 67
cross-fostering,
schizophrenia 281
crossing-over 54, 150, 153
cryptophasia 225
cultural transmission 322
cultural-familial retardation 30,
69, 73, 81
cyclic AMP 149
cycloheximide 111, 150–1
cystic fibrosis, family
reactions 325
incidence 60

Daphnia 139
Darwinian evolution 74
defecation 13, 14, see also
emotionality
deletion, chromosome 34
deme 202–3
dementia 275
dendrite 58
deoxyribonucleic acid
(DNA) 65, 79–80
dependent variable 114
depolarization, Paramecia 133
depression 195, see also
manic-depressive psychosis
deprivation experiment 22–3
deprivation model 295
development of
intelligence 251–7, 290

fluctuations 252
genetics 253-4
genotype-environment
correlation 255
in Down's syndrome 47-8
in Lesch-Nyhan
syndrome 63
Piaget 256
problem of
measurement 252
profile of change 254
development of
personality 269-73
development, and genetics 8
and maternal care 187-9
and sociobiology 323
infant 226-7
nervous system 162
racial differences 186
species differences 186
development-difference
controversy 81
developmental delay 195
diabetes 288
Diagnostic and statistical manual
of mental disorders (DSM) 10
diallel cross 105
Drosophila 106, 151
rat 198
diathesis-stress theory 279
dichorionic twins 228
diet 249
and phenylketonuria 60-1,
82
differential genetic loading 71
digit span 52, 302, 304
diploid cell, 34
directional dominance 100
and artificial selection 121
and evolution 137, 157-9,
200-1
and intelligence 241
and learning 156-7, 161,
185-6, 199
directional selection 157
disadvantaged children 302
dispersion of alleles 100
disruptive selection 157
divorce 325
dizygotic (DZ) twins 217-8
dogs, genotype-environment
interaction 166
inbreeding 166
selection 165
domestic animals 204

domestication, rodent 198
dominance 54, 95ff, 157, 233
ambidirectional 100
and assortative mating 236
and evolution 157
and intelligence 185, 236,
240-1
and maternal care 189
and rodent learning 185
behavioural 203
directional 100, 156
means 95ff
variances 101-2
domineering focus 147
dopamine 173
Down's syndrome 34, 44ff, 278
and maternal age 46, 180
early development 47
family background 49
intelligence and age 47-8
intelligence range 48
mosaicism 48
remedial programs 49, 50,
325
sex differences 50
translocation 46
trisomy 45
Drosophila, advantages 140
limitations 160
social mobility model 301
Drosophila melanogaster 86
courtship 121, 140-3
ethanol tolerance 155
gynandromorph 145-7
neurological mutants 143ff
shaker mutants 145
taxes 11-13
Drosophila melanogaster
mutants, *amnesiac* 149-50,
158, 168
aristaless 143
black 140-3
cabbage 150
cacophany 147
celibate 147
coitus interruptus 147
dropdead 144
dunce 149-50
ebony 140-2
Ether à go go 145
fruitless 147
hyperkinetic 145-6
Shaker 145
smellblind 149
stuck 147

extroversion-introversion 265–7, 269

F ratio test 222, 273
familiality 259
family background, in Down's syndrome 49
  in phenylketonuria 61–2
family of handicapped 324–5
family resemblance 114, 208–12, 219, 231, 233–4
family size 225
family stress 254
family studies in mental retardation 69–71
fast-2 132–3
fate-mapping 147
fearfulness, dogs 166
Fels longitudinal study 210, 253
fertility 298–300
fertility drug 218
fertilization 35, 43
fetal alcohol syndrome 29
  animal model 195–7
fixation 158
fixed action pattern 22
flagella 128–9
fluctuations in intelligence 252–4
focus, *Drosophila* 147
folic acid 72
foster children 250
founder effect 74–5
fragile X syndrome 52, 71, 81
fraying behaviour, mice 200
frontal lobe 40
fungus 127

G6PD deficiency 53, 73
galactosaemia 67
gamete 32, 35, 54
gender identity, in Turner's syndrome 40
gene 53
  dosage 95
  function 64ff
  mapping 53–4
  regulatory 8, 251
  structural 8
  structure 6
gene flow, wild mice 202
gene pool 105, 159
gene therapy 323
general intelligence g 16, 267–9, 264, 302, 311
  animal models 180–5

genetic analysis, human see human polygenic analysis
  means 94–101
  non-inbred animals 114ff
  variances 101–3
genetic architecture 90, 105, 124
  and different behaviours 242, 247
  and evolution 157–9
  and learning 156–7, 186
genetic background 160
genetic contamination 194
genetic counselling 77ff, 285, 291
genetic covariance 114
genetic diversity 204
genetic drift 75
  in artificial selection 122
genetic engineering 80, 323
genetic marker 275
genetic model 98, 124
genetic neurobiology see neurogenetics
genetic probe 79, 289
genetic variation, within and between families 220–1
genetics, and social mobility 247–9
  biometrical see polygenic inheritance
  change with age 254
  intelligence see intelligence, genetics
  Mendelian 53ff
  molecular 6, 289
  personality 265–8
  quantitative see polygenic inheritance
genotype 3, 55, 86, 89–90
genotype-environment correlation 216, 237, 255
  and personality 269
  and school achievement 264
genotype-environment interaction 49, 73, 125, 237, 320
  and heritability 110
  compensatory education 318
  degree of arousal 109
  dogs 166
  mice 106–10
geotaxis, and learning 154
  correlated response 12
  Drosophila 11–13
  selection lines 12, 152–4

transfusion syndrome,
twins 227–8, 232
translocation, chromosome 34,
46
triple test cross 199, 203
triplets, and
phenylketonuria 59
trisomy 33, 35, 46
Turner's syndrome 35, 37ff, 43,
324
  personality 40
  physical features 37, 39
  range of intelligence 37
  specific cognitive
  deficits 39–40, 260
twin-family method 236,
259–60
twinning, and maternal age 35
  identical (MZ) 217–18
  non-identical (DZ) 217–18
twins, ability differences 225
  and Down's syndrome 47
  and Huntington's disease 57
  and neuroses 9
  and schizophrenia 9, 10
  birthweight 270
  Burt's sample 223–4
  catch-up 256
  competition 229
  development of
  intelligence 252–5
  discordance for
  schizophrenia 283
  environmental
  divergence 254–5, 269, 273
  genotype-environment
  correlation 238
  genotype-environment
  interaction 237
  incidence 217–9
  language 225
  method 217–30
  mirror-imaging 263
  mortality 226
  parental impressions 274
  parental treatment 230
  personality 266
  physical differences 225
  placentation 227–8
  prematurity 226–7
  psychopathology 275
  racial studies 307
  reading disability 263
  reared apart 222–224, 235,
  250, 326

  representative of
  population 225
  retardation 226–7
  Siamese 275
  statistical
  resemblance 219–221, 232
  stuttering 226, 229
  suicide 278
  temperament 271–4
  transfusion syndrome 227–8
  232
  uniqueness 233
  variability compared to
  singleton 256
  zygosity 217–8, 228, 230
tyrosinase 173
tyrosine 62

unc-5 135
unipolar psychosis see
  manic-depressive psychosis

validity 8
variance 87ff
  genetic analysis 101–3, 110
variation, ecological 67, 155,
  194–5
ventricle (brain) 56, 57, 66
verbal ability, assortative
  mating 236
  in syndromes 40, 48, 51, 72, 81
  maternal effects 260
  twins 225, 255, 257
  twins, placentation 228
verbal IQ 310, 313
vibration, *Drosophila* 121,
  140–1
virus 282
visual cortex 172
visual system, *Daphnia* 139
visual-motor ability, in
  galactosaemia 67
  in phenylketonuria 62
visual-perceptual ability 48
vocabulary 257, 264

water maze, mice 170
Wechsler intelligence tests see
  intelligence tests
weight, mice 188–9
  twins 222, 225–7
wheat 89
White flight 314
wild rodents 124, 198–204, 205
wing length, *Drosophila* 121